OPTIMIZATION OF DIGITAL TRANSMISSION SYSTEMS

The Artech House Communication and Electronic Defense Library

Principles of Secure Communication Systems by Don J. Torrieri
Introduction to Electronic Warfare by D. Curtis Schleher
Electronic Intelligence: The Analysis of Radar Signals by Richard G. Wiley
Electronic Intelligence: The Interception of Radar Signals by Richard G. Wiley
Signal Theory and Random Processes by Harry Urkowitz
Signal Theory and Processing by Frederic de Coulon
Digital Signal Processing by Murat Kunt
Analysis and Synthesis of Logic Systems by Daniel Mange
Mathematical Methods of Information Transmission by Kurt Arbenz and Jean-Claude Martin
Advanced Mathematics for Practicing Engineers by Kurt Arbenz and Alfred Wohlhauser
Codes for Error-Control and Synchronization by Djimitri Wiggert
Machine Cryptography and Modern Cryptanalysis by Cipher A. Deavours and Louis Kruh
Microcomputer Tools for Communication Engineering by S.T. Li, J.C. Logan, J.W. Rockway, and D.W.S. Tam

OPTIMIZATION OF DIGITAL TRANSMISSION SYSTEMS

K. Tröndle and G. Söder

Artech House
Boston and London

Library of Congress Cataloging-in-Publication Data

Tröndle, K. (Karlheinz), 1939-
Digital transmission systems.

Translation of: Digitale Übertragunssysteme.
Bibliography: p.
Includes index.
1. Digital communications. I. Söder, Günter, 1946- . II. Title.
TK5103.7.T7513 1987 621.38'0413 87-1032
ISBN 0-89006-225-0

ARTECH HOUSE, INC.
685 Canton Street
Norwood, MA 02062

International Standard Book Number: 0-89006-225-0
Library of Congress Catalog Card Number: 87-1032

Translation from the German of *Digitale Übertragunssysteme*, copyright 1985 by Springer-Verlag, Heidelberg. Volume 14 of the series *Nachrichtentechnik* (Telecommunications) edited by H. Marko.

10 9 8 7 6 5 4 3 2 1

Preface

This book is a translation of the German textbook *Digitale Übertragungssysteme* which appeared in 1985 as a volume in the *Nachrichtentechnik* (Communications Technology) series published by Springer-Verlag, Berlin. The aim of the book is to describe baseband digital signal transmission and to make the reader familiar with modern transmission methods.

After the basic theories have been presented optimization methods and the resulting system designs are discussed. The subject matter, which is sometimes complicated, is clarified by numerous figures and examples, and the correlation between the important parameters is presented, together with their values.

This book is for the use of engineers and physicists who are involved in the design and development of digital transmission systems. It is designed to be of assistance to them in the solution of various types of theoretical and practical problems. However, because of its elementary approach, it is also suitable as a university text book and should enable students to familiarize themselves with the operation of and approach to the design of digital transmission systems.

Most of the processes described were investigated as part of a number of research projects undertaken at the Institute of Communications Technology, Technical University, Munich, and at the Military University, Munich, in which the authors have cooperated in the last few years. Therefore the fundamental results have been derived from a substantial body of material.

We thank the publisher of the *Nachrichtentechnik* series, Herr Professor Dr.-Ing. E.k. Hans Marko, for his constant interest in and extensive encouragement of our mutual research. As our tutor he acquainted us with the problems of digital transmission technology and gave us much valuable advice. Thanks are also due to our colleagues, assistants and undergraduate students who, through their research, have contributed to the basis of this book. In particular we wish to mention Herr Dr.-Ing. Udo Peters, who cooperated in the preparation of Chapter 6 and Section 7.4, and Herr Dr.-Ing. Erich Lutz, whose constructive criticism provided much stimulating encouragement. In addition we would like to thank Herr Dr.-Ing. Hans Dirndorfer (Section 8.1), Herr Dr.-Ing. Alfred Irber (Section 8.5) and Herr Dr.-Ing. Michael Dippold (Chapter 9) for allowing us to use the results of their research. Particular thanks are also due to Herr Dr.-Ing. Manfred Bertelsmeier who advised us with regard to English wording of the book and made many valuable comments.

Unfortunately, for technical reasons it was not possible to change the symbols and notation used in the German edition to be compatible with the English translation. Therefore a glossary, in which both the German and the English terminology are specified, is provided to simplify understanding of the abbreviations, superscripts and subscripts used.

Munich, October 1986

Gunter Söder
Karlheinz Tröndle

Contents

Chapter 1

Introduction

A digital transmission system became available as early as the middle of the 19th century through the introduction of telegraphy. The rapidly increasing demand for communication links led to a systematic examination of the physical and technical possibilities of this technology. New transmission media and continually advancing technology provided the basis for the emergence of new systems and larger networks, and allowed dynamic development of digital communication techniques—a process which continues unchanged to the present day. The introduction or planning of new services by the postal authorities (e.g. teletext, facsimile transmission, magnetic card systems, remote monitoring and control systems) and the effort towards automating the modern office (in which the storage, distribution and processing of information are of primary concern) have in recent years led directly to the wide-ranging consideration of reconfiguration of communication installations and systems.

The postal authorities are therefore aiming at the provision of an Integrated Services Digital Network (ISDN) which will give the subscriber extensive freedom of choice of transmitted information. In addition to voice transmission, computer system data transfer and transmission of photographs and documents in the automated office, consideration is now being given to the transmission of moving images. The channel capacity requirements for this are significantly higher. By suitable organization of the transmitted information (in packets, for example, and by employing multiplexing techniques) the user will be given access to a modern communications service which is distinguished by its multiplicity of applications and its flexibility. The conventional division of devices and techniques into modules for signal conditioning, transmission and switching has to be questioned, however, because the performance of all these three functions can in part be undertaken by single units. Also, very sophisticated techniques have become practicable as a consequence of the technological advances which have been made during the past few decades. These appear, for example, in the emergence of new digital communication systems and local area networks (LANs).

The aim of this book is to introduce the reader to the basic theory of digital transmission technology, and to explain design and optimization. The main topic is digital baseband transmission, which also illustrates the basics of modulated-carrier digital signal transmission. Direct application of the baseband system occurs in transmission by coaxial cable and symmetric lines, but digital transmission by optical fibers can also, in system terms, be regarded as equivalent to baseband transmission.

After the individual system components have been described and the error probabilities of binary and multilevel systems have been calculated, channel encoding methods are treated. The function of these is to match the source signal to the transmission channel and, in addition, to enable lines and systems to be monitored.

The characteristics of systems with an optimum receiver (Viterbi detector) or a receiver with decision feedback equalization are discussed exhaustively, and the limits of digital transmission technology are explained. In particular, systems without intersymbol interference (Nyquist systems) are examined in detail because of their good transmission characteristics. As has been shown in some recent papers, a reduction of error probability or of unit complexity can be achieved by the amalgamation of coding and modulation ("codulation") into a single unit attached to the channel. The characteristic features of an appropriate theory aimed at the common optimization of encoding, signal shaping and recovery are explained. Finally, optical transmission systems are treated; these differ from electrical systems because of the signal-dependent, and thus also time-dependent, noise.

At the start of each chapter the contents are listed and the appropriate assumptions are stated. The individual chapters are largely self-contained, so that the reader seeking specific information can refer to single chapters.

Chapter 2

Components of a Digital Transmission System

Contents In this chapter the individual components of a digital transmission system—source, transmitter, transmission channel, equalizer, detector and timing recovery—are described, and models for their signal transmitting behavior are given.

Assumptions Only systems with stationary noise over which the digital signal is transmitted unmodulated (baseband system) are examined. The characteristics of optical digital systems are dealt with in Chapter 9. Further, it is assumed that the digital source is ergodic and that the transmission is uncoded. The detector is assumed to be an optimum memoryless threshold device.

2.1 Principle of digital transmission

Figure 2.1 shows the block diagram of a digital transmission system in its simplest form. The digital source emits the **source symbol sequence** $\langle q_\nu \rangle$ as a signal which is physically represented by the source signal $q(t)$ and is to be transmitted to a distant signal sink. An error-free transmission results if the **sink symbol sequence** $\langle v_\nu \rangle$ is completely identical with the source symbol sequence, and this is only possible with an ideal transmission link. In real channels **symbol errors** ($v_\nu \neq q_\nu$) are unavoidable because of linear and nonlinear distortions as well as the intruding disturbances. The mean probability of occurrence of such symbol errors is designated the **mean symbol error probability**. This measure is the most important criterion for the assessment of digital transmission systems, and is derived in Chapter 3.

First, however, a description of the digital transmission system and its signals will be given. This system consists of the **transmitter**, the **transmission channel** (medium) and the **receiver** which contains the **equalizer**, the **detector** and the **timing recovery device** (TRD).

Figure 2.2 shows the signal time histories at various points in the block diagram for a typical example. The discrete-value time-sampled **source signal** $q(t)$ can only

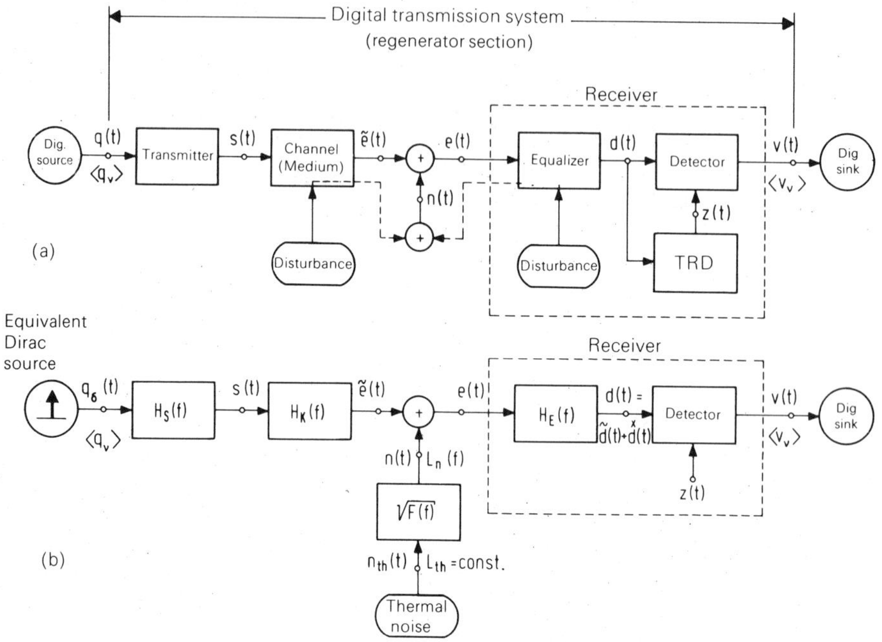

Fig. 2.1 (a) Block diagram and (b) equivalent block diagram of a digital baseband transmission system.

take on the instantaneous values 0 V and 1 V, and represents the specified binary source sequence $\langle q_v \rangle$ (Fig. 2.2(a)). The transmitter produces from this the **transmitted signal** $s(t)$ shown in Fig. 2.2(b) with the instantaneous values ± 3 V.

In transmission through the medium, this digital signal is distorted and has interference superimposed upon it. The final signal in the medium is the **received signal** $e(t)$, which consists of the useful received signal $\tilde{e}(t)$ and the (received) interference $n(t)$. The time history of the received signal $\tilde{e}(t)$ in Fig. 2.2(c) shows that the distortion and interference imposed by the medium are so strong that the transmitted signal can no longer be recognized.

The purpose of the equalizer is to remove, as far as is necessary and possible, the amplitude and phase distortion imposed by the transmission medium, and at the same time to limit the interference and noise power at the detector input. Full equalization to the original signal $s(t)$ is not possible because of this necessary noise power limitation.

The equalizer output signal is the **detector input signal** $d(t)$ (Fig. 2.2(d)), which for brevity will also be referred to as the **detection signal**. This consists of the useful detector input signal (useful detection signal) $\tilde{d}(t)$ and the noise component $\overset{\times}{d}(t)$ and, in contrast with the received signal, permits the transmitted signal to be recognized again owing to the equalization (amplification of the components with higher signal frequency).

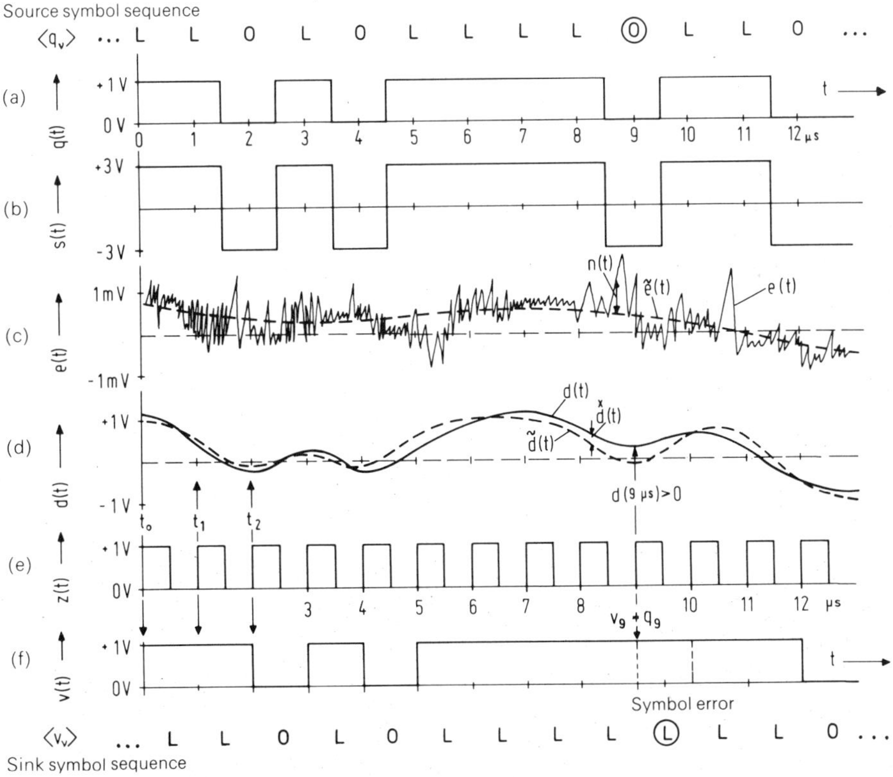

Fig. 2.2 Time histories for the digital baseband transmission.

Finally the detector input signal $d(t)$ is amplitude and time regenerated with the aid of the detector. To achieve this, the detector needs timing information, which is supplied to it by the TRD in the form of the **clock signal** $z(t)$ (Fig. 2.2(e)). At all detection time instants t_ν, which are denoted in this example by the leading edge of the clock pulses, the detector determines the instantaneous value of its input signal. Depending on whether this value is smaller or greater than zero, the output voltage is set to 0 V or +1 V accordingly and is kept constant until the start of the next timing pulse (Fig. 2.2(f)).

The output signal from the detector is the **sink signal** $v(t)$ which, analogous to the source signal $q(t)$, is represented by the sink symbol sequence $\langle v_\nu \rangle$. As long as the distortions and interferences do not exceed a certain limiting value, we have $\langle v_\nu \rangle = \langle q_\nu \rangle$ and the sink signal $v(t)$ is an identical but delayed version of the source signal $q(t)$ owing to the transit time through the transmission system. In this example a symbol error can be seen at $t = 9\ \mu s$; the source symbol $q_9 = \mathrm{O}$ has been corrupted in transmission to give the corresponding sink symbol $v_9 = \mathrm{L}$. The cause of the symbol error is that the distortion and interference imposed by the transmission medium has raised the detection signal above the 0 V threshold.

2.2 Digital source

In the following, we consider a digital communication source which can emit any of M different symbols $q_1, \ldots, q_M$. M is called the **source symbol set size** or the **level number**. $\{q_\mu\}$ is the **source symbol set**, where the subscript $\mu = 1, \ldots, M$. The appropriate level number M distinguishes between the **binary source** ($M = 2$) and **multilevel sources** ($M > 2$) (Tabie 2.1).

Table 2.1 Relationship between source symbols and amplitude coefficients

Source	*Number of levels* M	*Source symbol set* $\{q_\mu\}$	*Amplitude coefficients* a_μ *Unipolar*	*Bipolar*
Binary	2	{O; L}	$\{0; 1\}$	$\{-1; +1\}$
Ternary	3	$\{-; \mathrm{O}; +\}$	$\{0; \frac{1}{2}; 1\}$	$\{-1; 0; +1\}$
Quaternary	4	$\{\ominus; -; +; \oplus\}$	$\{0; \frac{1}{3}; \frac{2}{3}; 1\}$	$\{-1; -\frac{1}{3}; +\frac{1}{3}; +1\}$

An ensemble of selections made from the digital source symbol set at equal time intervals T is the temporal **source symbol sequence** $\langle q_\nu \rangle$ which represents the source message. In the following, the subscript ν always denotes a temporal sequence $\langle\ \rangle$, where ν runs from $-\infty$ to $+\infty$. The length of time which is available for the transmission of a source symbol is defined as the **symbol duration** T. $1/T$ is the **symbol rate** and has the unit baud which is defined as symbols per second.

To describe the statistical characteristics of the digital source, the **symbol probability** is used:

$$p_{\nu\mu} = P(q_\nu = q_\mu) \tag{2.1}$$

$p_{\nu\mu}$ is the probability that the νth symbol of the sequence $\langle q_\nu \rangle$ is equal to the μth symbol of the source symbol set $\{q_\mu\}$. If statistical relationships exist between consecutive symbols of the sequence, the occurrence of the symbol q_ν will be dependent on the preceding symbols $q_{\nu-1}, q_{\nu-2}, \ldots$, and accordingly the probability $p_{\nu\mu}$ will depend on ν.

If the symbol $q_\nu = q_\mu$, its **information content** is defined as follows:

$$I_\nu = \log_2\left(\frac{1}{p_{\nu\mu}}\right) \tag{2.2}$$

where $\log_2$ denotes logarithm to the base 2. In numerical analysis the unit "bit" (binary digit) is used. The smaller the probability is, the greater is the information content of the symbol. For example, if in a given text the letter b occurs with the probability 1/32, its information content amounts to 5 bits.

By averaging all the terms of the infinite sequence $\langle q_\nu \rangle$, we obtain the **entropy** (units, bits) of the digital source [1.8]:

$$H_q = \overline{I_\nu} = \lim_{N \to \infty} \frac{1}{2N + 1} \sum_{\nu = -N}^{N} I_\nu \tag{2.3}$$

The entropy H_q thus corresponds to the mean information content of a symbol. If the source symbols are statistically independent of each other, the symbol probability $p_{\nu\mu}$ is independent of ν. With **statistically independent symbol probability** p_μ we have

$$p_{\nu\mu} = P(q_\nu = q_\mu) = p_\mu \qquad \text{for all } \nu \tag{2.4}$$

and in this special case the entropy becomes

$$H_q = \sum_{\mu = 1}^{M} p_\mu \log_2\left(\frac{1}{p_\mu}\right) \tag{2.5}$$

For example, for a binary source ($M = 2$) with statistically independent source symbols with the symbol probabilities $p_1 = 0.25$ and $p_2 = 1 - p_1 = 0.75$, the entropy has the value $H_q = 0.81$ bit.

H_q is a maximum if the M symbol probabilities $p_\mu = 1/M$ are of equal magnitude. The maximum value $H_{q,\max}$ of the entropy is the **decision content** of the source:

$$H_{q,\max} = \log_2 M \tag{2.6}$$

For $M = 2$, $H_{q,\max} = 1$ bit. With a redundant source the entropy is smaller than the decision content. The **relative redundancy** of the source

$$r_q = \frac{H_{q,\max} - H_q}{H_{q,\max}} \qquad 0 \leqslant r_q \leqslant 1 \tag{2.7}$$

is defined for quantitative determination of the statistical dependence. A digital source is termed nonredundant ($r_q = 0$) if the source symbols are statistically independent of each other and the M possible source symbols occur with the same probability $p_\mu = 1/M$.

The entropy H_q is the mean information content of a source symbol. However, in the assessment of communication systems, the period of time T required for the transmission of a symbol is also of great significance. Thus the **information rate** (units, bits s^{-1}) is defined as

$$\phi = \frac{H_q}{T} \tag{2.8}$$

which defines the mean information content transmitted per unit time. A valid analog for the maximum information rate, which is called the (equivalent) **bit rate**, is

$$R = \frac{H_{q,\max}}{T} = \frac{\log_2 M}{T} \tag{2.9}$$

The term "maximum information rate" stems from information theory. In the following the designation "bit rate", which is more usual in communication technology, is used.

With a redundant source the information rate ϕ is a factor $1 - r_q$ smaller than the bit rate R, which is independent of r_q. When the source is binary ($\log_2 M = 1$), the bit rate R (with units bits per second) and the symbol rate $1/T$ (with the unit baud) have the same numerical value. With a multilevel source the bit rate defines how many symbols an equivalent binary source has to emit in unit time so that it possesses the same decision content as the M-level source under consideration.

2.3 Transmitter

The function of the transmitter is to produce from the source signal $q(t)$ a transmittable signal that completely represents the source information and is matched to the characteristics of the transmission channel, the interference and the receiving equipment. In addition, it supplies the necessary transmitter power.

2.3.1 Transmitter signal parameters

Each of the M possible source symbols q_μ ($\mu = 1, \ldots, M$) is represented by a **transmitter pulse** $a_\mu g_s(t)$ which differs from the others only in respect of the pulse amplitude as the shape $g_s(t)$ of all the transmitted pulses is the same. $g_s(t)$ is defined as the **basic transmitter pulse** which is to be optimized according to the spectral characteristics of the channel and the receiver.

The dimensionless **amplitude coefficients** a_μ can be assigned, in principle arbitrarily but unambiguously, to the source symbols q_μ. **Unipolar** ($0 \leqslant a_\mu \leqslant 1$) and **bipolar** ($-1 \leqslant a_\mu \leqslant +1$) amplitude coefficients or transmitter signals are distinguished by means of the range of values of a_μ. In general it is expedient to select equal step differences between adjacent amplitude values. In that case the possible amplitude coefficients ($\mu = 1, \ldots, M$) are as follows:

$$a_\mu = \begin{cases} \dfrac{\mu - 1}{M - 1} & \text{unipolar signals} \\[2ex] \dfrac{2\mu - M - 1}{M - 1} & \text{bipolar signals} \end{cases} \tag{2.10}$$

The unipolar and bipolar amplitude coefficients for the level numbers $M = 2$, $M = 3$ and $M = 4$ are given in Table 2.1.

If we consider further a time-independent **base component** (dc bias) s_0, which is necessary with optical transmitters for example, the transmitter signal can be written

$$s(t) = s_0 + \sum_{\nu=-\infty}^{+\infty} a_\nu \, g_s(t - \nu T) \tag{2.11}$$

Example Figure 2.3 shows a unipolar binary signal ($M = 2$) and a bipolar quaternary signal ($M = 4$) in which the basic transmitter pulse is of the $\cos^2$ form. The maximum value s_{max} and the minimum value s_{min} of the transmitter signal are also superimposed on this figure. The difference defines the **modulation range**:

$$\Delta s = s_{max} - s_{min} \tag{2.12}$$

For the same basic transmitter pulse, the modulation range of the bipolar signal is twice that of the unipolar signal.

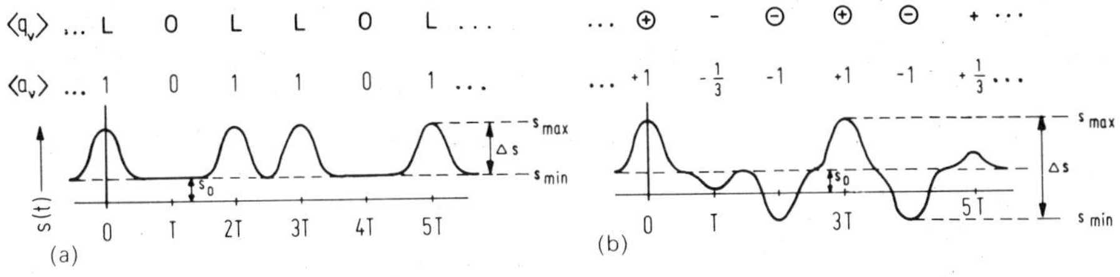

Fig. 2.3 Source symbol sequence $\langle q_\nu \rangle$ and transmitter signal $s(t)$ with the $\cos^2$ form of the basic transmitter pulse; (a) binary unipolar; (b) quaternary bipolar.

The following parameters are used to describe the basic transmitter pulse (Fig. 2.4(a)).

(a) The **transmitter pulse amplitude** $\hat{g}_s$ is the maximum value of the basic transmitter pulse.

(b) The **equivalent transmitter pulse duration** Δt_s is defined as the width of a rectangular pulse with the same amplitude and area as the basic transmitter pulse:

$$\Delta t_s = \frac{1}{\hat{g}_s} \int_{-\infty}^{+\infty} g_s(t) \, \mathrm{d}t \tag{2.13}$$

(c) The **absolute transmitter pulse duration** T_s is the length of time for which the basic transmitter pulse deviates from zero.

In all cases $\Delta t_s \leqslant T_s$; the terms are equal in the special case of the rectangular transmitter pulse. The basic transmitter pulse can be further distinguished as RZ or NRZ (see Fig. 2.6 below). In an **RZ pulse** (return-to-zero pulse) the absolute transmitter pulse duration T_s is smaller than the symbol duration T. In contrast, $T_s \geqslant T$ for an **NRZ pulse** (non-return-to-zero pulse).

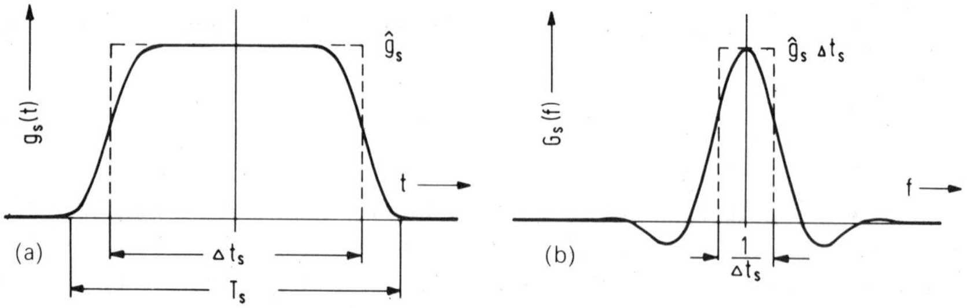

Fig. 2.4 (a) The basic transmitter pulse with cosine form leading and trailing edges and (b) the corresponding spectrum.

2.3.2 Equivalent block diagram for the transmitter

The basic transmitter pulse spectrum is the Fourier transform of $g_s(t)$ (Fig. 2.4(b)):

$$G_s(f) = \int_{-\infty}^{+\infty} g_s(t) \exp(-\mathrm{j}2\pi ft)\,\mathrm{d}t \quad \text{abbreviated to } G_s(f) \bullet\!\!-\!\!\!-\!\!\circ\, g_s(t) \tag{2.14}$$

In Chapters 7–9 it proves to be appropriate for system optimization to introduce the transmitter pulse-shaper frequency response (**transmitter frequency response**) as the normalized transmitter pulse spectrum:

$$H_S(f) = \frac{G_s(f)}{\hat{g}_s T} \tag{2.15}$$

If a Dirac impulse $\hat{g}_s T \delta(t)$ is applied to the input of the transmitter pulse shaper, the output pulse is $g_s(t)$ and its spectrum is $G_s(f)$. $\delta(x)$ is the Dirac (unit impulse) function as defined in the Appendix, Table A1.

Without loss of generality, in the following the maximum of the basic transmitter pulse is assumed to occur at time $t = 0$, i.e. $\hat{g}_s = g_s(0)$. With the relationship

$$g_s(0) = \int_{-\infty}^{+\infty} G_s(f)\,\mathrm{d}f$$

for the transmitter frequency response, which is obtained from system theory, we obtain the condition

$$T \int_{-\infty}^{+\infty} H_S(f)\,\mathrm{d}f = 1 \tag{2.16}$$

In the block diagram in Fig. 2.1 the transmitter can be replaced by the transmitter frequency response $H_S(f)$ if at the same time the equivalent Dirac source with the output signal

$$q_\delta(t) = \sum_{\nu=-\infty}^{+\infty} a_\nu \hat{g}_s T\delta(t - \nu T) \tag{2.17}$$

is used instead of the digital source. The transmitter signal $s(t)$ is identical in the two cases illustrated in Figs 2.1(a) and 2.1(b). However, the second representation offers the advantage that the transmitter can be described by the linear network $H_S(f)$. In general this is not possible for the block diagram shown in Fig. 2.1(a).

2.3.3 Spectral characteristics of the transmitter signal

As $s(t)$ represents a stochastic signal, the **power spectral density** $L_s(f)$, which is often abbreviated to **power spectrum**, is used to define its spectral characteristics. This quantity can be calculated from the **autocorrelation function** (ACF) of the transmitter signal, for which

$$l_s(\tau) = \lim_{T_0\to\infty} \frac{1}{2T_0} \int_{-T_0}^{T_0} s(t)s(t+\tau)\,\mathrm{d}t \tag{2.18}$$

According to the Wiener–Khintchine theorem the power spectrum $L_s(f)$ is the Fourier transform of the ACF:

$$L_s(f) = \int_{-\infty}^{+\infty} l_s(\tau)\exp(-\mathrm{j}2\pi f\tau)\,\mathrm{d}\tau \tag{2.19}$$

These definitions apply to an ergodic transmitter signal such that all statistical characteristics of the stochastic process $\{s(t)\}$ can be determined from a single "representative" function (sample) $s(t)$ of the process. In many applications this requirement is not strictly satisfied. The signals of these examples, however, can in general be considered as cyclo stationary processes, for which the above definition of the ACF is equally valid [1.21].

The ACF has the dimensions of power normalized to unit resistance (e.g. V^2 or A^2) and is a measure of the linear statistical dependence of the instantaneous value of the power-limited transmitter signal $s(t)$. Thus we form the time-averaged value of all products of the signal values, which are separated by time τ. By placing $\tau = 0$ in the definition of the ACF, we obtain the **mean transmitter signal power**:

$$S_s = \lim_{T_0\to\infty} \frac{1}{2T_0} \int_{-T_0}^{T_0} s^2(t)\,\mathrm{d}t = l_s(0) = \int_{-\infty}^{+\infty} L_s(f)\,\mathrm{d}f \tag{2.20}$$

To calculate the ACF, the discrete ACF $l_a(\lambda)$ of the amplitude coefficients and the energy ACF $l^{\bullet}_{g_s}(\tau)$ of the basic transmitter pulse are required. The **discrete autocorrelation function of the amplitude coefficients** is defined as follows:

$$l_a(\lambda) = \overline{a_\nu a_{\nu+\lambda}} = \lim_{N\to\infty} \frac{1}{2N+1} \sum_{\nu=-N}^{N} a_\nu a_{\nu+\lambda} \tag{2.21}$$

$l_a(\lambda)$ is dimensionless and is an expression of the linear statistical relationship between the amplitude coefficients a_ν and $a_{\nu+\lambda}$ or the corresponding source symbols q_ν and $q_{\nu+\lambda}$. For statistically independent source symbols (amplitude coefficients) it is a consequence of this definition that, with the mean value (first moment) $\overline{a_\nu}$ and the mean squared value (second moment) $\overline{a_\nu^2}$,

$$l_a(\lambda) = \begin{cases} \overline{a_\nu^2} & \text{for } \lambda = 0 \\ \overline{a_\nu}^2 & \text{for } \lambda \neq 0 \end{cases} \tag{2.22}$$

The following applies for the **energy ACF** of the basic transmitter pulse $g_s(t)$:

$$\dot{l}_{g_s}(\tau) = \int_{-\infty}^{+\infty} g_s(t) g_s(t+\tau)\,\mathrm{d}t \tag{2.23}$$

$\dot{l}_{g_s}(\tau)$ has the dimensions of normalized energy (e.g. $V^2 s$) and can be calculated from the **energy spectrum**:

$$\dot{L}_{g_s}(f) = |G_s(f)|^2 \tag{2.24}$$

The dot over the symbols for the energy ACF and the energy spectrum is to emphasize the difference from the corresponding parameters of the power-limited signals.

Analogous to eqn (2.19) is the energy ACF $\dot{l}_{g_s}(\tau)$, which is the inverse Fourier transform of the energy spectrum:

$$\dot{l}_{g_s}(\tau) = \int_{-\infty}^{+\infty} \dot{L}_{g_s}(f) \exp(\mathrm{j}2\pi f\tau)\,\mathrm{d}f \tag{2.25}$$

Figure 2.5 shows the relationship between the pulse, the spectrum, the energy ACF and the energy spectrum for the example of a rectangular pulse.

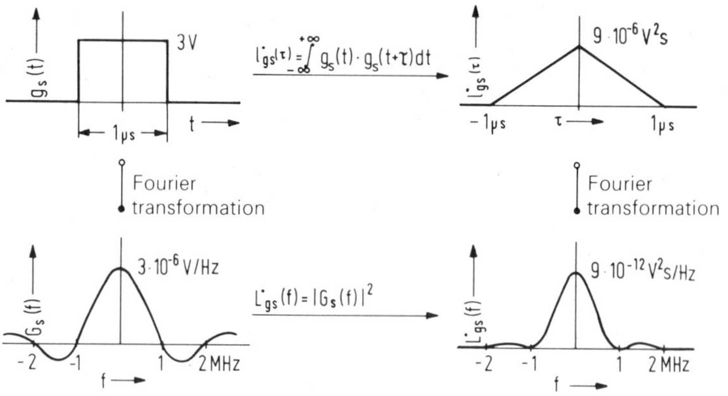

Fig. 2.5 Rectangular basic transmitter pulse and the spectrum, energy ACF and energy spectrum ($\hat{g}_s = 3$ V; $T_s = 1$ μs).

From eqn (2.11) and definitions (2.18), (2.21) and (2.23) we obtain for the ACF of the stochastic transmitter signal $s(t)$ [1.12]

$$l_s(\tau) = s_0{}^2 + 2\overline{a_v} s_0 \hat{g}_s \frac{\Delta t_s}{T} + \sum_{\lambda=-\infty}^{+\infty} \frac{1}{T} l_a(\lambda) \dot{l}_{g_s}(\tau - \lambda T) \tag{2.26}$$

If the base component s_0 is zero, the first two terms are both zero and therefore

$$l_s(\tau) = \sum_{\lambda=-\infty}^{+\infty} \frac{1}{T} l_a(\lambda) \dot{l}_{g_s}(\tau - \lambda T) \tag{2.27}$$

The ACF $l_s(\tau)$ of the transmitter signal is thus determined by two parameters, i.e. by the basic transmitter pulse, whose effect is defined by its energy ACF, and the statistical dependence of the transmitted symbol sequence, which is identified by the discrete ACF $l_a(\lambda)$.

By Fourier transformation of the ACF $l_s(\tau)$, we obtain the power spectrum for the condition $s_0 = 0$:

$$L_s(f) = \sum_{\lambda=-\infty}^{+\infty} \frac{1}{T} l_a(\lambda) |G_s(f)|^2 \cos(2\pi f \lambda T) \tag{2.28}$$

Proof From eqn (2.27) and the Wiener–Khintchine theorem,

$$L_s(f) = \int_{-\infty}^{+\infty} \sum_{\lambda=-\infty}^{+\infty} \frac{1}{T} l_a(\lambda) \dot{l}_{g_s}(\tau - \lambda T) \exp(-j2\pi f \tau) \, d\tau$$

By the substitution $\tau' = \tau - \lambda T$ and interchange of integration and summation, it follows that

$$L_s(f) = \sum_{\lambda=-\infty}^{+\infty} \frac{1}{T} l_a(\lambda) \exp(-j2\pi f \lambda T) \int_{-\infty}^{+\infty} \dot{l}_{g_s}(\tau') \exp(-j2\pi f \tau') \, d\tau'$$

In this equation the integral is equal to $|G_s(f)|^2$ (see eqn (2.24) and eqn (2.25)). Considering in addition to this the symmetry $l_a(-\lambda) = l_a(\lambda)$, we obtain the result shown in eqn (2.28). □

Now the **power spectrum of the amplitude coefficients** is defined as follows:

$$L_a(f) = \sum_{\lambda=-\infty}^{+\infty} l_a(\lambda) \exp(-j2\pi f \lambda T) = \sum_{\lambda=-\infty}^{+\infty} l_a(\lambda) \cos(2\pi f \lambda T) \tag{2.29}$$

Therefore the power spectrum of the transmitter signal can be written (assuming that $s_0 = 0$)

$$L_s(f) = \frac{1}{T} L_a(f) |G_s(f)|^2 \tag{2.30}$$

$L_s(f)$ can thus be represented as the product of two functions. $L_a(f)$ is dimensionless and describes the spectral shaping of the transmitter signal according to the statistical relationships of the source. When the source is nonredundant

$L_a(f)$ = constant, i.e. it is frequency independent. $|G_s(f)|^2$ takes account of the spectral shaping by the basic transmitter pulse $g_s(t)$. The narrower the transmitter pulses are, the wider is the power spectrum $L_s(f)$.

2.3.4 Spectral characteristics of the non-redundant transmitter signals

With a nonredundant source the discrete ACF of the amplitude coefficients a_ν is fully defined by the mean linear and mean squared values (see eqn (2.22)). With the possible amplitude coefficients a_μ from eqn (2.10) we obtain for the mean value

$$\overline{a_\nu} = \frac{1}{M}\sum_{\mu=1}^{M} a_\mu = \begin{cases} \frac{1}{2} & \text{for unipolar signals} \\ 0 & \text{for bipolar signals} \end{cases} \tag{2.31}$$

and by analogy for the mean squared value

$$\overline{a_\nu^2} = \frac{1}{M}\sum_{\mu=1}^{M} a_\mu^2 = \begin{cases} \dfrac{2M-1}{6(M-1)} & \text{for unipolar signals} \\ \dfrac{M+1}{3(M-1)} & \text{for bipolar signals} \end{cases} \tag{2.32}$$

On substituting the values in eqn (2.26) or (2.28), the equations in Table 2.2 are produced.

Table 2.2 Autocorrelation function, power spectrum and mean power of a nonredundant transmitter signal ($s_0 = 0$)

	Unipolar	*Bipolar*
$l_s(\tau)$	$\dfrac{M+1}{12T(M-1)}\dot{l}_{g_s}(\tau) + \dfrac{1}{4T}\sum_{\lambda=-\infty}^{+\infty} \dot{l}_{g_s}(\tau-\lambda T)$	$\dfrac{M+1}{3T(M-1)}\dot{l}_{g_s}(\tau)$
$L_s(f)$	$\dfrac{M+1}{12T(M-1)}\lvert G_s(f)\rvert^2 + \dfrac{1}{4T^2}\sum_{\lambda=-\infty}^{+\infty} \lvert G_s(f)\rvert^2 \delta\left(f-\dfrac{\lambda}{T}\right)$	$\dfrac{M+1}{3T(M-1)}\lvert G_s(f)\rvert^2$
S_s	$\dfrac{2M-1}{6T(M-1)}\dot{l}_{g_s}(0) + \dfrac{1}{2T}\sum_{\lambda=1}^{\infty} \dot{l}_{g_s}(\lambda T)$	$\dfrac{M+1}{3T(M-1)}\dot{l}_{g_s}(0)$

Example A rectangular transmitter pulse of duration T_s produces a triangular energy ACF (see eqn (2.23) and Fig. 2.5):

$$\dot{l}_{g_s}(\tau) = \hat{g}_s^2 T_s \wedge\left(\frac{\tau}{T_s}\right) \tag{2.33}$$

($\wedge(x)$ is given in Table A1). Thus from eqn (2.25) we obtain for the energy spectrum

$$\dot{L}_{g_s}(f) = |G_s(f)|^2 = \hat{g}_s^2 T_s^2 \,\mathrm{si}^2(\pi f T_s) \tag{2.34}$$

(si(x) is given in Table A1). Figure 2.6 shows a part of the signal time history, the ACF and the power spectrum for unipolar and bipolar nonredundant RZ and NRZ binary signals. The power spectrum of the unipolar RZ signal has a continuous component as well as discrete lines at all multiples of the symbol rate $1/T$ which can be used for time recovery. The weightings of the discrete lines are proportional to the continuous component of the power spectrum. The power spectrum of the unipolar NRZ signal, however, possesses in addition to the continuous component only one discrete line at frequency $f = 0$, i.e. a direct (dc) component.

A triangular ACF and continuous power spectrum of finite density always occur for a bipolar nonredundant rectangular signal. Here discrete lines can only be achieved by nonlinear methods, e.g. full-wave rectification (see Section 2.7).

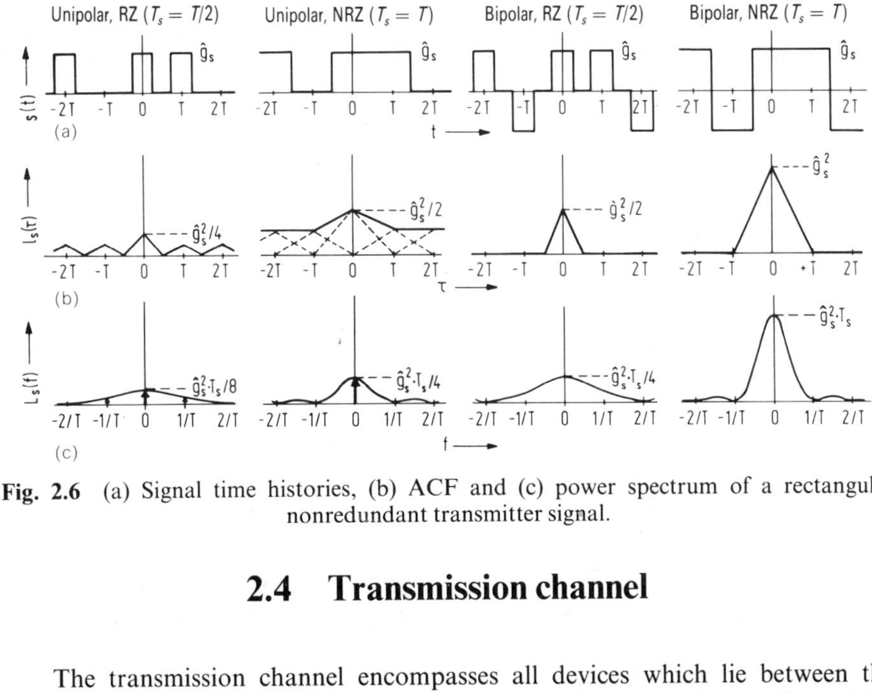

Fig. 2.6 (a) Signal time histories, (b) ACF and (c) power spectrum of a rectangular nonredundant transmitter signal.

2.4 Transmission channel

The transmission channel encompasses all devices which lie between the transmitter and the receiver. Its main component is the transmission medium which may be, for example, a symmetric pair cable, a coaxial cable, an optical waveguide or a radio relay section. In addition the transmission channel contains various items which are necessary for operational reasons (e.g. the power supply, lightning protection and fault-location devices).

In the most general case the transmission characteristics of the channel are time, frequency, amplitude and temperature dependent. Therefore the transmitted signal is distorted during transmission via the channel. Additive disturbances (e.g. thermal noise disturbances, pulse disturbances and cross-talk interference) are also imposed on the useful signal [2.2, 2.26].

2.4.1 Channel frequency response and its parameters

It can be assumed that the distortions are primarily caused by the frequency dependence of the transmission channel (linear distortions), so that its transmission characteristics are fully described by the **channel frequency response**:

$$H_K(f) = \exp\{-a_K(f)\} \exp\{-jb_K(f)\} \tag{2.35}$$

$a_K(f)$ is the **attenuation constant** and has the (pseudo-)units nepers (Np); $b_K(f)$ is the **phase constant.**

Many transmission media possess a low-pass characteristic, i.e. the amplitude response $|H_K(f)|$ decreases with increasing frequency (Fig. 2.7(a)). If the channel contains additional coupling devices, however, e.g. transformers or capacitors, then direct signal transmission is not possible*. These channels are referred to as **band-pass channels** (Fig. 2.7(c)). It is shown in Section 8.1.4 that the distorting influence of the lower band limit can be ignored if direct signal restoration techniques are employed at the receiver. Therefore a low-pass channel is assumed in the following.

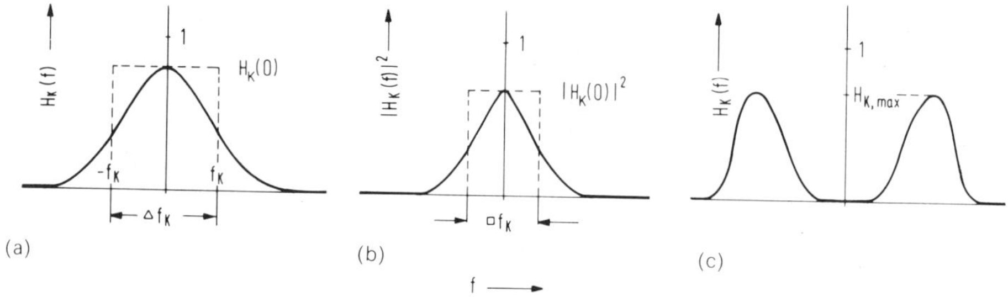

Fig. 2.7 (a) Frequency response and (b) squared absolute value of a low-pass channel; (c) frequency response of a band-pass channel.

An important parameter of the low-pass channel is the **direct signal transmission factor** $H_K(0)$ which is related to the **direct signal attenuation value** a_0 as follows:

$$|H_K(0)| = \exp\{-a_0(\text{Np})\} = 10^{-a_0/20\,\text{dB}} \tag{2.36}$$

The transmission bandwidth of a low-pass channel is described according to the application by the following parameters.

(a) The **system theoretic bandwidth** (Fig. 2.7(a)):

$$\Delta f_K = \frac{1}{H_K(0)} \int_{-\infty}^{+\infty} H_K(f)\,df \tag{2.37}$$

*In the following, by analogy with direct current or direct voltage, the term "direct signal" will be used to describe a signal having a spectral component only at zero frequency.

(b) The **channel cut-off frequency** (Fig. 2.7(a)):

$$f_K = \frac{1}{2}\Delta f_K = \frac{1}{H_K(0)}\int_0^{+\infty} H_K(f)\,df \tag{2.38}$$

(c) The **(equivalent) noise bandwidth** (Fig. 2.7(b)):

$$\Box f_K = \frac{1}{|H_K(0)|^2}\int_{-\infty}^{+\infty} |H_K(f)|^2\,df \tag{2.39}$$

For band-pass systems, $H_K(0)$ in eqns (2.37)–(2.39) has to be substituted by a suitable reference value $H_K(f_0)$, e.g. by the maximum value $H_{K,max}(f)$.

2.4.2 Pulse response and useful received signal

The **pulse response** $h_K(t)$ is the inverse Fourier transform of the channel frequency response and has the unit reciprocal second:

$$h_K(t) = \int_{-\infty}^{+\infty} H_K(f)\exp(j2\pi ft)\,df \tag{2.40}$$

If the transmitter signal $s(t)$ is applied to the channel input then, according to convolution theory [1.17], the output signal under noise-free assumptions (the useful received signal) is

$$\tilde{e}(t) = s(t) * h_K(t) = \int_{-\infty}^{+\infty} s(\tau)h_K(t-\tau)\,d\tau \tag{2.41}$$

The **basic received pulse** $g_e(t)$ is the transmission channel's response to a single basic transmitter pulse $g_s(t)$ at its input and is calculated as the convolution product

$$g_e(t) = g_s(t) * h_K(t) = \int_{-\infty}^{+\infty} g_s(\tau)h_K(t-\tau)\,d\tau \tag{2.42}$$

Consequently we obtain using eqn (2.11) the useful received signal

$$\tilde{e}(t) = H_K(0)s_0 + \sum_{\nu=-\infty}^{+\infty} a_\nu g_e(t-\nu T) \tag{2.43}$$

The power spectrum is generally used to describe the spectral characteristics of the useful received signal. This can be calculated from the transmitter power spectrum and the channel frequency response (see ref. 1.21 and eqn (2.30)):

$$L_{\tilde{e}}(f) = L_s(f)|H_K(f)|^2 = \frac{1}{T}L_a(f)|G_s(f)|^2|H_K(f)|^2 \tag{2.44}$$

The first part of this equation is generally valid, but the second part is only valid under the precondition $s_0 = 0$. The **useful received power** is calculated by analogy with eqn (2.20):

$$S_e = \overline{\tilde{e}^2(t)} = \int_{-\infty}^{+\infty} L_{\tilde{e}}(f)\,\mathrm{d}f = \frac{1}{T}\int_{-\infty}^{+\infty} L_a(f)|G_s(f)|^2|H_K(f)|^2\,\mathrm{d}f \qquad (2.45)$$

It can be seen from this relationship that the useful received power S_e is dependent not only on the channel frequency response but also on the basic transmitter pulse $g_s(t) \circ\!\!-\!\!\bullet\, G_s(f)$ and on the characteristics of the source, i.e. on $L_a(f)$.

2.4.3 Transmission channel disturbances

During transmission over the channel, additive disturbances are superimposed on the useful signal. These are caused by a multitude of noise sources, which may be distributed along the whole of the transmitter–receiver link. In a linear channel it is possible to replace this large number of noise sources with a single source which emits a noise signal $n(t)$ and is effective at the channel output (Fig. 2.1(a)). This location is the most suitable for the consideration of noise because the useful signal level is at its lowest, and consequently the noise is most strongly noticeable.

The noise signal $n(t)$ cannot usually be specified explicitly. Therefore it has to be described by its statistical characteristics, i.e. by its **probability density function** (PDF) $f_n(n)$ and its **noise power spectrum** $L_n(f)$. For the theoretical examination of digital transmission systems it can often be assumed that the disturbances approximate to a Gaussian distribution. For the PDF of a **Gaussian noise signal** with **root mean square** (rms) (standard deviation) σ_n

$$f_n(n) = \frac{1}{(2\pi)^{1/2}\sigma_n}\exp\left\{-\left(\frac{n^2}{2\sigma_n{}^2}\right)\right\} \qquad (2.46)$$

A fundamental disturbance present in every communication system is **thermal noise**, since every impedance R_{th} at an absolute temperature θ yields, under no-load conditions, a noise signal with the frequency-independent noise power density

$$L_{th,L} = 2k_B\theta R_{th} = \text{constant} \qquad \text{for } |f| \leqslant 6000\,\text{GHz} \qquad (2.47)$$

(Boltzmann's constant $k_B = 1.38 \times 10^{-23}\,\mathrm{W\,s\,K^{-1}}$). As thermal noise disturbances contain all frequency components in equal amounts, this is referred to as **white noise**. In the following, impedance matching is assumed. Thus the effective noise signal is halved, so that the effective **thermal noise power density** amounts to only a quarter of its no-load value:

$$L_{th} = \tfrac{1}{2}k_B\theta R_{th} \qquad \text{for } |f| \leqslant 6000\,\text{GHz} \qquad (2.48)$$

At room temperature ($\theta = 290$ K) we obtain for this the approximate value

$$L_{th} \approx 2 \times 10^{-21} R_{th}\ \mathrm{V^2\,Hz^{-1}\,\Omega^{-1}} \qquad (2.49)$$

The effect of the other sources of disturbances is considered using the **spectral noise factor** $F(f) \geqslant 1$, so that for the power spectrum of the total effective noise signal $n(t)$ (Fig. 2.1(b))

$$L_n(f) = F(f)L_{\text{th}} = \tfrac{1}{2}F(f)k_B\theta R_{\text{th}} \tag{2.50}$$

The receiver's input impedance also produces noise, so that $F(f) \geqslant 2$ [1.19]. Possible deviations from the assumed impedance matching can similarly be represented by increasing the noise factor $F(f)$.

Note The physical noise power density is independent of the value of the impedance considered, and amounts to $2k_B\theta$ with units watts per hertz in open circuit. However, we measure the noise signal, as usual in communication practice, on the impedance R_{th}, thus yielding eqn (2.47) where the noise power density now has the units volts squared per hertz. Equations (2.47)–(2.50) can therefore be used only if the signals considered are voltages.

2.4.4 Coaxial cable systems

As a special case let us now consider a digital system for **coaxial cables**. Because of the good shielding of the cable from external disturbances, e.g. pulse disturbances and cross-talk interference, the dominant source of disturbance here is thermal noise, where the noise power spectrum $L_n(f)$ can be considered as frequency independent.

A good approximation to the frequency response of a coaxial cable in the frequency range of most interest [2.36] is

$$H_K(f) = \exp(-\alpha_0 l)\exp\{-(\alpha_1 + j\beta_1)fl\}\exp\{-(\alpha_2 + j\beta_2)f^{1/2}l\} \tag{2.51}$$

where l is the coaxial cable length; α_0, α_1, α_2 and β_1, β_2 are specific **attenuation and phase coefficients** for the cable based on a length of 1 km and are given in Table 2.3 for conventional types of cable. The term

$$H_K(0) = \exp(-a_0) = \exp(-\alpha_0 l) \tag{2.52}$$

arising from ohmic losses causes only frequency-independent attenuation and no signal distortion. The values in Table 2.3 show further that $H_K(0) \approx 1$ for a 1.2 mm/4.4 mm or a 2.6 mm/9.5 mm coaxial pair cable ("small" or "normal" coaxial cable) with a length of up to several kilometers. The frequency-proportional attenuation component $\alpha_1 fl$ attributable to shunt losses is only noticeable at very high frequencies, in contrast with the component $\alpha_2 f^{1/2}l$, and is therefore neglected in the following. The frequency-proportional phase $\beta_1 fl$ leads to only a signal delay $\beta_1 l/2\pi$ in transmission time and no phase distortion. Hence the frequency response of a coaxial cable is principally determined by the influence of the constants α_1 and β_2 (**skin effect**), and in the frequency range of most interest (small coaxial cable,

200 kHz to 250 MHz; normal coaxial cable, 200 kHz to 400 MHz) an approximation is

$$H_K(f) \approx \exp(-\alpha_2 f^{1/2} l - j\beta f^{1/2} l) = \exp\{-\alpha_2 l (2jf)^{1/2}\} \qquad (2.53)$$

It is noticeable here that the values for α_2 (in nepers) and β_2 (in radians) are the same (see Table 2.3). Figure 2.8 shows that the attenuation constant $a_K(f)$ for small and normal coaxial cable is adequately approximated by eqn (2.53).

Table 2.3 Kilometric attenuation and phase coefficients

Cable type	α_0 (Np km^{-1} (dB km^{-1}))	α_1 (Np km^{-1} MHz^{-1} (dB km^{-1} MHz^{-1}))	α_2 (Np km^{-1} MHz$^{-1/2}$ (dB km^{-1} MHz$^{-1/2}$))	β_1 (rad km^{-1} MHz^{-1})	β_2 (rad km^{-1} MHz$^{-1/2}$)
Small coaxial cable (1.2 mm/4.4 mm)	0.00783 (0.068)	0.00044 (0.0039)	0.5984 (5.20)	22.18	0.5984
Normal coaxial cable (2.6 mm/9.5 mm)	0.00162 (0.014)	0.00043 (0.0038)	0.2722 (2.36)	21.78	0.2722

These values can be calculated from the geometric dimensions of the cable and have been experimentally verified by Wellhausen [2.35] for various cables. They are valid for a temperature of 290 K and frequencies greater than 200 kHz. Below this frequency the values only provide a coarse approximation.

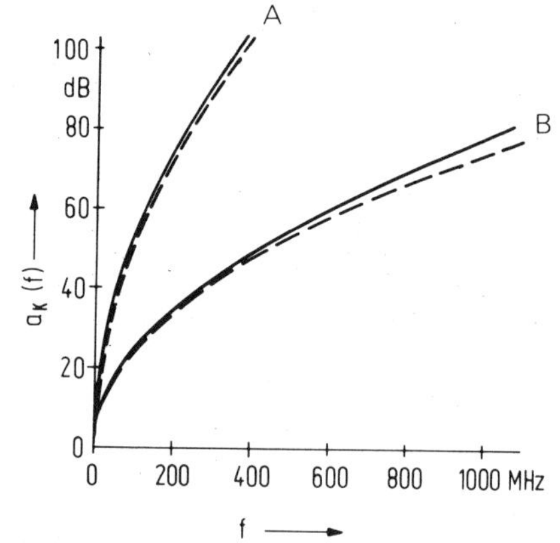

Fig. 2.8 Attenuation constant $a_K(f)$ for small coaxial cable (1.2 mm/4.4 mm) (curves A) and normal coaxial cable (2.6 mm/9.5 mm) (curves B) for a cable length of 1 km in each case: ——, exact; - - -, approximation.

The attenuation value at a bit rate of 50% is often used as a coaxial cable parameter. This is denoted subsequently as the **characteristic attenuation value** a_*:

$$a_* = a_K\left(f = \frac{R}{2}\right) \approx a_2 l\left(\frac{R}{2}\right)^{1/2} \quad (2.54)$$

Consequently the approximation to the coaxial cable frequency response is

$$H_K(f) \approx \exp\left\{-a_*\left(\frac{2f}{R}\right)^{1/2} - ja_*\left(\frac{2f}{R}\right)^{1/2}\right\} = \exp\left\{-2a_*\left(\frac{jf}{R}\right)^{1/2}\right\} \quad (2.55)$$

The introduction of the characteristic attenuation value a_* allows a uniform treatment of systems with different cables and various bit rates. Figure 2.9 presents the characteristic attenuation value a_* versus the bit rate R for various cable lengths l. The scale on the left applies to normal coaxial cable (2.6 mm/9.5 mm) and that on the right applies to small coaxial cable (1.2 mm/4.4 mm).

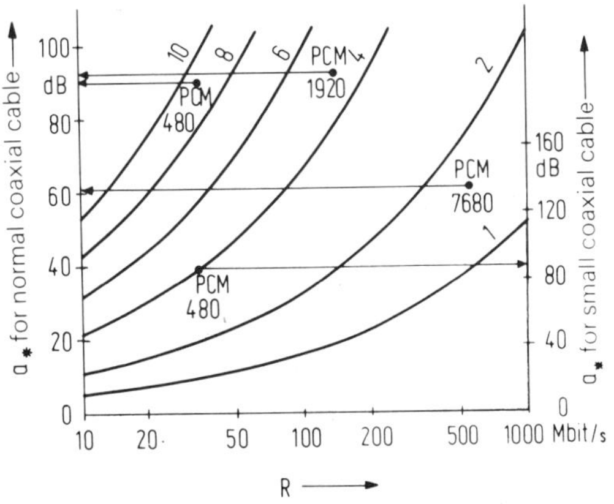

Fig. 2.9 Characteristic attenuation value a_* as a function of the bit rate R. The numbers on the curves give the cable length l in kilometers.

Table 2.4 presents a summary of the pulse code modulation (PCM) systems proposed by the Comité Consultatif International sur les Téléphones et Télécommunications. It shows that the systems of hierarchical levels 3–5 are designed for a characteristic attenuation value of 60–100 dB (Fig. 2.9).

The pulse response of the coaxial cable is obtained from the inverse Fourier transformation of eqn (2.55):

$$h_K(t) = \frac{a_*}{\pi(2Rt^3)^{1/2}} \exp\left(-\frac{a_*^2}{2\pi Rt}\right) \quad (2.56)$$

where a_* is in nepers. With a rectangular basic transmitter pulse we obtain the basic received pulse

$$g_e(t) = \hat{g}_s \operatorname{rec}\left(\frac{t}{T_s}\right) * h_K(t) = 2\hat{g}_s\left(\phi\left[\frac{a_* T^{1/2}}{\{\pi(t - T_s/2)\}^{1/2}}\right] - \phi\left[\frac{a_* T^{1/2}}{\{\pi(t + T_s/2)\}^{1/2}}\right]\right) \quad (2.57)$$

Table 2.4 Pulse code modulation systems for hierarchical stages 3–5

Hierarchical levels	*Bit rate* (Mbit s^{-1})	*Length* (km)	*Transmission medium*	a_* (dB)
3, PCM 480	34	9.30	Normal coaxial cable	90.5
	34	4.00	Small coaxial cable	86.2
4, PCM 1920	140	4.65	Normal coaxial cable	91.8
5, PCM 7680	560	1.55	Normal coaxial cable	61.2

where $\phi(x)$ denotes the Gaussian error integral shown in Table A1. For $a_* \geqslant 40$ dB, we obtain $g_e(t) \approx Th_K(t)$. Figure 2.10 makes it clear that the basic received pulse becomes lower and wider with increasing cable attenuation. The pulse maximum occurs at time $t = a_*^2/3\pi R$ and is approximately

$$\hat{g}_e \approx \frac{1.453RT}{a_*^2}\hat{g}_s \tag{2.58}$$

where a_* is in nepers.

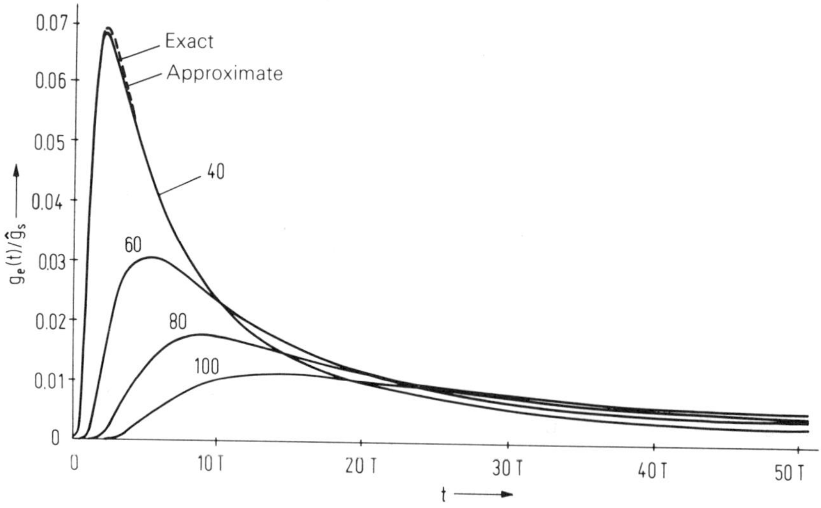

Fig. 2.10 Response of a coaxial cable with characteristic attenuation value a_* to a rectangular input pulse ($T_s = T$). The numbers on the curves give the value of a_* in decibels.

2.5 Equalizer (receiver filter)

The received signal $e(t)$ is unsuitable for immediate detection because, as a result of channel attenuation, it is very weak and there are large distortions (Fig. 2.2(c)). In addition the noise of the whole transmission frequency range is present. For these reasons the received signal must be filtered and equalized. Consequently the equalizer fulfils the following functions: (a) amplification of the received signal; (b) equalization of the useful received signal (reduction of the linear distortions); (c) limitation of the noise power. Complete equalization to the original transmitted signal is not possible, however, because of the necessary noise power limitation.

In the following a linear equalizer is assumed, so that it can be fully described by the **equalizer frequency response** $H_E(f) \bullet\!\!-\!\!\circ\, h_E(t)$ (Fig. 2.1(b)). The **direct signal transmission factor** $H_E(0)$ and the **equivalent noise bandwidth** $\square f_E$ are used for equalizer parameters by analogy with the transmission channel (see eqn (2.36) and eqn (2.39)).

2.5.1 Effect of the equalizer on the useful signal

The output signal of the equalizer is the detector input signal $d(t)$ which, by analogy with the received signal, is made up of a useful component $\tilde{d}(t)$ and a noise component $\check{d}(t)$:

$$d(t) = \tilde{d}(t) + \check{d}(t) \tag{2.59}$$

Firstly consider the effect of the equalizer on the useful signal. For clarification the **pulse shaper frequency response** is defined as

$$H_I(f) = H_K(f) H_E(f) \tag{2.60}$$

with the associated pulse response $h_I(t) \circ\!\!-\!\!\bullet\, H_I(f)$. $H_I(f)$ is the frequency response between the transmitter and the equalizer output which determines the distortion of the transmitted signal. Therefore, for the useful detection signal

$$\tilde{d}(t) = \tilde{e}(t) * h_E(t) = s(t) * h_I(t) \tag{2.61}$$

The **basic detection pulse** is defined, by analogy with eqn (2.42), as the response of the pulse shaper to a single basic transmitted pulse:

$$g_d(t) = g_s(t) * h_I(t) = \int_{-\infty}^{+\infty} G_s(f) H_I(f) \exp(\mathrm{j}2\pi f t)\, \mathrm{d}f \tag{2.62}$$

Therefore from eqns (2.11) and (2.61) we obtain for the useful detection signal

$$\tilde{d}(t) = H_I(0) s_0 + \sum_{v=-\infty}^{+\infty} a_v g_d(t - vT) \tag{2.63}$$

By analogy with eqn (2.44) the power spectrum of the useful detection signal is

$$L_{\tilde{d}}(f) = L_{\tilde{e}}(f) |H_E(f)|^2 = L_s(f) |H_I(f)|^2 \tag{2.64}$$

Integrating over $L_{\tilde{d}}(f)$ we obtain the **useful detection power** S_d. As an example, for a nonredundant bipolar transmitted signal this yields (see Table 2.2)

$$S_d = \frac{M+1}{3(M-1)} \frac{1}{T} \int_{-\infty}^{+\infty} |G_s(f)|^2 |H_I(f)|^2 \, df \tag{2.65}$$

Some pulse shapers which are easily described by system theory are summarized in Table A3 in the Appendix. In addition to the frequency response $H_I(f)$ and the pulse response $h_I(t)$, the **step response**

$$c_I(t) = \_\!\ulcorner(t) * h_I(t) = \int_{-\infty}^{t} h_I(\tau) \, d\tau \tag{2.66}$$

the one-sided cut-off frequency f_I and the equivalent noise bandwidth $\square f_I$ of the pulse shaper are given. These are determined by analogy with eqns (2.38) and (2.39). Pulse shapers 1–8 are defined by a single parameter, e.g. the cut-off frequency f_I. For the trapezoidal low-pass filter (row 9) and the cosine roll-off low-pass filter (row 10) the slope of the edges is another parameter additional to the cut-off frequency. The **roll-off factor**, which is defined as

$$r_I = \frac{f_2 - f_1}{f_2 + f_1} \qquad 0 \leqslant r_I \leqslant 1 \tag{2.67}$$

is generally used for a quantitative definition of the steepness of the edges. Frequencies f_1 and f_2 in (2.67) are defined in the sketches in Table A3.

2.5.2 Detector input noise signal and noise power

The noise component $\overset{\times}{d}(t)$ of the detector input signal emerges from the received noise signal $n(t)$ after linear filtering by the equalizer frequency response $H_E(f)$. Thus, from eqns (2.50) and (2.60), we obtain for the power spectrum of the detector input noise signal

$$L_{\overset{\times}{d}}(f) = L_n(f)|H_E(f)|^2 = L_n(f)\frac{|H_I(f)|^2}{|H_K(f)|^2} \tag{2.68}$$

from which the detector input noise power can be calculated by integration:

$$N_d = \int_{-\infty}^{+\infty} L_{\overset{\times}{d}}(f) \, df = \int_{-\infty}^{+\infty} L_n(f)|H_E(f)|^2 \, df \tag{2.69}$$

Both these equations make it clear that, from a given frequency, $|H_I(f)|$ must decay more rapidly than $|H_K(f)|$ as otherwise an infinitely large detector input noise power would result. With spectral noise factor $F(f)$ we have

$$N_d = L_{th} \int_{-\infty}^{+\infty} F(f)|H_E(f)|^2 \, df \tag{2.70}$$

N_d is a minimum when $F(f) = 1$, which is the smallest possible value. The ratio of noise powers at the outputs of noisy and noise-free equalizers is defined as the **average noise factor** or the **noise figure**:

$$F = \frac{N_d\{F(f)\}}{N_d\{F(f) = 1\}} = \frac{L_{\text{th}} \int_{-\infty}^{+\infty} F(f)|H_{\text{E}}(f)|^2 \, \mathrm{d}f}{L_{\text{th}} \int_{-\infty}^{+\infty} |H_{\text{E}}(f)|^2 \, \mathrm{d}f} \tag{2.71}$$

Thus the detector input noise power of eqn (2.70) can also be written

$$N_d = FL_{\text{th}} \int_{-\infty}^{+\infty} |H_{\text{E}}(f)|^2 \, \mathrm{d}f = FL_{\text{th}} \Box f_{\text{E}} |H_{\text{E}}(0)|^2 \tag{2.72}$$

where $\Box f_{\text{E}}$ is the equivalent noise bandwidth of the low-pass type of equalizer (see eqn (2.39)). Further, the ratio of the useful power to the noise power obtained from eqn (2.65) and eqn (2.72) respectively is defined as the **detector input signal-to-noise power ratio**:

$$\rho_d = \frac{S_d}{N_d} \tag{2.73}$$

To calculate the error probability, both the spectral distribution and the PDF $f_{\check{d}}(\check{d})$ of the detection noise signal have to be specified. If the (received) noise signal $n(t)$ has a Gaussian distribution, the detector input noise signal $\check{d}(t)$ also has a Gaussian distribution and has the PDF

$$f_{\check{d}}(\check{d}) = \frac{1}{(2\pi N_d)^{1/2}} \exp\left(-\frac{\check{d}^2}{2N_d}\right) \tag{2.74}$$

2.5.3 Equalizer with Gaussian pulse shaper

The effect of the equalizer on the useful and noise signals is now illustrated by means of an example. In this it is assumed that the channel frequency response $H_{\text{K}}(f)$ and the pulse shaper frequency response $H_{\text{I}}(f)$ can be approximated by a first-order low-pass filter and a Gaussian low-pass filter respectively (see Table A3, rows 1 and 7). Thus, from eqn (2.60), the absolute value of the equalizer frequency response is

$$|H_{\text{E}}(f)| = \frac{|H_{\text{I}}(f)|}{|H_{\text{K}}(f)|} = \left[1 + \left(\frac{\pi f}{2f_{\text{K}}}\right)^2 \exp\left\{-\pi\left(\frac{f}{2f_{\text{I}}}\right)^2\right\}\right]^{1/2} \tag{2.75}$$

f_{K} and f_{I} are the cut-off frequencies of the channel and pulse shaper according to eqn (2.38). $|H_E(f)|$ is shown in Fig. 2.11(a). The larger the ratio f_I/f_K is, the more marked is the increase in the equalizer frequency response and the larger is the detection noise power. For white noise with noise factor F we obtain

$$N_d = \sqrt{2} f_I F L_{th} \left(1 + \frac{\pi^2 f_I^2}{4 f_K^2}\right) \tag{2.76}$$

The basic detector input pulse $g_d(t)$ is independent of the channel cut-off frequency f_K and can be calculated using eqn (2.62). With rectangular transmitter pulses of amplitude $\hat{g}_s$ and duration T_s

$$g_d(t) = g_s(t) * h_I(t) = \hat{g}_s \operatorname{rec}\left(\frac{t}{T_s}\right) * 2f_I \exp\{-\pi(2f_I t)^2\} \tag{2.77}$$

With the step response $c_I(t)$ of the Gaussian low-pass filter as in Table A3 we obtain the basic detection pulse

$$\begin{aligned} g_d(t) &= c_I\left(t + \frac{T_s}{2}\right) - c_I\left(t - \frac{T_s}{2}\right) \\ &= \hat{g}_s\left[\phi\left\{2(2\pi)^{1/2} f_I\left(t + \frac{T_s}{2}\right)\right\} - \phi\left\{2(2\pi)^{1/2} f_I\left(t - \frac{T_s}{2}\right)\right\}\right] \end{aligned} \tag{2.78}$$

Here $\phi(x)$ is the Gaussian error integral given in Table A1. Figures 2.11(b) and 2.11(c) make it clear that $g_d(t)$ becomes wider and the pulse amplitude

$$\hat{g}_d = g_d(0) = \hat{g}_s[2\phi\{(2\pi)^{1/2} f_I T_s\} - 1] \tag{2.79}$$

becomes smaller as $f_I T_s$ decreases.

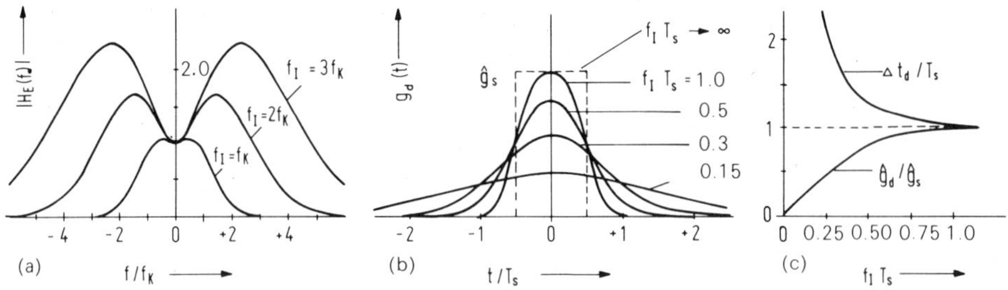

Fig. 2.11 (a) Equalizer frequency response with a first-order low-pass channel and a Gaussian pulse shaper; (b) basic detector input pulse with a rectangular transmitter pulse and a Gaussian pulse shaper; (c) amplitude and equivalent pulse width of the basic detector input pulse.

2.6 Detector

The digital receiver requires a device to extract from the detection signal $d(t)$ containing noise and intersymbol interference a signal that reproduces the source information as clearly as possible. This device for amplitude and time recovery is referred to as the **detector**. There is a distinction between detectors with and without memory. A memoryless detector requires only one sample value of the detector input signal to detect a symbol. In comparison, several (adjacent) sample values are extracted for a detector with a memory (see Chapters 5 and 6).

The simplest type of detector is the **memoryless threshold detector**. It consists of a threshold device for amplitude regeneration and a sampling device for time regeneration of the digital signal. For this the detector requires information about the clock of the transmitted digital signal, which is fed to it by the TRD in the form of the clock signal $z(t)$. The clock signal determines the **detection time** t_ν at which the detector samples its input signal $d(t)$. In the example shown in Fig. 2.2 the times of detection are fixed by the leading edge of the rectangular clock signal. In general it has to be assumed, however, that the times t_ν of detection do not follow each other equidistantly. This is then referred to as a jittering clock signal (Fig. 2.12(a)). However, in the following an ideal clock signal is assumed so that, as shown in Fig. 2.12(b),

$$t_\nu = T_D + \nu T \tag{2.80}$$

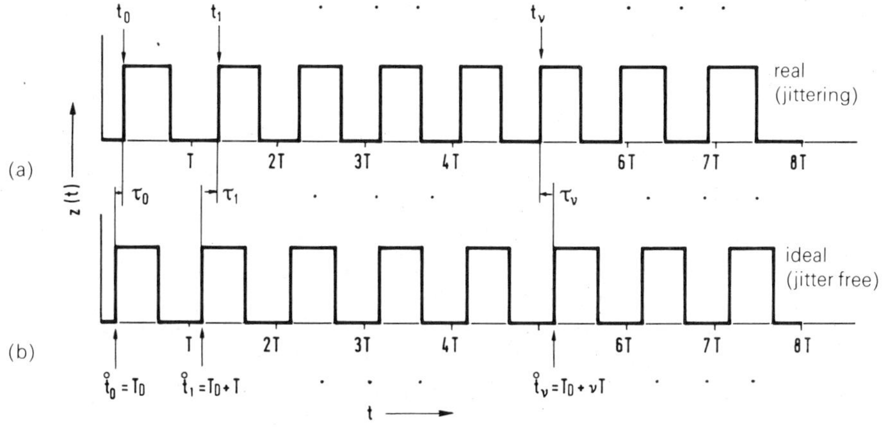

Fig. 2.12 Definition of the detection times for (a) a jittering clock signal and (b) an ideal clock signal.

T_D is the time at which the symbol v_0 is detected. It indicates the (constant) shift of the detection time t_ν relative to the time reference νT. If it is assumed that the basic detection pulse $g_d(t)$ is a maximum at time $t = 0$, i.e. that $\hat{g}_d = g_d(0)$, then T_D defines the time difference between the occurrence of the pulse maximum and pulse detection. $T_D < 0$ signifies that the approaching pulses are detected before their maxima are reached.

The operation of the threshold device can be described using its characteristic curve $v(t) = f\{d(t)\}$ (Fig. 2.13). For a binary system there is only a single threshold value E, which for bipolar transmission is generally chosen to be zero. If the **detection sample value**

$$d_\nu = d(t_\nu) = \tilde{d}_\nu + \check{d}_\nu \tag{2.81}$$

at the detection time t_ν is greater than the threshold value, the detector output signal $v(t)$ is set equal to $a_2 v_{\max}$ until the next clock pulse. This then corresponds to the sink symbol $v_\nu = $ L. However, if $d_\nu < E$, then $v(t) = a_1 v_{\max}$ and the sink symbol $v_\nu = $ O (see Fig. 2.2(f)). Here a_1 and a_2 are the possible amplitude coefficients defined in Section 2.3.

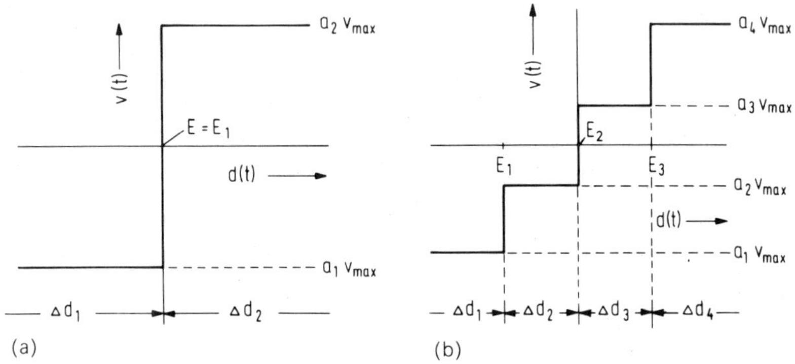

Fig. 2.13 Characteristic curve of the threshold device: (a) binary; (b) quaternary.

In an M-level system the overall range of the detection signal $d(t)$ is divided into M increments $\Delta d_1, \ldots, \Delta d_\mu, \ldots, \Delta d_M$, which results in the time and value discrete M-level sink signal $v(t)$ (Fig. 2.13(b)). Its possible instantaneous values are $a_\mu v_{\max}$. The **decision value** or **threshold value** E_μ ($\mu = 1, \ldots, M - 1$) is the limit value which separates the increments Δd_μ and $\Delta d_{\mu+1}$.

If the detection sample value $d_\nu = d(t_\nu)$ lies between the threshold values $E_{\mu-1}$ and E_μ (thus in the interval Δd_μ) the sink signal $v(t)$ is put equal to $a_\mu v_{\max}$ in the time interval $t_\nu < t < t_{\nu+1}$, which corresponds to the sink symbol $v_\nu = v_\mu$.

2.7 Timing recovery device

The purpose of the TRD is to produce a suitable clock signal $z(t)$ from the detection signal (or from its zero-crossings). Therefore a possible means of timing recovery is to use a **tank circuit** (band-pass $H_{BP}(f)$) with center frequency $1/T$

(Fig. 2.14(a)). When a signal $x(t)$ with power spectral density $L_x(f)$ is applied to the input terminal of the tank circuit, the output signal $y(t)$ has a power spectrum given by

$$L_y(f) = L_x(f)|H_{BP}(f)|^2 \tag{2.82}$$

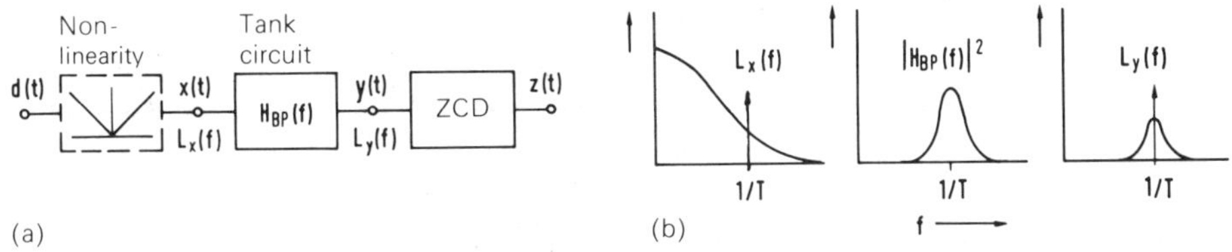

Fig. 2.14 (a) TRD and (b) corresponding power spectra.

If $L_x(f)$ contains a discrete component with symbol rate $1/T$, this discrete component is also contained in the power spectrum $L_y(f)$ of the output signal. Hence the output signal

$$y(t) = \hat{y}\cos\left(2\pi\frac{t}{T}\right) + \check{y}(t) \tag{2.83}$$

is a cosine oscillation with a symbol rate $1/T$ if the noise $\check{y}(t)$ is neglected. $\check{y}(t)$ arises from the continuous part of the power spectrum and causes a shift of the zero-crossings of $y(t)$. This noise component becomes larger as the figure of merit of the tank circuit becomes smaller, i.e. as the bandwidth of $H_{BP}(f)$ is increased. The subsequent zero-crossing detector (ZCD) detects when the signal $y(t)$ passes through zero and produces from this the rectangular clock signal $z(t)$. If $\check{y}(t) = 0$, all the rectangular pulses have equal width and the clock signal is ideal.

This method of timing recovery is thus only applicable if the power spectrum $L_x(f)$ of the input signal $x(t)$ exhibits a discrete line at the symbol rate $1/T$ or multiples thereof, as is the case for unipolar RZ signals for example. However, this periodic component is not present with nonredundant bipolar signals (see Fig. 2.6). Here the discrete lines have to be produced by a nonlinearity (e.g. two-way rectification or quadratic characteristic).

With the present state of technology, however, a timing recovery device with a **phase-locked loop** (PLL) is most frequently used (Fig. 2.15(a)). This contains a ZCD at both input and output, a phase comparator (PC), a clock filter $H_T(f)$ and a voltage-controlled oscillator (VCO).

The purpose of the PC is to compare the position of the zero-crossings of the detection signal $d(t)$ and the clock signal $z(t)$. The output signal $w(t)$ of the PC can thus be considered as a sequence of small rectangular pulses at times νT, where the pulse amplitudes are proportional to the time between the observed zero-crossings (Fig. 2.15(c)). If the detector input signal $d(t)$ has no zero-crossings in the time interval $(\nu - 1)T$ to νT, then $w(\nu T) = 0$.

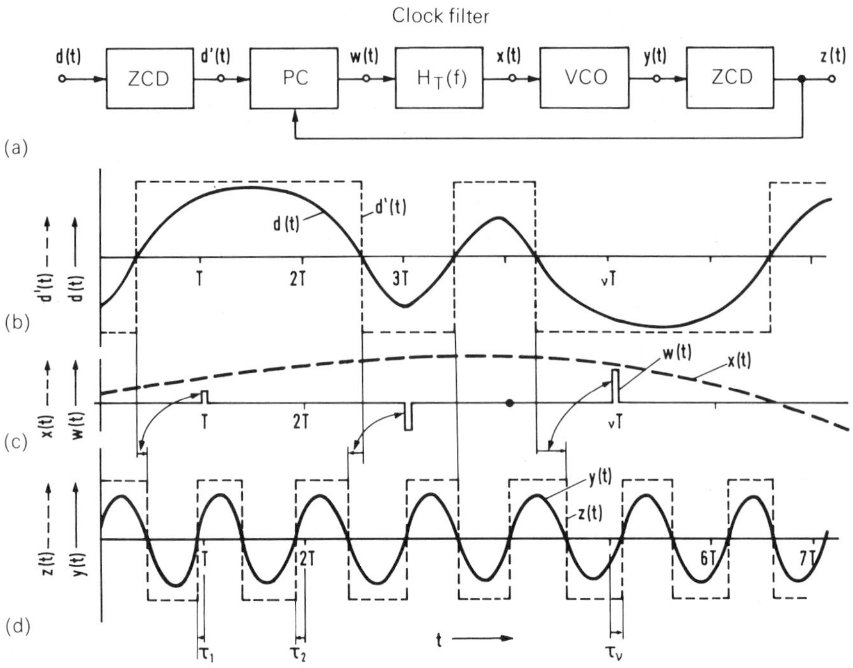

Fig. 2.15 (a) TRD with PLL; (b), (c), (d) corresponding signals.

The output signal of the PC is passed through the clock filter whose frequency response $H_T(f)$ exhibits low-pass filter characteristics and whose cut-off frequency is very much lower than the clock frequency $1/T$. Thus the control signal $x(t)$ for the VCO is a very low frequency signal whose instantaneous value at time νT depends on the shift of the preceding zero-crossing (Fig. 2.15(c)). Thus the low-pass filter works as an averager over the phase shift between the signals $d(t)$ and $z(t)$.

If the signals $d(t)$ and $y(t)$ are of equal phase, the control signal $x(t)$ is zero and the VCO oscillates at its resonant frequency, which in the ideal case is equal to the symbol rate $1/T$. Otherwise, the output signal of the VCO is

$$y(t) = \hat{y} \cos\left\{2\pi \frac{t}{T} + \varphi_T(t)\right\} \tag{2.84}$$

$\varphi_T(t)$ denotes the phase of the clock oscillation, which is controlled by the signal $x(t)$:

$$\varphi_T(t) = 2\pi \int_{-\infty}^{t} \{\Delta f_{\mathrm{VCO}} + V_{\mathrm{VCO}} x(t')\} \, \mathrm{d}t' \tag{2.85}$$

Δf_{VCO} is the detuning of the VCO and V_{VCO} is its sensitivity (in Hz V^{-1}). The following approximate relationship exists between the phase position $\varphi_T(t)$ and the clock jitter value τ_ν (time deviation):

$$\tau_\nu \approx -\frac{\varphi_T(\nu T)}{2\pi} T \tag{2.86}$$

The **clock jitter value** τ_ν defines the deviation of the detection instant t_ν from the actual value $\mathring{t}_\nu = T_D + \nu T$ (see Fig. 2.12). For a jitter-free clock signal $\tau_\nu = 0$ for all values of ν. In general, however, it must be assumed that the clock signal is not ideal (jittering), in which case the error probability is higher; this can lead to difficulties in data transfer.

The origin of phase jitter and its accumulation within a regenerator chain has been examined by a number of workers [2.5, 2.17, 2.20, 2.28, 2.31, 2.32] who have shown that the effective value of the clock jitter with conventional regenerator section lengths can in general be kept small by using a clock recovery device with a PLL. Therefore in the following an ideal clock signal is assumed.

Chapter 3

Error Probability of a Digital Transmission System

Contents In Sections 3.1 and 3.2 the error probability for a given binary sequence is determined and the eye pattern is discussed. Next the mean error probability for binary and multilevel signals is calculated (Sections 3.3 and 3.4). Finally various approximations for the mean error probability are stated and compared in Section 3.5.

Assumptions The provisions of Chapter 2 remain valid. In addition, if not explicitly stated otherwise, a nonredundant source, equidistant amplitude coefficients (eqn (2.10)) and signal-independent Gaussian noise are assumed. The base component of the transmitted signal is set to zero.

3.1 Error probability of a given binary sequence

3.1.1 Definition of the error probability

To calculate error probability, a transmitted binary source symbol sequence $\langle q_\nu \rangle$ is considered first. The probability that the νth symbol of the sequence is corrupted is the **symbol error probability** (SEP):

$$p_{S_\nu} = P(v_\nu \neq q_\nu) \tag{3.1}$$

If a memoryless threshold detector as described in Section 2.6 is assumed, the SEP p_{S_ν} is defined by the detection sample value $d_\nu = \tilde{d}_\nu + \overset{\times}{d}_\nu$ and the decision value E. Thus for a binary threshold detector (Fig. 2.13(a))

$$p_{S_\nu} = \begin{cases} P(d_\nu > E) = P(\overset{\times}{d}_\nu > E - \tilde{d}_\nu) & \text{when } q_\nu = \text{O} \\ P(d_\nu < E) = P(\overset{\times}{d}_\nu < E - \tilde{d}_\nu) & \text{when } q_\nu = \text{L} \end{cases} \tag{3.2}$$

Example Figure 3.1(a) shows the bipolar useful detection signal $\tilde{d}(t)$ corresponding to the source symbol sequence ... O L O L L L ..., for which a Gaussian basic detection pulse is assumed. Detection occurs at equispaced detection times t_ν

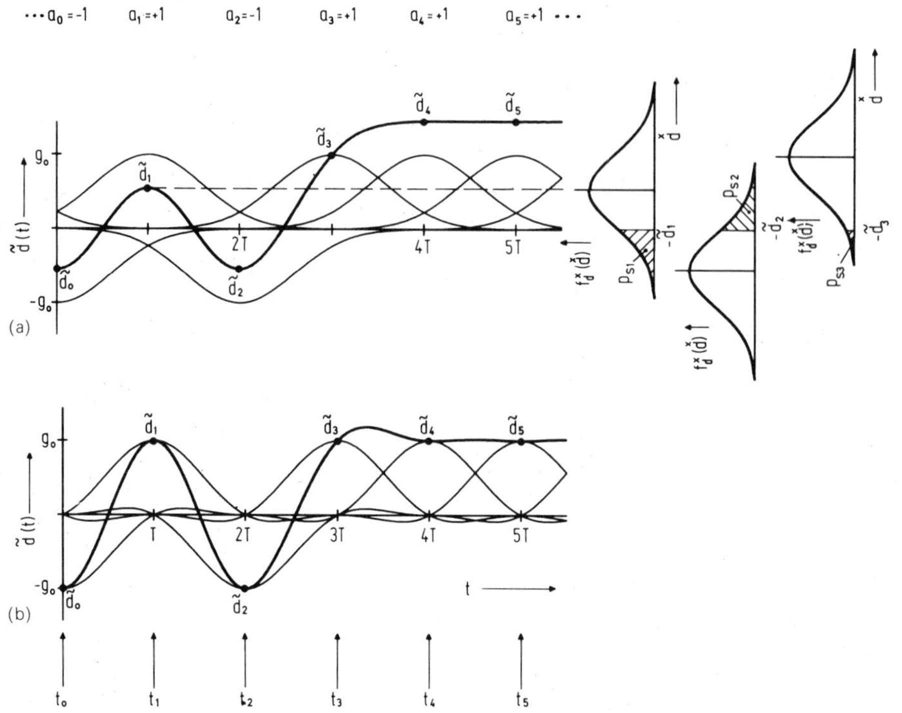

Fig. 3.1 Calculation of the error probability p_{S_ν} with (a) a Gaussian or (b) an intersymbol-interference-free basic detection pulse.

$= \nu T$, i.e. $T_{\mathrm{D}} = 0$. A decision value of $E = 0$ is used. From eqn (3.2) we obtain for the probability that the symbol $q_1 = \mathrm{L}$ is corrupted

$$p_{\mathrm{S}_1} = P(v_1 = \mathrm{O}) = P(\check{d}_1 < -\tilde{d}_1) \tag{3.3}$$

Therefore a symbol error occurs if the noise component $\check{d}_1$ of the sample value is of opposite sign to and larger magnitude than the useful component $\tilde{d}_1$. With the probability density function (PDF) $f_{\check{d}}(\check{d})$ of the stationary detection noise signal $\check{d}(t)$ given by eqn (2.74),

$$p_{\mathrm{S}_1} = \int_{-\infty}^{-\tilde{d}_1} f_{\check{d}}(\check{d})\,\mathrm{d}\check{d} \tag{3.4}$$

The SEP p_{S_1} thus corresponds to the shaded area below the PDF $f_{\check{d}}(\check{d})$ in Fig. 3.1(a).

By analogy we obtain for the probability that the symbol $q_2 = \mathrm{O}$ is falsely detected (cf. eqn (3.2))

$$p_{S_2} = P(v_2 = \mathrm{L}) = P(\check{d}_2 > -\tilde{d}_2) = \int_{-\tilde{d}_2}^{+\infty} f_{\check{d}}(\check{d})\,\mathrm{d}\check{d} \tag{3.5}$$

As in this example $\tilde{d}_2 = -\tilde{d}_1$ and a symmetric PDF of the noise signal is assumed, the SEP $p_{S_2} = p_{S_1}$. In contrast, the probabilities p_{S_3}, p_{S_4} and p_{S_5} are very much smaller because the useful components $\tilde{d}_3$, $\tilde{d}_4$ and $\tilde{d}_5$ have a larger separation from the decision value $E = 0$.

To generalize this example for the case $E \neq 0$ it is appropriate to introduce the **detection distance** D_ν as the separation of the useful detection sample value $\tilde{d}_\nu$ from the decision value E:

$$D_\nu = |\tilde{d}_\nu - E| \tag{3.6}$$

Equations (3.4) and (3.5) can then be combined as follows. If the detection noise signal $\check{d}(t)$ is symmetrically distributed around zero, i.e. if $f_{\check{d}}(-\check{d}) = f_{\check{d}}(\check{d})$, then the SEP is given by

$$p_{S_\nu} = \int_{D_\nu}^{+\infty} f_{\check{d}}(\check{d})\,\mathrm{d}\check{d} \tag{3.7}$$

In this equation it is assumed that the threshold detector at least makes the correct decision when there is no noise. It means that the useful detection sample values $\tilde{d}_\nu$ must fulfil the following condition:

$$\tilde{d}_\nu < E \text{ when } q_\nu = \mathrm{O} \qquad \text{or} \qquad \tilde{d}_\nu > E \text{ when } q_\nu = \mathrm{L} \tag{3.8}$$

If the detection noise signal $\check{d}(t)$ has a Gaussian distribution, eqn (2.74) yields

$$p_{S_\nu} = Q\left(\frac{D_\nu}{N_d^{\;1/2}}\right) \tag{3.9}$$

Here $Q(x)$ is the **complementary Gaussian error integral** (Table A1) and N_d is the detection noise power as defined by eqn (2.69). The values given in Table A2 in the Appendix show that $Q(x)$ approaches zero asymptotically as x increases. Therefore even a slight increase in the detection distance D_ν leads to a marked reduction in the SEP p_{S_ν} (see Fig. 3.1(a)).

3.1.2 Intersymbol interference

The useful detection sample value $\tilde{d}_\nu$ and hence the detection distance D_ν are in general strongly affected by neighboring pulses. Depending on which symbols the neighboring pulses represent, D_ν can be increased or reduced relative to its original value. This is called **intersymbol interference**. The decaying trailing edges of the preceding pulses are defined as the **postcursors**, and the ascending leading edges of the following pulses are defined as the **precursors**.

For clarity, in the following we consider only the detection time $t_0 = T_D$ at which the symbol v_0 is detected. The useful detection sample value $\tilde{d}_0$ is completely specified by the detection pulse $g_d(t)$, the detection time T_D and the amplitude coefficients a_ν (see eqn (2.63)):

$$\tilde{d}_0 = \tilde{d}(T_D) = \sum_{\nu=-\infty}^{+\infty} a_\nu g_d(T_D - \nu T) \tag{3.10}$$

Here it is assumed that the base component s_0 of the transmitter signal is zero.

For brevity, in the following the instantaneous values of the basic detection pulse at times $T_D + \nu T$ are denoted as the **basic detection pulse values**:

$$g_\nu = g_d(T_D + \nu T) \tag{3.11}$$

$g_0 = g_d(T_D)$ is the **main value** of the basic detection pulse (Fig. 3.2). Values with negative ν denote the precursors of the basic detection pulse, and those with positive ν denote the postcursors.

Separating the summation in eqn (3.10) we obtain

$$\tilde{d}_0 = a_0 g_0 + \sum_{\nu=1}^{\infty} a_{-\nu} g_\nu + \sum_{\nu=1}^{\infty} a_\nu g_{-\nu} \tag{3.12}$$

$a_0 g_0$ is the original value of the useful detection signal, which would be the result without precursors and postcursors. The first summation describes the effect of the preceding pulses, whose decaying components (postcursors) are still present at the detection time T_D. The second summation takes into account that the detection of symbol v_0 is also affected by the increasing components of the following pulses (precursors).

It is assumed in the following that the basic detection pulse only has v

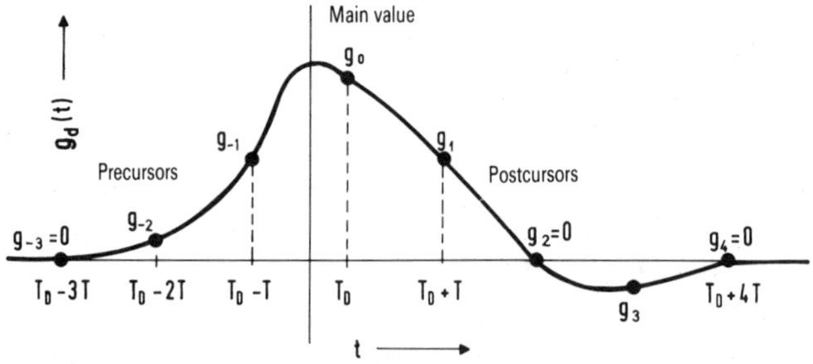

Fig. 3.2 Definition of the basic detection pulse values g_ν ($v = 2, n = 3$).

precursors and n postcursors; therefore

$$g_\nu \approx 0 \quad \text{for} \quad \nu < -v \text{ and } \nu > n \tag{3.13}$$

With this limitation the useful detection sample value can be written

$$\tilde{d}_0 = a_0 g_0 + \sum_{\nu=1}^{n} a_{-\nu} g_\nu + \sum_{\nu=1}^{v} a_\nu g_{-\nu} \tag{3.14}$$

If the basic detection pulse values g_ν are identically zero ($v = n = 0$) for all $\nu \neq 0$, $\tilde{d}_0 = a_0 g_0$. This means that in this special case the leading and trailing edges of the adjacent pulses do not interfere with the symbol detection. Such systems are referred to as **intersymbol-interference-free systems** (see Section 8.3). They are also often called **Nyquist systems.**

Figure 3.1(b) shows a sample from a binary bipolar useful detection signal $\tilde{d}(t)$ with a basic detection pulse of the type $(\sin x)/x$. At the moments of detection νT, $\tilde{d}(t) = \pm g_0$, i.e. the system is free of intersymbol interference. Except at these equally spaced times, intersymbol interference occurs.

3.2 Eye pattern for binary signals

The **eye pattern** or **eye diagram** is the sum of all the superimposed segments of a digital signal whose duration is an integer multiple of the symbol width T. This diagram has some similarity to an eye—hence its name. It can, for example, be presented on an oscilloscope which is triggered by the clock signal $z(t)$. It is assumed in the following that the digital signal contains all permissible symbol sequences and that the noise is negligibly small. In this case the eye pattern furnishes a statement of the intersymbol interference present in the digital signal, and it is completely determined by the level number M and the basic detection pulse $g_d(t)$.

The eye diagram can be produced for any digital signal, e.g. the transmitted or the received signal. The eye pattern of the detector input signal is of particular significance because it is an important expression of the transmission quality of the digital communication system.

3.2.1 Characteristics of the eye pattern

Figure 3.3 shows the eye pattern of the binary bipolar useful detection signal $\tilde{d}(t)$ for various basic detection pulses. A nonredundant source is assumed, so that all possible symbol sequences are indeed permissible. It should be noted that the time scales for the eye pattern and the basic pulse are different.

Since the useful detection signal $\tilde{d}(t)$ is from $t = -\infty$ to $t = +\infty$, an infinite number of lines are superimposed in the eye pattern. However, if the basic pulse $g_d(t)$ differs from zero only within the time interval mT, many lines coincide and in the

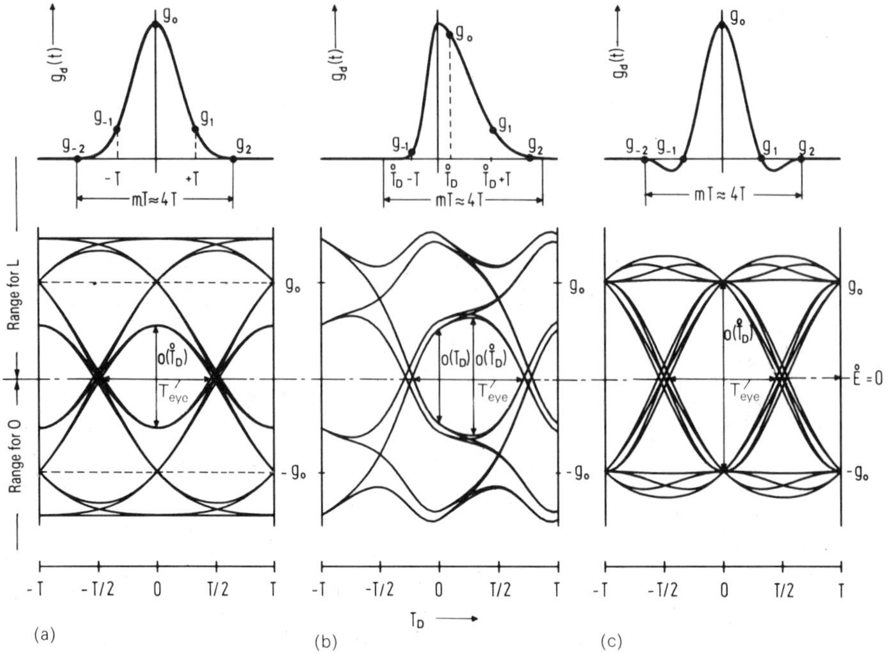

Fig. 3.3 Basic detection pulse and binary eye pattern for (a) a symmetric, (b) a non-symmetric and (c) an intersymbol-interference-free basic detection pulse.

binary eye pattern only 2^m discrete lines can be distinguished in the time period T. For the basic pulse considered in Fig. 3.3, $m = 4$, and so 16 eye lines can be identified. The figure shows further that the eye pattern of a bipolar nonredundant signal is always symmetrical about the decision value $E = 0$ (Fig. 3.4(a)). With redundant and/or unipolar signals this symmetry is not always present. If the basic detection pulse is symmetrical, i.e. $g_d(-t) = g_d(t)$, the eye pattern is also symmetrical about the ordinate $t = 0$.

In the following we consider the individual lines $\tilde{d}_i(t)$ of the eye pattern in Fig. 3.4. The subscript i is an index over all identifiable eye pattern lines. If $g_d(t)$ is limited to the time segment mT, for a binary signal

$$1 \leqslant i \leqslant 2^m \tag{3.15}$$

In general it can be deduced from this that, in the absence of noise, each arbitrary symbol sequence $\langle q_v \rangle_i$ can be transmitted without error (**bit sequence independence**). This means that for all values of i there is at least one time T_D at which the following condition is fulfilled:

$$\begin{aligned} \tilde{d}_i(T_D) &< E \text{ when } q_0^{(i)} = \mathrm{O} \\ \tilde{d}_i(T_D) &> E \text{ when } q_0^{(i)} = L \end{aligned} \tag{3.16}$$

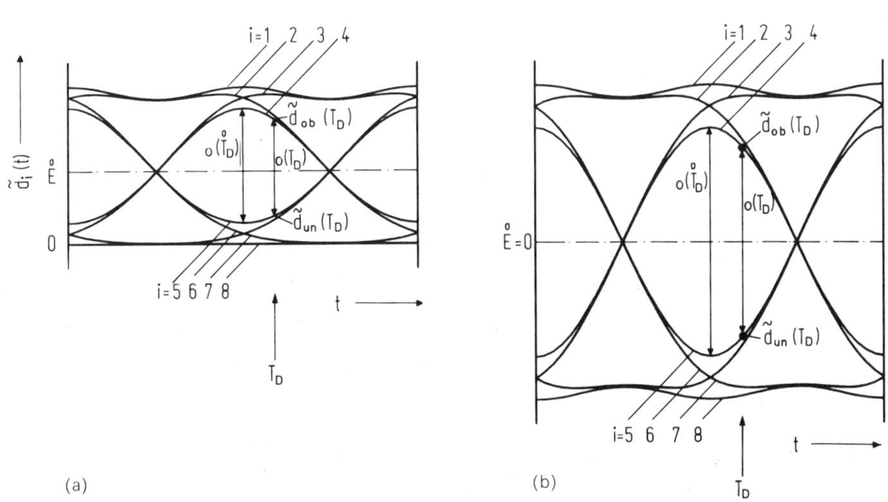

Fig. 3.4 Calculation of the vertical eye opening: (a) unipolar; (b) bipolar.

In this case we refer to an "open" eye.

The individual eye lines $\tilde{d}_i(t)$ are segments of the useful detection signal $\tilde{d}(t)$ and the $\langle q_v \rangle_i$ are the corresponding symbol sequences. The detection distance D_i is the separation of the eye line $\tilde{d}_i(t)$ from the threshold value E at time T_D:

$$D_i = |\tilde{d}_i(T_D) - E| \tag{3.17}$$

The SEP p_{S_i} of the ith eye line is the probability that the "middle" symbol of the sequence $\langle q_v \rangle_i$ (i.e. the symbol to be detected) is corrupted:

$$p_{S_i} = P(v_0{}^{(i)} \neq q_0{}^{(i)}) \quad i = 1, \ldots, 2^m \tag{3.18}$$

By analogy with eqn (3.9) we obtain for Gaussian noise and an open eye

$$p_{S_i} = \int_{D_i}^{+\infty} f_{\check{d}}(\check{d})\,\mathrm{d}\check{d} = Q\left(\frac{D_i}{N_d{}^{1/2}}\right) \tag{3.19}$$

It should be noted that the index i used here is different from the variable v which appears in definition (3.1).

3.2.2 Eye opening and worst-case error probability

Of particular interest are the symbol sequences with the worst (smallest) detection distance

$$D_U = \min_i D_i = \min_i |\tilde{d}_i(T_D) - E| \tag{3.20}$$

as in this case the probability for false detection is greatest. D_U is defined as the **worst-case detection distance**, and the corresponding symbol sequences are the **worst-case symbol sequences** (worst-case patterns). In general, for the error probability of the worst-case symbol sequence $\langle q_\nu \rangle_U$,

$$p_U = \int_{D_U}^{+\infty} f_{\check{d}}(\check{d})\, d\check{d} \tag{3.21}$$

If the noise signal is Gaussian, the worst-case error probability can be calculated by analogy with eqn (3.9) from

$$p_U = Q\left(\frac{D_U}{N_d^{1/2}}\right) = Q(\rho_U{}^{1/2}) \tag{3.22}$$

Here ρ_U is the **worst-case signal-to-noise ratio**:

$$\rho_U = \frac{D_U{}^2}{N_d} \tag{3.23}$$

The SEP p_{S_i} of any arbitrary symbol sequence $\langle q_\nu \rangle_i$ is less than or equal to p_U. The significance of the worst-case SEP is that it can be used as the upper limit for the mean SEP, and in general it is a good approximation to it (see Section 3.5).

In the eye pattern the worst-case symbol sequence results in the innermost eye line, which is the inner limit of the eye opening. The **vertical eye opening** $o(T_D)$ is the distance between the two innermost eye lines at the detection time T_D.

The upper boundary of the eye opening is denoted by $\tilde{d}_{up}(T_D)$ and the lower boundary by $\tilde{d}_{lo}(T_D)$, and so corresponding to Fig. 3.4 we obtain

$$o(T_D) = \begin{cases} \tilde{d}_{up}(T_D) - \tilde{d}_{lo}(T_D) & \text{for } \tilde{d}_{up}(T_D) > \tilde{d}_{lo}(T_D) \\ 0 & \text{otherwise} \end{cases} \tag{3.24}$$

It is often appropriate to relate the vertical eye opening to the transmitted signal and pulse-shaper characteristics. Thus the **normalized eye opening** for a low-pass type of pulse shaper ($|H_I(0)| > 0$) is defined as follows:

$$o_{norm}(T_D) = \frac{o(T_D)}{\Delta s |H_I(0)|} \tag{3.25}$$

Δs is the modulation range of the transmitted signal as shown in Fig. 2.3 and $H_I(0)$ is the direct signal transmission factor of the pulse shaper.

The normalization chosen here guarantees that $o_{norm}(T_D)$ has a value between zero and unity. The lower limit zero thus corresponds to a closed eye at $T_D = 0$. However, for a binary Nyquist system there is at least one optimum detection time T_D, for which the normalized eye opening has the maximum value of unity.

It is assumed in the following that the threshold is located at

$$E = \mathring{E} = \tfrac{1}{2}\{\tilde{d}_{lo}(T_D) + \tilde{d}_{up}(T_D)\} \tag{3.26}$$

This value is optimum for signal-independent disturbances with symmetrical PDF; the optimality is indicated by superscript ∘. For this precondition, all detection distances are equal for the upper and lower worst-case symbol sequences and

$$D_U = \tfrac{1}{2} o(T_D) \tag{3.27}$$

Thus for the worst-case signal-to-noise ratio defined in (3.23) with optimum decision value and Gaussian noise we obtain

$$\rho_U = \frac{o(T_D)^2}{4N_d} \tag{3.28}$$

The **optimum detection time** $\mathring{T}_D$ should be selected so that the vertical eye opening has the largest possible value. For the maximum eye opening we obtain

$$o(\mathring{T}_D) = \max_{T_D} \{o(T_D)\} \tag{3.29}$$

If the basic detection pulse $g_d(t)$ is symmetrical, the optimum detection time $\mathring{T}_D$ is zero (see Figs 3.3(a) and 3.3(c)). However, for an asymmetrical detection pulse $\mathring{T}_D$ is generally nonzero (see Fig. 3.3(b)).

In addition to the vertical eye opening, the **temporal eye opening** T_{eye} is also an important parameter of digital systems. T_{eye} defines the maximum horizontal separation of the innermost eye lines, where $T_{eye} \leqslant T$ (see Fig. 3.3). The larger the temporal eye opening is, the smaller is the probability that an unstable clock will lead to an erroneous decision.

3.2.3 Vertical eye opening for nonredundant signals

All previous findings in this section are generally valid even for redundant sources. Nonredundant signals are considered exclusively in the following. In this case, for a worst-case symbol sequence all precursors and postcursors contribute to diminution of the detection pulse. Thus the upper boundary line $\tilde{d}_{up}(T_D)$ of the eye opening results when the symbol considered is $q_0 = \mathrm{L}$ and all leading and trailing edges of the neighboring pulses reduce the detection pulse considered (Fig. 3.4). For a bipolar system, all the precursor and postcursor absolute values have to be subtracted from the main value g_0. With a unipolar binary system, however, the amplitude coefficients possess the values zero and unity, so that only an amplitude coefficient $a_\nu = 1$ in combination with a corresponding basic pulse value $g_{-\nu} < 0$ causes a reduction in $\tilde{d}_{up}(T_D)$. Thus

$$\tilde{d}_{up}(T_D) = \begin{cases} g_0 - \sum\limits_{\nu \neq 0} |g_\nu| & \text{for bipolar signals} \\ g_0 - \sum\limits_{\substack{\nu \neq 0 \\ g_\nu < 0}} |g_\nu| & \text{for unipolar signals} \end{cases} \tag{3.30}$$

Here $g_\nu = g_d(T_D + \nu T)$ denotes the basic detection pulse value as defined in (3.11). By analogy the signal value for the lower boundary line ($q_0 = 0$) is

$$\tilde{d}_{lo}(T_D) = \begin{cases} -g_0 + \sum\limits_{\nu \neq 0} |g_\nu| & \text{for bipolar signals} \\ \sum\limits_{\substack{\nu \neq 0 \\ g_\nu > 0}} |g_\nu| & \text{for unipolar signals} \end{cases} \tag{3.31}$$

Substituting these two equations in eqn (3.24) we obtain for the vertical eye opening of a nonredundant binary system

$$o(T_D) = \begin{cases} 2\left(g_0 - \sum\limits_{\nu \neq 0} |g_\nu|\right) & \text{for bipolar signals} \\ g_0 - \sum\limits_{\nu \neq 0} |g_\nu| & \text{for unipolar signals} \end{cases} \tag{3.32}$$

If the sum of all precursors and postcursors is larger than the main value g_0, the eye is closed and, by definition, the vertical eye opening $o(T_D) = 0$.

Figures 3.4(a) and 3.4(b) show clearly that the eye patterns for unipolar and bipolar systems can be different. However, the vertical eye opening in each case is identical except for a factor of 2. As the modulation range Δs for bipolar signals is similarly twice that of unipolar signals, the normalized eye openings as defined by (3.25) have the same value. While the optimum threshold value is $\mathring{E} = 0$ for a bipolar system, for unipolar nonredundant signals eqn (3.26) yields

$$\mathring{E} = \frac{1}{2} \sum_{\nu = -\infty}^{+\infty} g_\nu \tag{3.33}$$

It is further obtained from eqns (3.30) and (3.31) that for the symbols of a worst-case binary sequence $\langle q_\nu \rangle_U$

$$q_\nu = \begin{cases} q_0 & \text{when } g_{-\nu} < 0 \\ \overline{q_0} & \text{when } g_{-\nu} > 0 \end{cases} \qquad \nu \neq 0 \tag{3.34}$$

$\overline{q_0}$ denotes the complementary symbol of q_0. If $g_{-\nu} = 0$, the symbol q_ν of a worst-case symbol sequence has no effect and can be both O and L.

Example With a Gaussian basic detection pulse (see Fig. 3.3(a)) all pulse values g_ν are positive. Therefore the two sequences ...O O L O O... and ...L L O L L... are both worst cases. With the Nyquist pulse of Fig. 3.3(c), all basic pulse values are identically zero for $\nu \neq 0$. For this reason, all symbol sequences here are worst case and they possess the same error probability.

These examples show that the inverse sequence of a worst-case symbol sequence is itself a worst-case symbol sequence. The worst-case symbol sequences $\langle q_\nu \rangle_U$ are determined by the signs of the basic detection pulse values $g_\nu = g_d(T_D + \nu T)$. It thus emerges that the worst-case symbol sequences depend not only on the basic pulse $g_d(t)$ but also on the detection time T_D. Therefore for different values of T_D there can also be different worst-case symbol sequences. In general, the eye opening $o(T_D)$ is thus determined by different worst-case symbol sequences.

3.3 The mean error probability of a binary system

An important criterion for assessment of digital transmission systems is the **mean error probability**:

$$p_{\mathrm{M}} = \overline{P(v_\nu \neq q_\nu)} = \lim_{N\to\infty} \frac{1}{2N+1} \sum_{\nu=-N}^{+N} p_{\mathrm{S}_\nu} \tag{3.35}$$

Calculation according to this definition corresponds to a time average across the sequence $\langle q_\nu \rangle$ (see Fig. 3.1). However, it is often more advantageous to deal with the average corresponding to Fig. 3.4 across all the distinguishable symbol sequences $\langle q_\nu \rangle_i$ in the eye pattern. For an average across all i we obtain the mean SEP

$$p_{\mathrm{M}} = \sum_i p_i p_{\mathrm{S}_i} \tag{3.36}$$

Here p_i is the **probability of occurrence** of the ith eye pattern line $\tilde{d}_i(t)$ and p_{S_i} is the corresponding SEP according to definition (3.18). If the basic detector input pulse $g_d(t)$ has only v precursors and n postcursors, the individual eye lines $\tilde{d}_i(T_{\mathrm{D}})$ at the detection time T_{D} can have at the most 2^{n+v+1} different values. For a nonredundant source all possible eye lines occur with equal probability $p_i = 1/2^{n+v+1}$, and the mean SEP is

$$p_{\mathrm{M}} = 2^{-(n+v+1)} \sum_{i=1}^{2^{n+v+1}} p_{\mathrm{S}_i} \tag{3.37}$$

This equation shows that, in contrast with calculation by means of a time averaging according to definition (3.35), the computing cost is considerably reduced as long as the number of precursors and postcursors to be considered is kept small. With optimum decision value $\mathring{E}$ the computing cost can be further halved if the symmetry of the eye pattern is used in the calculation. Therefore in the case of the Gaussian basic detection pulse in Fig. 3.3(a) only four different probabilities need to be considered. For the intersymbol-interference-free system in Fig. 3.3(c) $n = v = 0$ and therefore all symbol sequences are corrupted with equal probability:

$$p_{\mathrm{S}_i} = p_{\mathrm{M}} = Q\left(\frac{g_0}{N_d^{\,1/2}}\right) \quad \text{for all } i \tag{3.38}$$

Here a bipolar useful signal and a Gaussian noise signal are assumed. The detection value and the detection time are optimal.

Now consider further the error probability of a **regenerative transmission system**. For long-distance digital transmission (**long-haul systems**), intermediate amplifiers (**regenerators**) have to be inserted in the transmission link at regular intervals. In these regenerators the received signal is equalized, detected and transmitted along the following section of the transmission link with its original amplitude, shape and correct timing. Figure 2.1 is a block diagram representing each section of the transmission link.

In contrast with an analog system, in the case of a digital system there is no accumulation of noise but an increase in the mean error probability. Assuming that the digital system consists of K equally designed regenerator sections and that the mean SEP of each individual section is p_M, we obtain for the mean error probability of the whole system

$$p_K \leqslant K p_M \tag{3.39}$$

If p_M is relatively small, which can be assumed for long-haul systems, the equals sign is valid to a close approximation. However, it is possible with a larger error probability that a previous symbol error is removed by a wrong decision in a later regenerator section. In addition, with nonideal clock recovery there is an accumulation of the phase jitter originating in the regenerators and hence the error probability in a regenerative transmission system is slightly increased.

3.4 Error probability of a multilevel transmission system

There is only one threshold value E in a binary system. Accordingly, each of the binary symbols O and L can be corrupted in only one way, i.e. O to L or L to O. However, in a quaternary system ($M = 4$) with the symbol set $\ominus$, $-$, $+$, $\oplus$ and the $M - 1 = 3$ threshold values E_1, E_2 and E_3, the inner symbols $-$ and $+$ can be corrupted in both directions (see Fig. 2.13(b)). Correspondingly the error probability calculation for a multilevel transmission system differs from the computation procedures for the binary systems discussed above.

3.4.1 Multilevel Nyquist systems

To make this basic difference as clear as possible, the intersymbol-interference-free detector input pulse shown in Fig. 3.5(c) and a bipolar quaternary transmitted signal with the possible amplitude coefficients ± 1 and $\pm 1/3$ are considered first. As no intersymbol interference occurs, at the optimum detection time $\mathring{T}_D = 0$ all eye pattern lines $\tilde{d}_i(t)$ intersect at the four points $\pm g_0$ and $\pm g_0/3$. Therefore $M - 1 = 3$ eyes can be seen in the eye pattern for a quaternary system. In it the vertical eye opening defined by eqn (3.24) is the same for all eyes. Thus

$$o(\mathring{T}_D = 0) = \tfrac{2}{3} g_0 \tag{3.40}$$

for the maximum eye opening at the optimum detection time ($\mathring{T}_D = 0$). It is assumed in the following that the $M - 1$ threshold values are symmetrically placed about zero ($E_1 = -E$; $E_2 = 0$; $E_3 = E$) and that the detection noise signal $\check{d}(t)$ is distributed symmetrically. First the error probability for the symbol ⊖ is calculated. The useful detection sample value is equal to $-g_0$ in this case, and by analogy with eqn (3.5) we obtain

$$p_{\ominus} = P\{\overset{\times}{d}(T_{\mathrm{D}}) > g_0 - E\} = \int_{g_0 - E}^{+\infty} f_{\overset{\times}{d}}(\overset{\times}{d})\,\mathrm{d}\overset{\times}{d} \tag{3.41}$$

In the negative direction the noise signal can be arbitrarily large without the occurrence of an erroneous decision. However, for the SEP of the inner symbol − we have

$$p_{-} = P\left\{\overset{\times}{d}(T_{\mathrm{D}}) < \frac{g_0}{3} - E\right\} + P\left\{\overset{\times}{d}(T_{\mathrm{D}}) > \frac{g_0}{3}\right\} \tag{3.42}$$

As a consequence of the symmetrical characteristics assumed above $p_{\oplus} = p_{\ominus}$ and $p_{+} = p_{-}$. It therefore follows that for the mean error probability of the quaternary

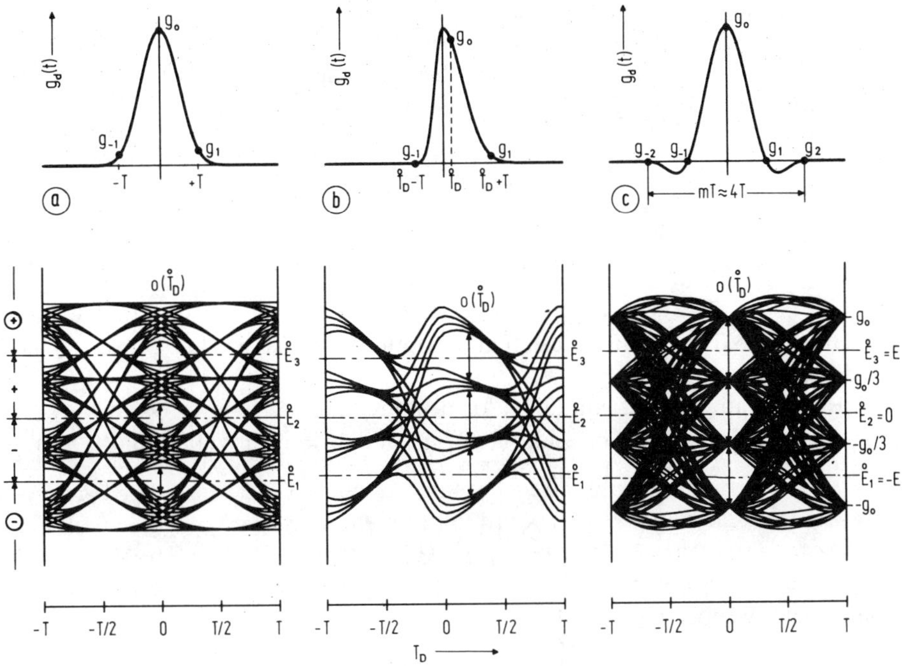

Fig. 3.5 Basic detection pulse and quaternary eye pattern for (a) a symmetric, (b) a non-symmetric and (c) an intersymbol-interference-free basic detection pulse.

Nyquist system, with equally probable source symbols,

$$p_{\mathrm{M}} = \frac{1}{4}(p_{\ominus} + p_{-} + p_{+} + p_{\oplus}) = \frac{1}{2}\int_{g_0 - E}^{+\infty} f_{\check{d}}(\check{d})\,\mathrm{d}\check{d} + \frac{1}{2}\int_{E - g_0/3}^{+\infty} f_{\check{d}}(\check{d})\,\mathrm{d}\check{d}$$

$$+\frac{1}{2}\int_{g_0/3}^{+\infty} f_{\check{d}}(\check{d})\,\mathrm{d}\check{d} \qquad (3.43)$$

p_{M} depends very strongly on the decision value. The optimum decision value $\mathring{E}$ certainly lies between $g_0/3$ and g_0. If the PDF $f_{\check{d}}(\check{d})$ of the noise signal decreases monotonically, which can in general be assumed, the mean error probability is a minimum for $E = 2g_0/3$, and for disturbances of Gaussian distribution

$$p_{\mathrm{M}} = \frac{3}{2}\int_{g_0/3}^{+\infty} f_{\check{d}}(\check{d})\,\mathrm{d}\check{d} = \frac{3}{2}Q\left(\frac{g_0}{3N_d^{1/2}}\right) \qquad (3.44)$$

This result shows that, with a nonredundant transmitter signal and signal-independent noise, the optimum decision value lies midway between the upper and lower eye pattern boundaries. With the possible amplitude coefficients a_μ defined by eqn (2.10), it follows that the **optimum decision value** of an M-level system with monotonically decreasing PDF of the noise signal is

$$\mathring{E}_\mu = \begin{cases} \dfrac{2\mu - 1}{2(M-1)}\sum_v g_v & \text{for unipolar signals} \\ \dfrac{2\mu - M}{M-1} g_0 & \text{for bipolar signals} \end{cases} \qquad (3.45)$$

Here $\mu = 1, \ldots, M-1$ and $g_0 = g_d(T_{\mathrm{D}})$, which is the main value of the basic detection pulse. Thus it can be seen that the optimum decision values also depend on the detection time T_{D}.

By generalization of eqn (3.44) for a nonredundant M-level Nyquist system with signal-independent Gaussian noise, optimum decision values and optimum detection times, we obtain the mean SEP

$$p_{\mathrm{M}} = \frac{2(M-1)}{M} Q\left\{\frac{o(\mathring{T}_{\mathrm{D}})/2}{N_d^{1/2}}\right\} \qquad (3.46)$$

The vertical eye opening

$$o(\mathring{T}_{\mathrm{D}}) = \begin{cases} \dfrac{g_o}{M-1} & \text{for unipolar signals} \\ \dfrac{2g_o}{M-1} & \text{for bipolar signals} \end{cases} \qquad (3.47)$$

for an M-level Nyquist system is thus a factor of $M - 1$ smaller than that for the corresponding binary system.

3.4.2 Intersymbol interference effect on multilevel systems

If the basic detection pulse $g_d(t)$ shows intersymbol interference, the eye opening $o(T_D)$ is smaller than the value given by eqn (3.47). In this case the individual eye pattern lines $\tilde{d}_i(t)$ do not intersect at only M points as for a Nyquist system (Figs 3.5(a) and 3.5(b)).

To calculate the vertical eye opening, the worst-case symbol sequences $\langle q_v \rangle_U$ again have to be determined first. While (at least) two worst-case symbol sequences exist for a binary system, i.e. the upper and lower eye boundary lines, for an M-level transmission system there are at least $2(M - 1)$ such worst-case sequences.

For example, let us consider the upper eye of Fig. 3.5(a). Since all the precursors and postcursors are greater than or equal to zero, the upper boundary line of the upper eye arises from the sequence ... $\ominus \ominus \oplus \ominus \ominus$ ..., and for a bipolar transmitter signal

$$\tilde{d}_{up}(T_D) = g_0 - \sum_{v \neq 0} g_v \tag{3.48}$$

The lower boundary line $\tilde{d}_{lo}(t)$ relates to the symbol sequence ... $\oplus \oplus + \oplus \oplus$ ..., so that at the detection time T_D

$$\tilde{d}_{lo}(T_D) = \frac{g_0}{3} + \sum_{v \neq 0} g_v \tag{3.49}$$

Thus, by analogy with eqn (3.32), we obtain for the vertical eye opening

$$o(T_D) = \tilde{d}_{up}(T_D) - \tilde{d}_{lo}(T_D) = 2\left(\frac{g_0}{3} - \sum_{v \neq 0} g_v\right) \tag{3.50}$$

The central eye is bounded by the lines of the sequences ... $\ominus \ominus + \ominus \ominus$ ... and ... $\oplus \oplus - \oplus \oplus$ .... As the intersymbol interference originates from the same symbols, equally spaced amplitude coefficients produce the same eye opening. Thus it is shown that, for a nonredundant M-level transmission system, all $M - 1$ eyes have the same vertical eye opening even if the eye shapes are different. For a quaternary system the central eye is symmetrical about the decision value $\mathring{E}_2 = 0$, while both outer eyes are asymmetrical about their thresholds $\mathring{E}_1$ and $\mathring{E}_3$.

It is generally true for the eye pattern of a nonredundant bipolar multilevel system, if arbitrary basic detection pulse values g_v are assumed, that

$$o(T_D) = 2\left(\frac{g_0}{M - 1} - \sum_{v \neq 0} |g_v|\right) \tag{3.51}$$

Here it is assumed that the eye is open, i.e. that the sum of all precursors and postcursors fulfils the following condition:

$$\sum_{\nu \neq 0} |g_\nu| < \frac{g_0}{M-1} \tag{3.52}$$

Otherwise $o(T_D)$ is set to zero.

The result shows that intersymbol interference is more noticeable in a multilevel system than in a binary system. For illustration consider again the basic pulse shown in Fig. 3.3(a). For a level number $M = 2$ the normalized eye opening amounts to more than 35%. In contrast a closed eye would result for a four-level transmission because the sum of all precursors and postcursors is greater than $g_0/3$.

3.4.3 Mean error probability of a multilevel system

Calculation of the mean SEP of an M-level transmission system is again derived from the eye pattern. If the basic detection pulse has v precursors and n postcursors, there is a maximum of M^{n+v+1} different values of the useful detection sample value $\tilde{d}_i(T_D)$. With a nonredundant M-level source all these values have equal probability, so by using eqn (3.36) the mean error probability can be written

$$p_M = M^{-(n+v+1)} \sum_{i=1}^{M^{n+v+1}} p_{S_i} \tag{3.53}$$

p_{S_i} is the SEP, defined by eqn (3.18), for the ith symbol sequence, the calculation of which is more costly for a multilevel system than for a binary system. If the middle symbol of the sequence (the symbol to be detected) is an "inner" symbol ($q_0^{(i)} = q_\mu, \mu = 2, \ldots, M-1$), then the useful detection sample value $\tilde{d}_i(T_D)$ lies between the neighboring decision values $E_{\mu-1}$ and E_μ. Therefore two detection distances have to be calculated, i.e. the lower and the upper detection distances (see definition (3.17)):

$$\begin{aligned} D_{i,\mathrm{lo}} &= \tilde{d}_i(T_D) - E_{\mu-1} \\ D_{i,\mathrm{up}} &= E_\mu - \tilde{d}_i(T_D) \end{aligned} \tag{3.54}$$

Consequently, and by analogy with eqn (3.42), with Gaussian noise we have for the SEP of the ith symbol sequence

$$p_{S_i} = Q\left(\frac{D_{i,\mathrm{lo}}}{N_d^{1/2}}\right) + Q\left(\frac{D_{i,\mathrm{up}}}{N_d^{1/2}}\right) \tag{3.55}$$

If $q_0^{(i)}$ is an "outer" symbol (q_1 or q_M), only one of two decision values (E_1 or E_{M-1} respectively) can be exceeded. On equating the decision values $E_0 = -\infty$ and $E_M = +\infty$, the symbol error probabilities of the "outer" symbols can also be calculated from eqn (3.55).

3.5 Approximations for the mean error probability

If a large number of precursors and postcursors contribute to the intersymbol interference, precise calculation of the mean SEP from eqns (3.37) and (3.53) respectively has a very large computing cost associated with it. For example, for a basic detection pulse with five precursors and five postcursors ($n = v = 5$) we have to average over $2^{11} = 2048$ values to determine the mean SEP of a nonredundant binary system. When the symmetry with an optimum threshold is taken into account, the number of values reduces to $2^{10} = 1024$.

For multilevel systems the computing cost increases very rapidly with the number of levels M. For example, with $n = v = 5$ and $M = 4$ the calculation of $4^{11} = 4\,194\,304$ or $4^{10} = 1\,048\,576$ values is necessary. Therefore the development of approximations for the mean SEP is of particular importance here.

For the derivation of these approximations, the detection sample values $d_\nu = d(T_D + \nu T)$ are considered as random variables with the PDF $f_d(d_\nu)$, from which the mean SEP p_M can be calculated. The sample value d_ν consists of the useful component $\tilde{d}_\nu$ and the noise component $\overset{\times}{d}_\nu$, and with a signal-independent noise these components are statistically independent of each other. Therefore, for the PDF $f_d(d_\nu)$ we have [1.21]

$$f_d(d_\nu) = f_{\tilde{d}}(d_\nu) * f_{\overset{\times}{d}}(d_\nu) = \int_{-\infty}^{+\infty} f_{\overset{\times}{d}}(x) f_{\tilde{d}}(d_\nu - x)\,\mathrm{d}x \tag{3.56}$$

Consequently $f_d(d_\nu)$ is calculated as the convolution product of the two PDFs $f_{\tilde{d}}(\tilde{d}_\nu)$ and $f_{\overset{\times}{d}}(\overset{\times}{d}_\nu)$. If a stationary noise signal with Gaussian distribution is assumed, $f_{\overset{\times}{d}}(\overset{\times}{d}_\nu)$ is defined by one parameter, i.e. the detection noise power N_d. The difficulty of exact calculation of the mean SEP p_M lies in the fact that, in general, the PDF $f_{\tilde{d}}(\tilde{d}_\nu)$ of the useful sample value cannot be specified in closed form. Depending on the basic detection pulse values g_ν, the useful detection sample value $\tilde{d}_\nu$ has quite specific discrete values $\tilde{d}_i$. Therefore the PDF $f_{\tilde{d}}(\tilde{d}_\nu)$ is composed of a sum of Dirac functions (see the first row of Table 3.1). For a basic pulse with v precursors and n postcursors there are up to M^{n+v+1} such Dirac functions. If the M-level source considered is nonredundant, the weights of the Dirac functions are all the same and we obtain

$$f_{\tilde{d}}(\tilde{d}_\nu) = M^{-(n+v+1)} \sum_{i=1}^{M^{n+v+1}} \delta(\tilde{d}_\nu - \tilde{d}_i) \tag{3.57}$$

For symmetric pulse values ($g_{-\nu} = g_\nu$), several Dirac functions coincide and there are correspondingly fewer Dirac functions with different weights.

3.5.1 Summary of the approximations

The approximations for the mean SEP summarized in Table 3.1 differ in the approximation to the PDF $f_{\tilde{d}}(\tilde{d}_\nu)$. All the approximations specified are valid for nonredundant bipolar signals, a Gaussian distributed noise signal and optimum decision values defined by eqn (3.45).

In the **best-case mean error probability** p_G the disturbing effect of the intersymbol interference is not considered. In the special case of the Nyquist system $p_G = p_M$ (eqn (3.46)); otherwise p_G is a very inaccurate lower limit for the mean error probability p_M. It only serves as an estimate of how far p_M can be reduced, at most, if all intersymbol interferences (at equal detection noise power N_d) are completely removed (see Chapter 6).

The **worst-case symbol error probability** p_U is defined as the error probability for the worst-case symbol sequence (see definition (3.21)). p_U is never smaller than the mean SEP and is thus an upper limit for p_M. This approximation assumes that all sequences are corrupted with equal (i.e. the maximum) probability (see row 3 of Table 3.1). From eqns (3.22) and (3.27) we obtain for a nonredundant bipolar binary system

$$p_U = Q\left\{\frac{o(T_D)/2}{N_d^{1/2}}\right\} \quad \text{where} \quad \frac{o(T_D)}{2} = g_0 - g_U \tag{3.58}$$

Table 3.1 Approximations for the mean symbol error probability

Designation	PDF $f_{\tilde{d}}(\tilde{d}_v)$ for M = 2	PDF $f_{\tilde{d}}(\tilde{d}_v)$ for M = 4
Mean symbol error probability $p_M = M^{-(n+v+1)} \sum_{i=1}^{M^{n+v+1}} p_{Si}$	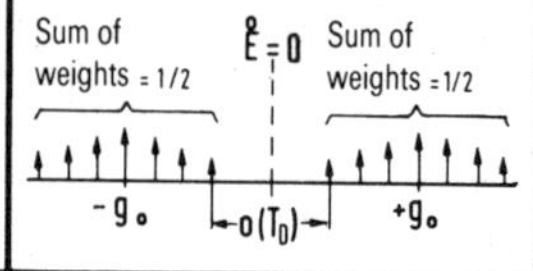	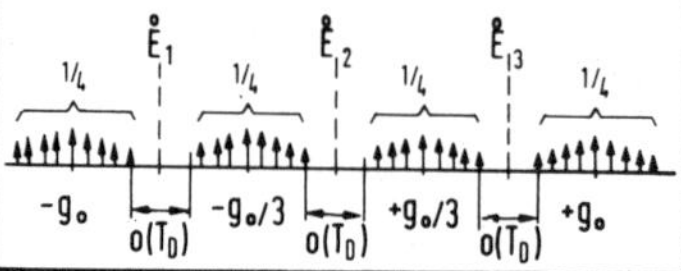
Best-case mean symbol error probability $p_G = \frac{2(M-1)}{M} \cdot Q\left(\frac{g_0/(M-1)}{\sqrt{N_d}}\right)$	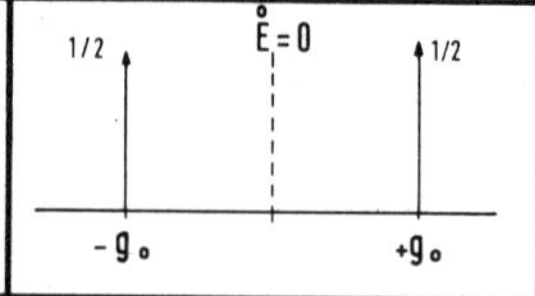	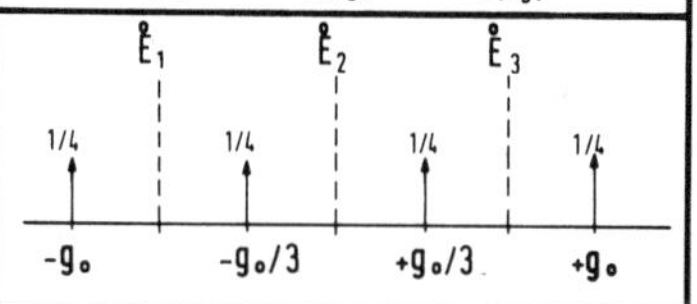
Worst-case symbol error probability $p_U = \gamma_{pU} \cdot Q\left(\frac{o(T_D)/2}{\sqrt{N_d}}\right)$	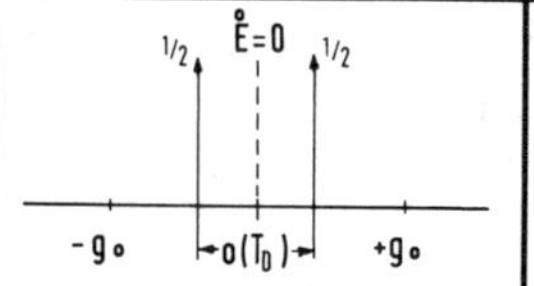	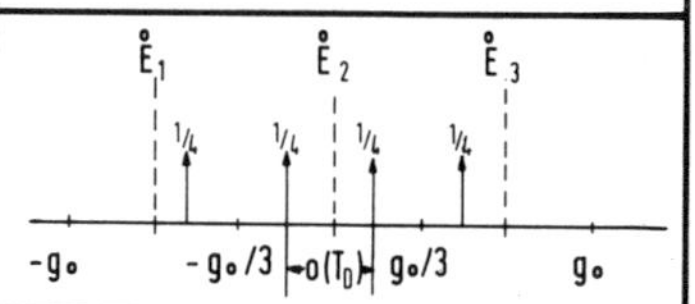
Mean squared error $p_{MSE} = Q\left(\frac{g_0/(M-1)}{(MSE)^{1/2}}\right)$	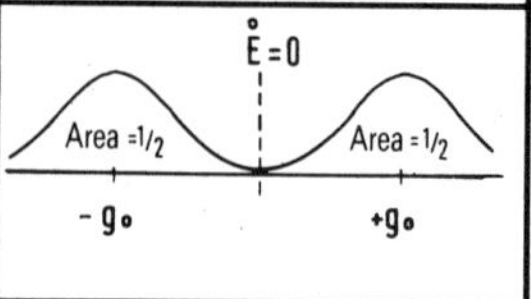	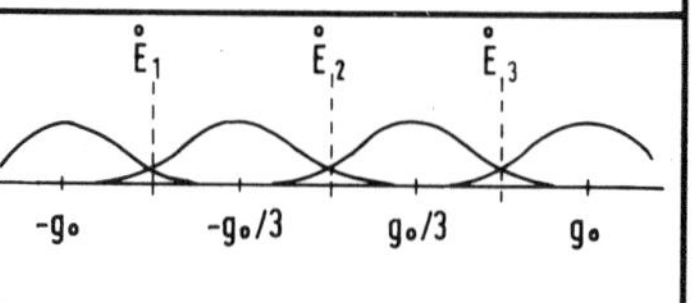
Glave's variance bound for M = 2 $p_{Glv} = \frac{\beta}{2} \gamma_{pU} \cdot Q\left(\frac{o(T_D)/2}{\sqrt{N_d}}\right) + (1-\beta) Q\left(\frac{g_0}{\sqrt{N_d}}\right)$	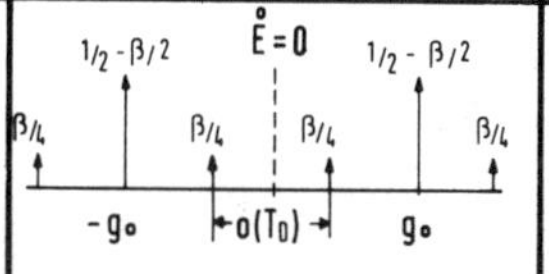	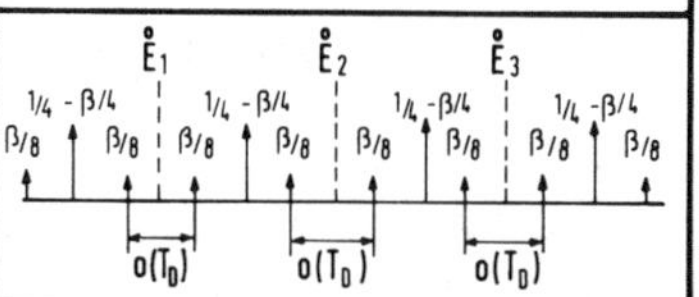

g_U is an abbreviation for the sum of the absolute values of all precursors and postcursors:

$$g_U = \sum_{v \neq 0} |g_v| \tag{3.59}$$

In contrast, for the worst-case SEP of an M-level system

$$p_U = Q\left\{\frac{1}{N_d^{1/2}}\left(\frac{g_0}{M-1} - g_U\right)\right\} + Q\left\{\frac{1}{N_d^{1/2}}\left(\frac{g_0}{M-1} + g_U\right)\right\} \tag{3.60}$$

Proof In a multilevel system the two "outer" symbols can only be corrupted in one direction whereas the "inner" symbols can be corrupted in both directions. Therefore the symbol to be detected in a worst-case symbol sequence is certainly an "inner" symbol, and from eqn (3.55)

(a) $$p_U = \max_i (p_{S_i}) = \max_i \left\{Q\left(\frac{D_{i,\mathrm{up}}}{N_d^{1/2}}\right) + Q\left(\frac{D_{i,\mathrm{lo}}}{N_d^{1/2}}\right)\right\}$$

Since the Q-function derivation decreases monotonically, the maximum value of p_U results for the case when either $D_{i,\mathrm{up}}$ or $D_{i,\mathrm{lo}}$ has the smallest possible value. With optimum decision values defined by eqn (3.45) we obtain from eqn (3.51)

(b) $$\min_i (D_{i,\mathrm{up}}) = \min_i (D_{i,\mathrm{lo}}) = \frac{o(T_D)}{2} = \frac{g_0}{M-1} - g_U$$

The summation

(c) $$D_{i,\mathrm{up}} + D_{i,\mathrm{lo}} = \frac{2g_0}{M-1}$$

is independent of the intersymbol interference. If $D_{i,\mathrm{up}}$ ($D_{i,\mathrm{lo}}$) has the smallest value as defined by (b), $D_{i,\mathrm{lo}}$ ($D_{i,\mathrm{up}}$) is a maximum (see Fig. 3.5):

(d) $$\max_i (D_{i,\mathrm{lo}}) = \max_i (D_{i,\mathrm{up}}) = \frac{g_0}{M-1} + g_U$$

The equation above emerges from this. □

The two equations for the worst-case error probability of the binary system and the multilevel system can be obtained by the introduction of the correction factor γ_{p_U} defined in row 3 of Table 3.1. This factor has a value of unity for the binary system. For a multilevel system it lies between 1 and 2 and is approximately given by

$$\gamma_{p_U} \approx 1 + \frac{g_0 - (M-1)g_U}{g_0 + (M-1)g_U} \exp\left\{-\frac{2g_0 g_U}{(M-1)N_d}\right\} \tag{3.61}$$

For a multilevel Nyquist system with optimum detection time ($M > 2$; $g_U = 0$) we have $\gamma_{p_U} = 2$. The larger the effect of intersymbol interference is, i.e. the larger the value of g_U (eqn (3.59)), the nearer is the value of γ_{p_U} to unity for $M > 2$.

The **mean squared error** (MSE) is often used as the optimization criterion for digital transmission systems. This is the mean value of the square of the deviation of the detection sample values d_ν from their original values $a_\nu g_0$:

$$\text{MSE} = \lim_{N\to\infty} \frac{1}{2N+1} \sum_{\nu=-N}^{+N} (d_\nu - a_\nu g_0)^2 = \overline{(d_\nu - a_\nu g_0)^2} \tag{3.62}$$

Splitting the squared term and using eqn (2.81) we obtain

$$\text{MSE} = \overline{\check{d}_\nu^{\,2}} + 2\overline{(\tilde{d}_\nu - a_\nu g_0)}\,\overline{\check{d}_\nu} + \overline{(\tilde{d}_\nu - a_\nu g_0)^2} \tag{3.63}$$

Here it is assumed that the useful and the noise signals are not correlated. The mean squared value of the noise sample values is identical with the detection noise power N_d. With zero mean noise, the middle term of eqn (3.63) vanishes. The last term corresponds to the "mean squared value of the intersymbol interferences", which can be calculated from eqn (3.12):

$$\overline{(\tilde{d}_\nu - a_\nu g_0)^2} = \overline{\left(\sum_{\nu\neq 0} a_\nu g_{-\nu}\right)^2} = \sum_{\nu\neq 0}\sum_{\kappa\neq 0} \overline{a_\nu a_\kappa}\, g_{-\nu} g_{-\kappa} \tag{3.64}$$

The mean value $\overline{a_\nu a_\kappa}$ is identical with the discrete autocorrelation function $l_a(\kappa - \nu)$ (see definition (2.21)). If a nonredundant bipolar signal is assumed, then for $\kappa \neq \nu$ the autocorrelation function $l_a(\kappa - \nu)$ is equal to zero and we obtain from eqn (2.32)

$$\text{MSE} = N_d + l_a(0) \sum_{\nu\neq 0} g_\nu^{\,2} = N_d + \frac{M+1}{3(M-1)} \sum_{\nu\neq 0} g_\nu^{\,2} \tag{3.65}$$

It is not usually possible to deduce the mean SEP p_{M} from the MSE. Nevertheless, an approximation for p_{M} is often derived from eqn (3.65) (see row 4 of Table 3.1). The approximation p_{MSE} implies that the disturbing effect of the intersymbol interference can be sufficiently accurately taken into account by an increase in the detection noise power N_d. However, this is only possible if the PDF $f_{\tilde{d}}(\tilde{d}_\nu)$ can be approximated by a sum of M Gaussian functions. The Gaussian approximation is allowed, for example, if the intersymbol interferences are attributed to many relatively small precursors and postcursors. In all other cases this approximation is not possible and causes the system optimization to produce an invalid result.

Other approximations given in the references for the mean error probability will only be mentioned briefly here. Saltzberg's [3.10] and Lugananni's [3.7] approximations are each based on the Chernoff inequality and are therefore often referred to as the **Chernoff bounds**. Both approximations give a significantly better estimate than the worst-case error probability only if the intersymbol interference effect is very large (approximately closed eye) and is the product of a large number of "disturbers" (basic pulse values). Calculation of both these approximations involves considerable computing cost and therefore they are suitable as optimization criteria to only a limited degree. The same is true for the approximation given by Shimbo and Celebiler [3.11] which is based on a Gram–Charlier series expansion. Finally we again mention the **variance bound**, which was derived by Glave [3.4]. Glave shows that for a given "total magnitude" g_{U} and a given "variance" σ^2 of the

$$\text{MSE} = \hat{d}_\nu^{\,2} + 2(\tilde{d}_\nu - \quad)\,\hat{d}_\nu + (\tilde{d}_\nu - \quad)^2 \tag{3.63}$$

intersymbol interferences, the PDF $f_{\tilde{d}}(\tilde{d}_\nu)$ can be approximated as shown in row 5 of Table 3.1. The resulting approximation p_{GIv} is an upper boundary for p_{M}. The β term is calculated from

$$\beta = \frac{\sigma^2}{g_{\mathrm{U}}{}^2} = \frac{M+1}{3(M-1)} \sum_{\nu \neq 0} g_\nu{}^2 \Big/ \Big(\sum_{\nu \neq 0} |g_\nu| \Big)^2 \qquad (3.66)$$

3.5.2 Comparison of the approximations

Figure 3.6 shows the mean SEP and the above approximations to it as a function of the mean detection signal-to-noise ratio $10 \lg \rho_d$ (see definition (2.73)). The basic detection pulse of Fig. 2.11, which was produced by rectangular NRZ transmitter pulses and a Gaussian pulse shaper with cut-off frequency f_{I}, was used as the basis for this figure. The characteristics of the basic detection pulse are that it exhibits no negative values and that the intersymbol interference is caused by only a few precursors and postcursors.

Figure 3.6(a) is valid for a narrow-band pulse shaper with cut-off frequency $0.3R$, so that large intersymbol interferences result from the basic detection pulse values $g_{-1} = g_1 = 0.4g_0$. $f_{\mathrm{I}} = 0.5R$ in Fig. 3.6(b), i.e. the detection pulse values are significantly smaller in this case ($g_{-1} = g_1 = 0.13g_0$). On the basis of these figures the following conclusions are possible.

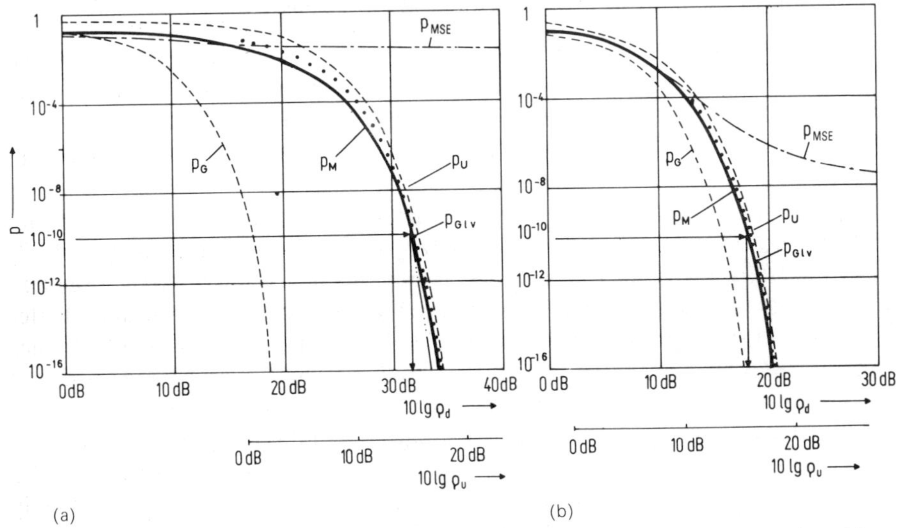

Fig. 3.6 Comparisons of approximations to the mean SEP for a nonredundant binary system: (a) $f_{\mathrm{I}} = 0.3R$; (b) $f_{\mathrm{I}} = 0.5R$.

(a) The mean SEP p_M depends strongly on the basic detection pulse considered. The larger the intersymbol interferences are, the larger the (mean) detection signal-to-noise ratio ρ_d has to be to prevent p_M exceeding a given value. For the cut-off frequency $f_I = 0.3R$, for example, because of the comparatively larger intersymbol interferences $10 \lg \rho_d$ must be greater than 32 dB for the value of p_M not to be greater than 10^{-10}. For $f_I = 0.5R$ about 18 dB is sufficient for the same SEP. Therefore the (mean) detection signal-to-noise ratio ρ_d is unsuitable for an optimization and comparison criterion.

(b) The smaller the intersymbol interferences are, the better the mean SEP p_M is represented by the defined approximations. For a binary system without intersymbol interference ($f_I \to \infty$), all the approximations are exact.

(c) When the intersymbol interference effects are large, the best-case mean SEP p_G is an inexact and too extreme lower boundary. The significance of this approximation is that it can be used to estimate the achievable improvement by complete compensation for all the precursors and postcursors. This requires a more complicated detector, however (see Chapter 6).

(d) With high noise power N_d and thus smaller detection signal-to-noise ratio $10 \lg \rho_d$, the approximation p_{MSE} derived from the MSE is a good measure of p_M but is not the upper bound [1.15]. With large detection signal-to-noise ratio, and hence small mean SEP, this approximation is very inaccurate because the assumed Gaussian distribution for the intersymbol interferences is not valid.

(e) The worst-case SEP p_U gives with open eyes a good reliable upper bound for the mean SEP, which is about an order of magnitude greater than p_M throughout the range. Glave's approximation p_{GIv} is in general only insignificantly more precise. As, in addition, p_U is very simple to calculate, this approximation is often applied as an optimization and comparison criterion for digital transmission systems.

Chapter 4

Coded Transmission Systems (Transmission Codes)

Contents In this chapter the effect of coding on digital transmission is described and discussed. First, in Section 4.1 the basic characteristics of a coded transmission are stated and valid parameters for all transmission codes are defined. Subsequently, in Sections 4.2 and 4.3 the special features of pseudo-multilevel codes and block codes are comparatively assessed.

Assumptions The premises of Chapters 2 and 3 remain valid. The digital source is assumed always to be nonredundant and binary.

4.1 Principles of coded transmission

The source symbol sequence $\langle q_v \rangle$ is often unsuitable for the transmission of data through the channel. For example, in the case of a band-pass channel neither a long O nor a long L sequence can be transmitted with sufficiently small error probability because of the lower band cut-off. Thus special measures have to be introduced at the transmitter which prevent the appearance of these long O and L sequences. For example, the source symbol sequence can be coded prior to transmission. By **coding** we mean the substitution of one or several source symbols by the corresponding code symbol or code word as defined by a fixed coding rule.

Figure 4.1 shows the block diagram for a digital transmission system with coding. The **coder** performs the task of converting the source symbol sequence $\langle q_v \rangle$ into the **coded symbol sequence** $\langle c_v \rangle$. The transmitter signal $s(t)$ that results is better matched to the characteristics of the transmission channel, the noise and the receiving equipment. The **decoder** is necessary to reconvert the coding performed at the transmitter. From the **regenerated symbol sequence** $\langle r_v \rangle$ it produces the sink symbol sequence $\langle v_v \rangle$. When no transmission errors occur $\langle r_v \rangle = \langle c_v \rangle$ and $\langle v_v \rangle = \langle q_v \rangle$.

For the remaining digital transmission system—consisting of transmitter pulse shaper, transmission channel, equalizer and detector—the statements made in Chapters 2 and 3 are valid. However, the level number M and the symbol duration

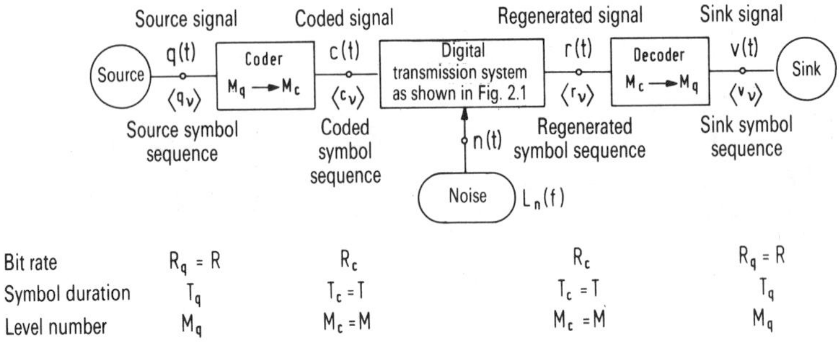

Fig. 4.1 Block diagram of a digital transmission system with coding.

no longer relate to the source signal but to the coded signal $c(t)$. Equally, the mean symbol error probability

$$p_M = \overline{P(r_\nu \neq c_\nu)} \tag{4.1}$$

relates to the coded symbol sequences $\langle c_\nu \rangle$ and $\langle r_\nu \rangle$ rather than as shown in eqn (3.35).

In the following sections we attempt to characterize the various **transmission codes (line codes)** by means of common parameters.

4.1.1 Symbol-wise and block coding

There is a basic difference between symbol-wise and block coding. With **symbol-wise coding** a coded symbol c_ν is produced from each incoming source symbol q_ν, which can depend on the N preceding symbols $q_{\nu-1}, q_{\nu-2}, \ldots, q_{\nu-N}$ as well as on the immediate symbol. N is the **order of the code**.

In the case of a binary source we often refer to **bit-wise coding**. It is typical of symbol-wise (or bit-wise) coding that the symbol durations T_c and T_q of the coded and source signals agree (Fig. 4.2(a)). In general no large time delays occur, which is unavoidable with block coding because of causality.

In **block coding** a block of m_q source symbols corresponds to a block of m_c coded symbols (Fig. 4.2(b)). For most **block codes** the block lengths m_q and m_c differ and so do the symbol durations of the coded and source signal, so that

$$m_q T_q = m_c T_c \tag{4.2}$$

Example Figure 4.2 shows a segment of the binary source signal $q(t)$ and the corresponding coded signal $c(t)$ for a ternary code in the cases of symbol-wise and block coding.

For the twinned binary code shown in Fig. 4.2(a) the following coding rule is an example of symbol-wise coding. If $q_\nu = q_{\nu-1}$ then the code symbol $c_\nu = 0$ is selected.

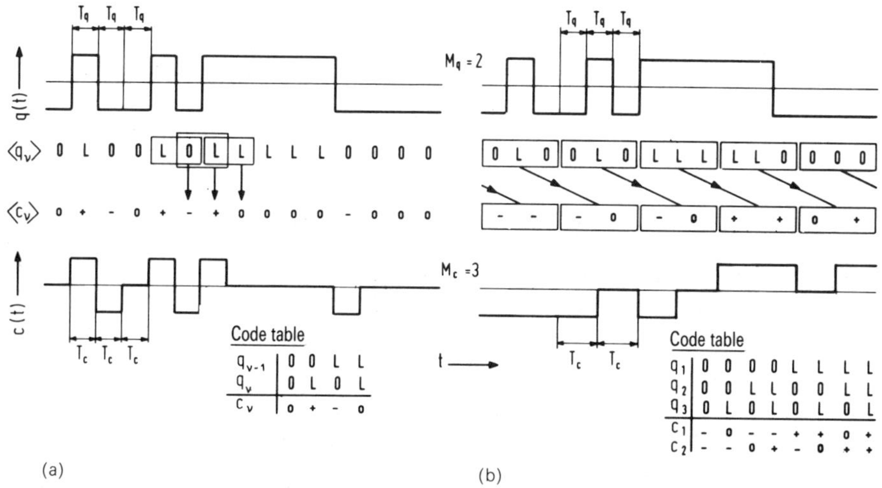

Fig. 4.2 Examples of symbol-wise and block coding: (a) twinned binary code ($N = 1$); (b) 3B2T code ($m_q = 3$; $m_c = 2$).

If the symbols q_ν and $q_{\nu-1}$ differ, however, the ternary symbol $c_\nu = +$ is substituted for the binary symbol $q_\nu = \mathrm{L}$ and the ternary symbol $-$ is substituted for the binary symbol O.

Figure 4.2(b) shows block coding with block lengths $m_q = 3$ and $m_c = 2$. Every three binary symbols are replaced by two ternary symbols (3B2T code), where the relationship between the binary and ternary blocks is defined by the code table shown. In contrast with symbol-wise coding, with block coding the coded signal symbol rate $1/T_c$ is about one-third less than the source signal symbol rate $1/T_q$, which is better suited to the spectral characteristics of most transmission channels.

When the twinned binary code is used as in Fig. 4.2(a) it is guaranteed by the coding rule that the coded symbol sequence $\langle c_\nu \rangle$ contains no long plus or minus sequences, so that the coded signal $c(t)$ can also be transmitted via a direct component-free channel. Similarly, with the 3B2T code illustrated here, the symbol 0 cannot occur more than twice in sequence.

4.1.2 Code redundancy

We now define the transmission code parameters relevant to Fig. 4.2. As well as the code level number M_c (or the **code symbol set size**), the ratio T_q/T_c of the symbol duration of the source to the symbol duration of the coded signal is of decisive significance. This ratio defines the extent by which the symbol rate can be reduced by the use of a code and corresponds to the bandwidth expansion ratio which is often used to describe modulated systems. The smaller the ratio T_q/T_c is, the better is the effect on the digital signal transmission, especially in the case of channels for which the attenuation increases markedly with frequency.

With level number M_c and symbol rate $1/T_c$ we obtain for the (equivalent) bit rate at the coder output

$$R_c = \frac{\log_2 M_c}{T_c} \tag{4.3}$$

With the corresponding bit rate $R_q = \log_2 M_q / T_q$ of the source signal, the **relative redundancy**, an important code parameter, can be calculated:

$$r_c = \frac{R_c - R_q}{R_c} = 1 - \frac{T_c \log_2 M_q}{T_q \log_2 M_c} \tag{4.4}$$

As an example, the ternary code for symbol-wise coding (Fig. 4.2(a)) has a relative redundancy of about 36.9%, whereas the 3B2T block code (Fig. 4.2(b)) shows only about 5.4%. Both codes result in a suitable spectral shape of the coded signal and hence a signal which is better matched to the transmission channel.

The redundancy is often also used for **error monitoring**. It is certain that the decoder will, for example, conclude that there is a transmission error if, in a coded sequence as shown in Fig. 4.2(a), two + symbols or two − symbols directly follow each other.

Note r_c is the relative redundancy of the code. For a nonredundant source ($r_q = 0$), which was assumed for this chapter, r_c is also equal to the relative redundancy of the coded signal $c(t)$. In comparison, the relative redundancy of the coded signal for a redundant source is

$$r_{\text{coded signal}} = 1 - (1 - r_c)(1 - r_q) \tag{4.5}$$

4.1.3 Spectral characteristics of a coded signal

The spectral shape of the transmitted signal $s(t)$ has a direct relationship to the statistics of the amplitude coefficients a_ν. In an analogous manner to Section 2.3, the probability density function $f_a(a_\nu)$, the discrete autocorrelation function (ACF) $l_a(\lambda)$ and the "power spectrum" $L_a(f)$ are used to describe these statistical characteristics.

First we consider the discrete ACF for a nonredundant bipolar M-level digital signal. According to eqns (2.22) and (2.31) $l_a(\lambda) = 0$ for $\lambda \neq 0$, whereas

$$l_a(0) = \overline{a_\nu^2} = \frac{M+1}{3(M-1)} \tag{4.6}$$

defines the mean squared value of the amplitude coefficients.

These equations are no longer valid in the case of a redundant signal. Also, at least some of the ACF values for $\lambda \neq 0$ differ from zero. Therefore the power spectrum $L_a(f)$ of the amplitude coefficients a_v, defined by eqn (2.29), is no longer independent of frequency but consists of a summation of cosine waves and is thus periodic with the symbol rate $1/T = 1/T_c$.

Example Figure 4.3(a) shows the discrete ACF of the amplitude coefficients a_v for the code symbol sequence shown in Fig. 4.2(a). In this sequence the symbol c_v depends only on the immediately preceding symbol c_{v-1}, and not on the previous symbols c_{v-2} and c_{v-3} (twinned binary code). Therefore $l_a(\lambda) = 0$ for $|\lambda| > 1$. The corresponding power spectrum $L_a(f)$ is illustrated in Fig. 4.3(b). As in this example the coding rule on which it is based precludes long + or − sequences, $L_a(0) = 0$.

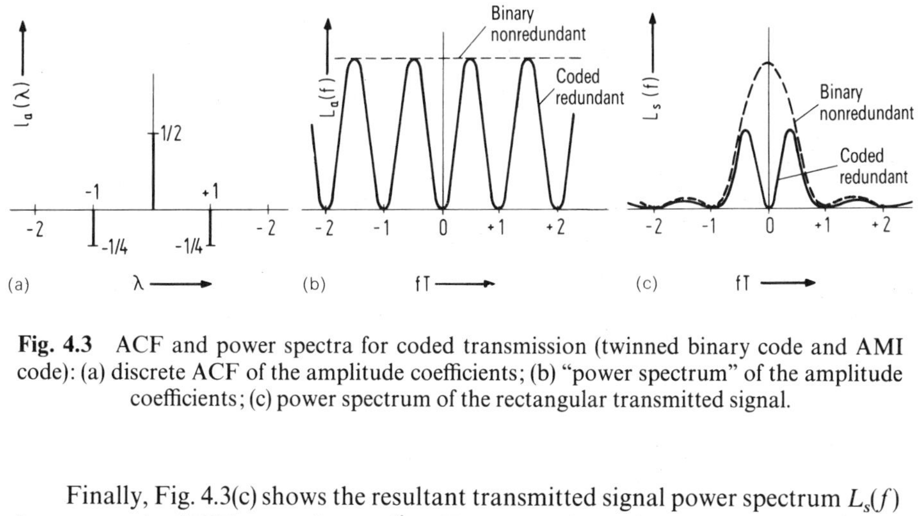

Fig. 4.3 ACF and power spectra for coded transmission (twinned binary code and AMI code): (a) discrete ACF of the amplitude coefficients; (b) "power spectrum" of the amplitude coefficients; (c) power spectrum of the rectangular transmitted signal.

Finally, Fig. 4.3(c) shows the resultant transmitted signal power spectrum $L_s(f)$ for rectangular NRZ transmitter pulses. For comparison, the power spectrum of the binary bipolar nonredundant transmitter signal, which for NRZ rectangular pulses is proportional to $\mathrm{si}^2(\pi f T)$, is also drawn. This example shows clearly that the "power spectrum"

$$L_a(f) = \frac{L_s(f)}{L^{\cdot}_{g_s}(f)} \tag{4.7}$$

reflects the spectral characteristics of the code but not those of the transmitter pulse shape. This value, which is often referred to in the references as the **normalized power spectrum**, provides particular statements about the direct component freedom of a coded signal. In the following a code is called **direct signal free** if $L_a(0) = 0$. This signifies not only that the code output signal contains no direct signal component, which would relate to a Dirac pulse $\delta(f)$ in the spectrum, but also that the number of positive and negative amplitude coefficients is the same for *every* arbitrary sample signal.

4.2 Codes for symbol-wise coding

4.2.1 Partial response coder block diagram

An important class of transmission code for symbol-wise coding is **partial response codes (pseudo-multilevel codes)** which produce a multilevel redundant coder signal of equal symbol rate from the binary source signal. Figure 4.4(a) shows the block diagram for such a coder. The level number M and the coding rules of the individual partial response codes depend on the order N and the coefficients $k_1, \ldots, k_N$. Codes with the level numbers $M = 3$, $M = 4$ and $M = 5$ are common. In the following the so-called **pseudo-ternary codes** ($M = 3$) are considered exclusively; the block diagram for these appears in Fig. 4.4(b). Here the coefficient k_N is ± 1 and all other coefficients are zero.

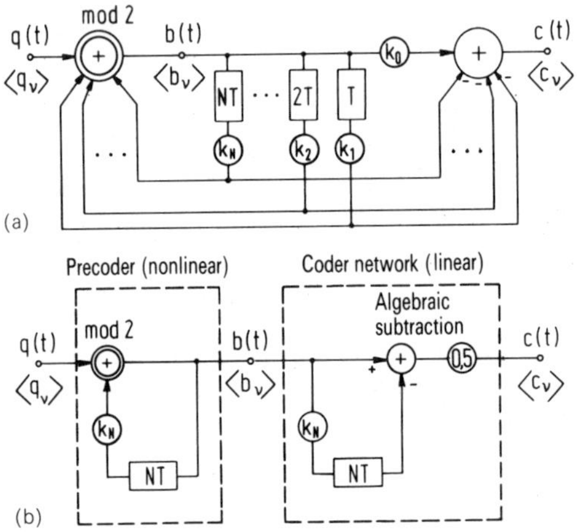

Fig. 4.4 Partial response coder block diagram: (a) general; (b) ternary.

The coder consists of a nonlinear **precoder** and a linear **coder network**. For illustration of these two components, the delay element NT and the weighting factor k_N are shown twice. The precoder, by means of a modulo 2 addition (antivalence) of the symbols q_ν and $b_{\nu-N}$, produces the binary precoded symbols b_ν which, like the source symbols q_ν, are statistically independent of each other. This precoding prevents error propagation due to a transmission error. In addition, it permits the simple realization of a decoder in the form of a two-way rectifier.

The actual conversion from binary into ternary is effected by the linear coding network using delay by NT and an (analog) subtraction, so that for the redundant ternary coder signal

$$c(t) = 0.5\{b(t) - k_N b(t - NT)\} \tag{4.8}$$

As both the precoded signal $b(t)$ and the coefficient k_N are either $+1$ or -1, the coded signal $c(t)$ can have the (normalized) ternary values of $+1, 0$ and -1.

The individual pseudo-ternary codes are distinguished by the parameters N and k_N.

The best-known type of pseudo-ternary code is the **AMI code** (from alternate mark inversion) with code parameters $N = 1$ and $k_N = 1$. This code is also known as the **first-order bipolar code**. Figure 4.5(b) shows the ternary coded signal that is produced by AMI coding of the binary signal $q(t)$ shown in Fig. 4.5(a). The figure demonstrates that with AMI code a binary O at the input is converted into a ternary 0 at the output. However, the binary symbol L is alternately coded as the ternary symbol + and −, so that two symbols of the same polarity (+ or −) never follow each other and thus the AMI code is direct signal free.

(a)
(b)
(c)
(d)
(e)
(f)

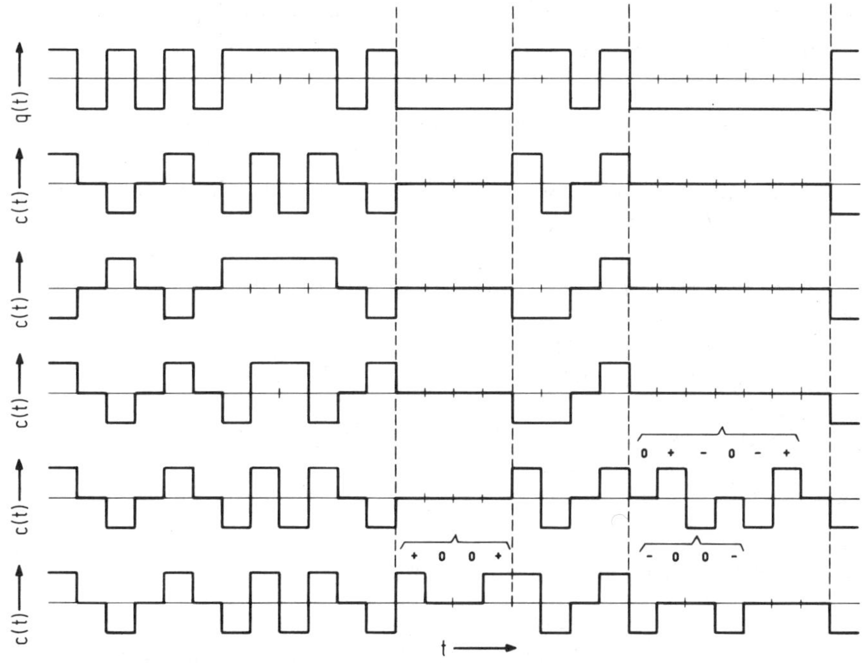

Fig. 4.5 (a) Source signal and (b)–(f) coded signal with ternary partial response coding: (b) AMI code; (c) duo-binary code; (d) second-order bipolar code; (e) B6ZS code; (f) HDB3 code.

By means of the simple AMI coding rule, the decoding can be achieved by a two-way rectifier. If a transmission error occurs, here as with the other partial response codes it does not lead to propagation of an error. Furthermore, the decoder can detect by a violation of the above coding rule when transmission errors have occurred.

If the coder does not include a precoder, the AMI code becomes the **twinned binary code** mentioned earlier (see Fig. 4.2(a)). This code is more difficult to decode and to monitor but otherwise is indistinguishable from the AMI code.

Figure 4.5(c) shows the coder signal for the **duo-binary code** ($N = 1, k_N = -1$). This is not direct signal free, i.e. there are arbitrary long + or − sequences in the coded signal also. However, symbol sequences extremely distorted by intersymbol interference, i.e. ... − + − ... or ... + − + ..., do not occur, and so the vertical and horizontal eye opening can be significantly greater than in the case of AMI code (see Section 8.1.2).

Finally, we should mention the **second-order bipolar code** ($N = 2, k_N = 1$). In some cases this is also referred to as the modified duo-binary code. In the second-order bipolar code, a binary O is converted into a ternary 0 as previously. However, the coding of the binary symbol L is not as simple as for the AMI code. Here up to two symbols of equal polarity + or − can follow each other (see Fig. 4.5(d)). However, as the number of symbols of the same polarity is limited here also, the second-order bipolar code is, like the AMI code, direct signal free.

4.2.2 Partial response code power spectrum

If a nonredundant binary signal is applied to the coder input, the precoder output signal $b(t)$ is also nonredundant. This means that the precoder has no effect on the power spectrum. The latter is due solely to the linear network with the **coder frequency response** $H_C(f)$ which can be determined directly from Fig. 4.4(b) by means of the shift theorem:

$$H_C(f) = \tfrac{1}{2}\{1 - k_N \exp(-j2\pi f N T)\} \tag{4.9}$$

It follows that the "power spectrum" of the amplitude coefficients a_v is

$$L_a(f) = |H_C(f)|^2 = \frac{1 + k_N{}^2}{4} - \frac{k_N}{2}\cos(2\pi f N T) \tag{4.10}$$

If we consider that, for all ternary partial response codes, $k_N = \pm 1$, then

$$L_a(f) = \tfrac{1}{2}\{1 - k_N \cos(2\pi f N T)\} \tag{4.11}$$

The amplitude and phase of the coder frequency response $H_C(f)$ and the "power spectrum" $L_a(f)$ of the three codes described above are shown in Fig. 4.6. The AMI code has no spectral components at multiples of the symbol rate $1/T$ but has a large component with frequency $1/2T$ which can be used for clock recovery. The AMI code, like the second-order bipolar code, is free of direct signals ($L_a(0) = 0$). These both exhibit zero values in the spectrum at multiples of half the symbol rate. In contrast, the duo-binary code is not direct signal free, and the power spectrum only vanishes at frequencies $\pm 1/2T$, $\pm 3/2T$ etc.

4.2.3 Error probability of partial response codes

Often only the power spectrum is considered when transmission codes are

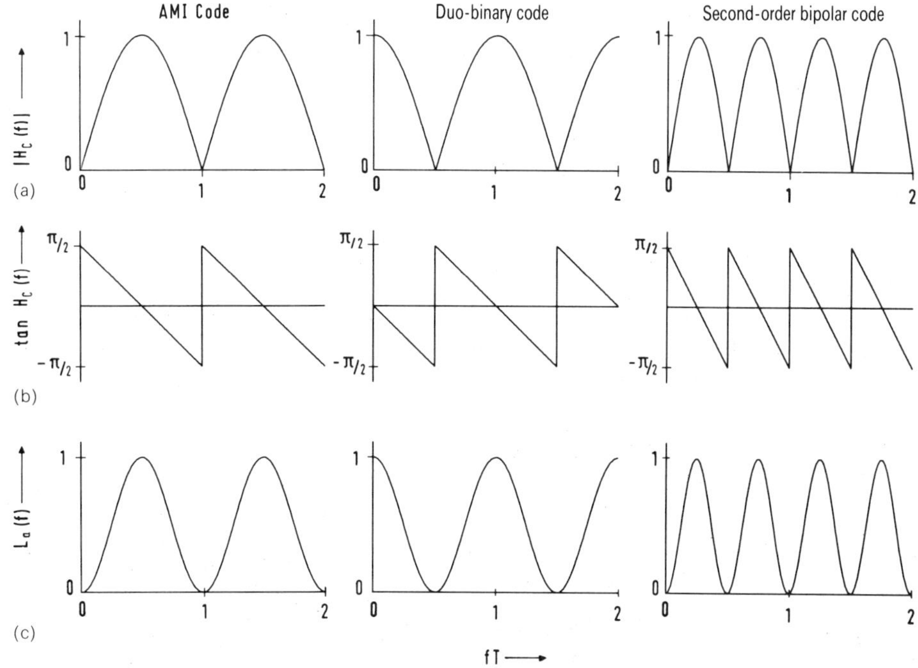

Fig. 4.6 Code frequency response $H_C(f)$ and "normalized power spectrum" $L_a(f)$ for the ternary partial response codes.

described and compared. However, for a more exact assessment the mean symbol error probability (eqn (3.35)) is the critical value. In contrast with uncoded transmission, it must be remembered that not all possible symbol sequences $\langle c_v \rangle_i$ are allowable. In addition, distinguishable lines in the eye pattern do not all appear with equal probability, i.e. the probability of occurrence p_i in eqn (3.36) in general depends on i.

If the transmission system allows direct signals ($H_I(0) \neq 0$) or if the transmission code is free from direct signals ($L_a(0) = 0$), the mean error probability p_M is approximated with sufficient accuracy by the worst-case symbol error probability

$$p_U = Q\left\{\frac{o(T_D)/2}{N_d^{\,1/2}}\right\} \tag{4.12}$$

for coded transmissions also ($p_M \approx p_U$). Here Gaussian noise and optimum decision values are assumed. In this section a given equalizer frequency response $H_E(f)$ is assumed, so that the detection noise power N_d in eqn (2.69) is independent of the transmission code under consideration. Thus the vertical eye opening $o(T_D)$ is an objective measure for comparison of the error probability of different transmission codes.

Figure 4.7(a) shows the eye pattern for a nonredundant binary system based on NRZ rectangular transmitted pulses and a Gaussian pulse shaper with cut-off frequency $f_I = 0.4R$. The basic detection pulse $g_d(t)$ shown in Fig. 2.11(b) is obtained. The normalized vertical eye opening can be calculated using eqn (3.32) and is approximately 37% for the optimum detection time $\mathring{T}_D = 0$.

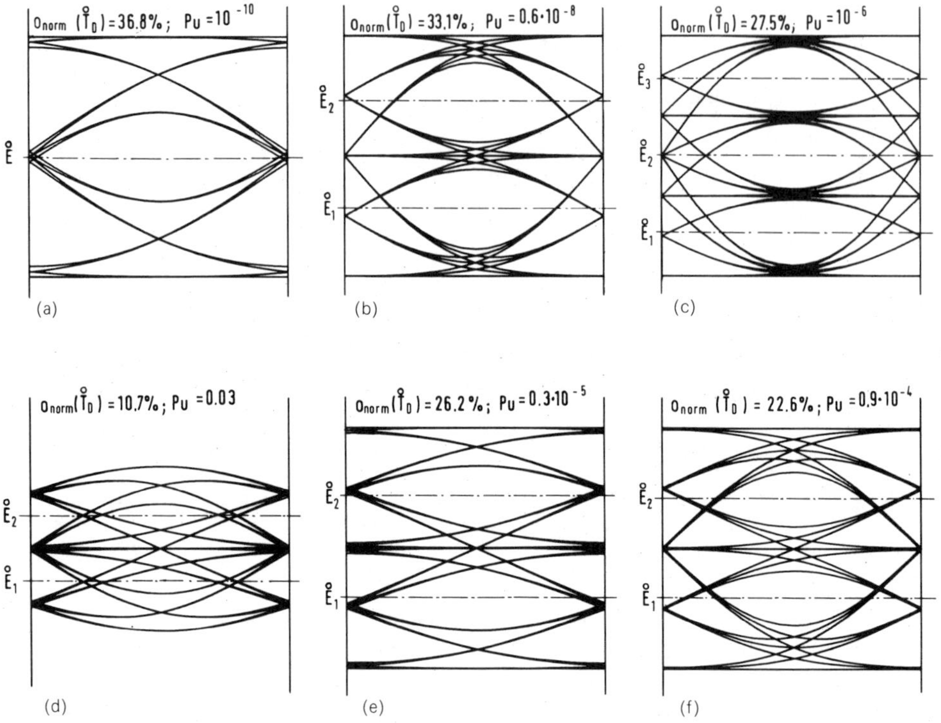

Fig. 4.7 Eye pattern for various codes with a Gaussian pulse shaper ($f_I = 0.4R$) and NRZ rectangular transmitter pulses: (a)–(c) nonredundant binary, ternary and quaternary code; (d) AMI code; (e) duo-binary code; (f) 4B3T code.

Since the symbol rate $1/T$ is not changed by the partial response coding, calculation of the eye opening for the coded system has to be derived from the same pulse $g_d(t)$. When using a transmission code, however, $o(T_D)$ cannot be specified in closed form if certain restrictions are not met. Therefore it is assumed here that the values of the basic detection pulse are symmetrical ($g_{-\nu} = g_\nu$) and nonnegative ($g_\nu \geq 0$). Further, it is assumed that

$$\ldots < g_{-2} < g_{-1} < g_0 > g_1 > g_2 \ldots \tag{4.13}$$

With this restriction the eye opening of a partial response coded transmission system can be calculated exactly. For example, if we consider that with AMI coding the symbols $+$ and $-$ are repeated alternately, the worst-case symbol sequence can be determined as in Section 3.2.

The upper boundary line $\tilde{d}_{up}(T_D)$ of the upper eye (see Fig. 4.7(d)) is caused by the sequence $...00 - + - 00....$ Accordingly we obtain at time $\mathring{T}_D = 0$

$$\tilde{d}_{up}(\mathring{T}_D = 0) = g_0 - 2g_1 \tag{4.14}$$

All other coded symbol sequences $\langle c_v \rangle_i$ whose middle symbol (the symbol to be detected) is $+$ result in a larger value because of the AMI coding rule and assumption (4.13). The lower boundary line $\tilde{d}_{lo}(T_D)$ of the upper eye is attributable to both sequences $...00 + \mathring{0}000...$ (for $T_D \leqslant 0$) and $...000\mathring{0} + 00...$ (for $T_D \geqslant 0$), so that $\tilde{d}_{lo}(\mathring{T}_D = 0) = g_1$. Therefore eqn (3.24) provides the maximum vertical eye opening for AMI coding as follows:

$$o(\mathring{T}_D = 0) = \tilde{d}_{up}(0) - \tilde{d}_{lo}(0) = g_0 - 3g_1 \tag{4.15}$$

A glance at Fig. 4.7(d) shows that the normalized eye opening for AMI coding is only 11% and thus is more than a factor of 3.5 smaller than that for the (nonredundant) binary code. If the detection noise power N_d is equal in both cases, as was assumed here, the worst-case error probability p_U is 10^{-10} for nonredundant binary coding and about 3% for AMI coding. The smaller eye opening (and therefore larger error probability) can be attributed to the ternary detection without simultaneous reduction of the symbol rate. The particularly unfavorable symbol sequences $... - + - ...$ and $... + - + ...$ with regard to the vertical eye opening are not excluded from transmission by the AMI coding rules, as is the case with other partial response codes.

Figure 4.7(e) shows, for example, the eye pattern for the duo-binary code. Within the assumptions applicable here, the upper eye is bounded by the following symbol sequences:

upper boundary $\quad ... - - - 0 + 0 - - - ... \qquad \tilde{d}_{up}(T_D = 0) = g_0 - 2\sum_{v=2}^{\infty} g_v$
$\qquad\qquad\qquad\qquad\qquad \uparrow$

lower boundary $\begin{cases} ... + + + + 00 + + + ... (T_D \leqslant 0) \\ \qquad\qquad\quad \uparrow \\ ... + + + 00 + + + + ... (T_D \geqslant 0) \end{cases} \qquad \tilde{d}_{lo}(T_D = 0) = g_1 + 2\sum_{v=2}^{\infty} g_v$
$\qquad\qquad\qquad\qquad\qquad\qquad \uparrow$

Therefore eqn (3.24) gives the vertical eye opening as

$$o(\mathring{T}_D = 0) = g_0 - g_1 - 4\sum_{v=2}^{\infty} g_v = 3(g_0 + g_1) - 2\hat{g}_s H_I(0) \tag{4.16}$$

It is considered here that for NRZ signals summation over the basic detection pulse values is equal to the detection signal value $\hat{g}_s|H_I(0)|$ for a long $+$ sequence:

$$\sum_{v=-\infty}^{+\infty} g_v = \sum_{v=-\infty}^{+\infty} g_d(T_D + vT) = \hat{g}_s H_I(0) \tag{4.17}$$

For the example considered we obtain with duo-binary coding a (normalized) eye opening of about 26% and a worst-case error probability p_U of approximately 0.3×10^{-5} (see Fig. 4.7(e)). Correspondingly, for the second-order bipolar code

$$o(\mathring{T}_D = 0) = g_0 - 2g_1 - 3g_2 \tag{4.18}$$

which corresponds to an eye opening of 18% and an error probability of 0.8×10^{-3} for the values assumed.

Note If the sample values g_ν of the basic detection pulse do not fulfil the assumptions of this section, other worst-case symbol sequences result. In these cases eqns (4.15), (4.16) and (4.18) are no longer valid. However, the vertical eye opening can be determined from similar considerations. The optimum decision value always lies halfway between the upper and lower bounds of the eye. Equation (3.45), which is only valid for nonredundant systems, is not applicable.

4.2.4 Modified AMI codes

Figure 4.5 has shown that with some partial response codes the occurrence of long + and long − sequences can be prevented. However, the occurrence of long zero sequences is quite possible, and no clock information is then transmitted for a long period.

Some codes have been developed to avoid this; they are based on the AMI code but are modified so that long zero sequences do not happen. A zero sequence which exceeds a certain length is coded by a ternary sequence which violates the AMI code rules and thus can be re-interpreted by the decoder as a zero sequence. Two of these modified AMI codes are the B6ZS and the HDB3 codes [4.6].

In the **B6ZS** code six consecutive 0 symbols are substituted by the symbol sequences

$$0 + - 0 - + \quad \text{or} \quad 0 - + 0 + -$$

depending on the polarity of the preceding symbols (see Fig. 4.5(e)). If no errors occur during transmission the decoder can recognize these symbol sequences, because they violate the AMI rules, and re-interpret them as six zeros.

In the **HDB3** code (from **h**igh **d**ensity **b**ipolar of order **3**), every four consecutive 0 symbols are represented by four symbols which violate AMI coding. There are four different substitution possibilities for the sequence 0000 depending on the polarity of the last symbol prior to this sequence and the polarity of the last symbol that has violated the AMI rules, and these are given in Table 4.1.

It can be seen from Fig. 4.5 that the maximum number of consecutive 0 symbols for the HDB3 code is three, whereas for the B6ZS code a maximum of five zeros can occur. The power spectra of both these codes differ only to an insignificant degree from that of the AMI code.

Table 4.1 HDB3 code

Polarity of the last code violation	*Polarity of preceding symbol*	
	+	−
+	0000 → −00−	0000 → 000−
−	0000 → 000+	0000 → +00+

4.3 Block codes

4.3.1 Nonredundant codes

Of the block codes, only the nonredundant codes and 4B3T codes will be considered in detail here.

In M-level nonredundant coding every $\log_2 M$ binary source symbols are substituted by one code symbol. The relationship between the source symbols and the code symbols is defined by a code table (Fig. 4.8(a)). If the source symbol sequence $\langle q_\nu \rangle$ is nonredundant, then this is also the case for the code symbol sequence $\langle c_\nu \rangle$, i.e. the individual code symbols c_ν are statistically independent of each other and all M possible symbols c_μ occur with equal probability $p_\mu = 1/M$ (see eqn (2.4)). Nonredundant coding is thus equivalent to an increase in the level number. Therefore the symbol rate can be reduced relative to binary transmission by a factor $1/(\log_2 M)$. According to eqn (2.29) the normalized power spectrum $L_a(f)$ is frequency independent. The eye opening and the error probability can be calculated as described in Section 3.4.

$q_{\nu-1} q_\nu$	00	0L	L0	LL
Dual code	⊖	-	+	⊕
Gray code	⊖	-	⊕	+

(a)

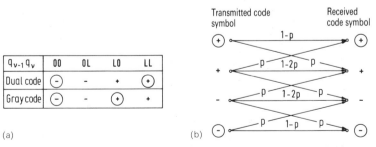

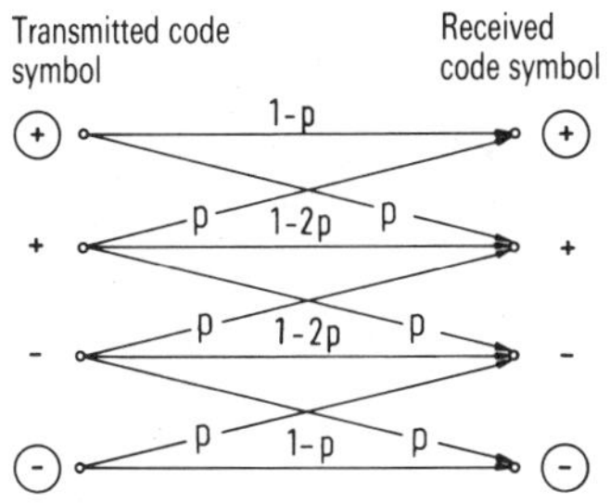

(b)

Fig 4.8 (a) Code tables for the four-level dual and Gray codes; (b) the quaternary channel model.

Figure 4.7(b) shows the eye pattern for the nonredundant ternary code ($M = 3$), again based on a Gaussian pulse shaper with cut-off frequency $f_I = 0.4R$. With nonredundant ternary coding the symbol duration is about a factor of $\log_2 3 \approx 1.58$ greater than that of uncoded binary transmission, so that the effect of intersymbol interference is significantly reduced. Therefore, in contrast with the binary system and despite the initial reduction in the eye opening by a factor of 2 owing to the ternary conversion, there is only a slight reduction of the eye to 33% and the error probability is increased by only an insignificant amount. With optimum pulse-shaper cut-off frequencies $\mathring{f}_I$ the nonredundant ternary code is even better than the binary code (see Section 8.1.2).

The mean and the worst-case symbol error probabilities p_M and p_U relate to the difference between the coded symbol sequences $\langle c_\nu \rangle$ and $\langle r_\nu \rangle$. However, to compare the individual codes it is necessary to derive the uncoded sequences $\langle q_\nu \rangle$ and $\langle v_\nu \rangle$. As the digital source is assumed to be binary, the **mean bit error probability** is defined as follows:

$$p_B = \overline{P(v_\nu \neq q_\nu)} \tag{4.19}$$

In the application of a transmission code the error probabilities p_M and p_B differ slightly, which is taken into consideration in the following by means of the **error propagation factor due to coding**:

$$\gamma_{EP,code} = \frac{p_B}{p_M} = \frac{\overline{P(v_\nu \neq q_\nu)}}{\overline{P(r_\nu \neq c_\nu)}} \tag{4.20}$$

Example With a nonredundant quaternary code ($M = 4$) every two binary symbols are substituted by a four-level code symbol. The two conversions shown in Fig. 4.8(a) are commonly used for this. Basically they differ in bit error probability p_B. We assume that the transmission characteristics of the quaternary transmission system can be described by the channel model shown in Fig. 4.8(b), which is only precisely possible with a Nyquist system. $P(\oplus|+)$ denotes, for example, the conditional probability that the transmitted symbol $+$ is corrupted during transmission to the symbol $\oplus$. The probabilities that a transmitted symbol is corrupted into a "neighboring" symbol will all be equal:

$$P(\oplus|+) = P(+|\oplus) = P(-|+) = \ldots = P(-|\ominus) = p \tag{4.21}$$

The probability that two or three decision values are transgressed will be negligibly small. Therefore, according to eqn (3.46), the mean symbol error probability p_M is $6p/4$.

Firstly let us consider the case of **dual coding**. If the coded symbol $+$ is corrupted by transmission into the symbol $\oplus$, this symbol error causes precisely one bit error. However, if the symbol $+$ is corrupted in the opposite direction, namely into the symbol $-$, this one symbol error corresponds to two bit errors. Therefore for the (mean) bit error probability with quaternary dual coding ($M = 4$)

$$p_{\mathrm{B}} = \frac{1/4}{2}\{P(+|\oplus) + P(\oplus|+) + 2P(-|+) + 2P(+|-) + P(\ominus|-) + P(-|\ominus)\}$$
$$= p$$
$$= \frac{2}{3}p_{\mathrm{M}} \quad (4.22)$$

The factor 1/4 specifies the probability of occurrence of the $M = 4$ equally probable code symbols. The factor 2 in the denominator accounts for the fact that two binary symbols are represented by only one quaternary symbol.

With **Gray coding**, however, every symbol error causes precisely one bit error. Therefore here the bit error probability p_{B} is only half as large as the mean symbol error probability p_{M}. It is generally valid for M-level nonredundant Gray coding and with model validity defined by Fig. 4.8(b) that

$$p_{\mathrm{B}} = \frac{p_{\mathrm{M}}}{\log_2 M} \quad (4.23)$$

4.3.2 4B3T codes

4B3T codes constitute a class of block coding in which a block of four binary symbols is coded into a block of three ternary symbols. Thus $T_q/T_c = 0.75$ (eqn (4.2)), and correspondingly the relative code redundancy from eqn (4.4) is

$$r_c = 1 - \frac{4\log_2 2}{3\log_2 3} \approx 15.9\% \quad (4.24)$$

Coding of the 16 possible binary blocks into the corresponding ternary blocks could in principle be done with a fixed code table. To improve the spectral characteristics of these codes further, however, code tables with two or more code alphabets are used in the common types of 4B3T codes. The selection of the appropriate code alphabet at any moment depends on the blocks coded previously.

For this reason the **running digital sum** (RDS) of the amplitude coefficients is calculated after each block. After the transmission of m (coded) blocks,

$$\mathrm{RDS}_m = \sum_{\nu=1}^{3m} a_\nu \quad (4.25)$$

The choice of code alphabet for coding the $(m+1)$th block depends on RDS_m. The code alphabets are selected so that the running digital sum does not exceed certain upper and lower limits, i.e. for every block

$$\mathrm{RDS}_{\min} \leqslant \mathrm{RDS}_m \leqslant \mathrm{RDS}_{\max} \quad (4.26)$$

In this way the freedom from direct (dc) signals of the 4B3T code is assured.

The individual 4B3T codes differ in the number and content of the code alphabets. These are summarized in Table 4.2 for the most important 4B3T codes, namely that defined by Jessop and Waters [4.11], the **MS43** code (from **m**onitored **s**um **4**B**3**T code) [4.6] and the **FOMOT** code (from **fo**ur-**mo**de **t**ernary) [4.23].

Table 4.2 4B3T code alphabets

Binary word	*Jessop–Waters* ($-2 \leqslant RDS_m \leqslant 3$) RDS_m		*MS43* ($-1 \leqslant RDS_m \leqslant 2$) RDS_m			*FOMOT* ($-1 \leqslant RDS_m \leqslant 2$) RDS_m			
	−2;−1;0	1; 2; 3	−1	0; +1	+2	−1	0	+1	+2
LLLO	0 + −			0 − +				+ − 0	
LLLL	− 0 +			− 0 +				+ 0 −	
OOOO	+ 0 −			− + 0				0 + −	
OOOL	− + 0			+ − 0				0 − +	
OOLO	0 − +			+ 0 −				− 0 +	
OOLL	+ − 0			0 + −				− + 0	
LOOO	+ + −	− − +	+ − +	+ − +	− − −	0 + +	0 + +	− − 0	− − 0
LOOL	− + +	+ − −	0 0 +	0 0 +	− − 0	+ 0 +	+ 0 +	− 0 −	− 0 −
LOLO	+ − +	− + −	0 + 0	0 + 0	− 0 −	+ + 0	+ + 0	0 − −	0 − −
LOLL	+ 0 0	− 0 0	+ 0 0	+ 0 0	0 − −	+ − +	+ − +	+ − +	− − −
LLOO	0 + 0	0 − 0	− + +	− + +	− − +	− + +	− 0 0	− + +	− 0 0
LLOL	0 0 +	0 0 −	+ + −	+ − −	+ − −	+ 0 0	+ − −	+ 0 0	+ − −
OLOO	+ + 0	− − 0	+ + 0	0 0 −	0 0 −	0 + 0	0 − 0	0 + 0	0 − 0
OLOL	0 + +	0 − −	+ 0 +	0 − 0	0 − 0	0 0 +	− − +	0 0 +	− − +
OLLO	+ 0 +	− 0 −	0 + +	− 0 0	− 0 0	+ + −	0 0 −	+ + −	0 0 −
OLLL	+ + +	− − −	+ + +	− + −	− + −	+ + +	− + −	− + −	− + −

The calculation of the power spectrum $L_a(f)$ of the amplitude coefficients is significantly more far-reaching for 4B3T codes than for the partial response codes discussed in Section 4.2. It is briefly outlined in the following. The transfer of the running digital sum from RDS_m to RDS_{m+1} can be described by a homogeneous stationary first-order Markov process. The associated transition probability $P(RDS_{m+1}|RDS_m)$ for the individual codes can be deduced from the respective code tables. The number of possible states (values for RDS_m) is hence $RDS_{max} - RDS_{min} + 1$. Thus a Markov diagram like those in Fig. 4.9 can be drawn for every 4B3T code. The numbers on the arrows denote the transition probabilities; the state probabilities $P(RDS_m)$ are given below the individual Markov diagrams.

The ACF $l_a(\lambda) = \overline{a_\nu a_{\nu+\lambda}}$ of the amplitude coefficients can be determined from these Markov diagrams. If we wish to do this analytically, a very high computing cost is involved [4.6]. It is simpler to simulate the Markov process shown in Fig. 4.9 with a digital computer and to determine the values of the ACF $l_a(\lambda)$ from this. The power spectrum $L_a(f)$ can be calculated from the ACF values by means of eqn (2.29).

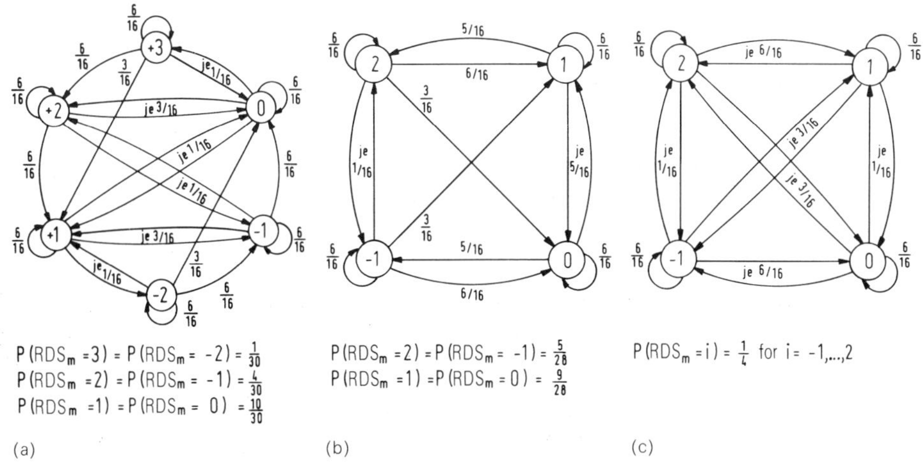

Fig. 4.9 Markov diagrams for the RDS_m "states" of 4B3T codes: (a) Jessop–Waters; (b) MS43; (c) FOMOT.

Figure 4.10 shows the plot of the "normalized power spectrum" $L_a(f)$ for the 4B3T codes compared with that for the AMI code. It is apparent that at low frequencies the 4B3T code power spectrum is greater than that of the AMI code.

Finally let us again consider the vertical eye opening. For a low-pass-type pulse shaper ($H_1(0) \neq 0$) we obtain approximately the same expression as for the nonredundant ternary code (cf. eqn (3.51)):

$$o(\mathring{T}_D = 0) \geqslant g_0 - 2 \sum_{v \neq 0} |g_v| \tag{4.27}$$

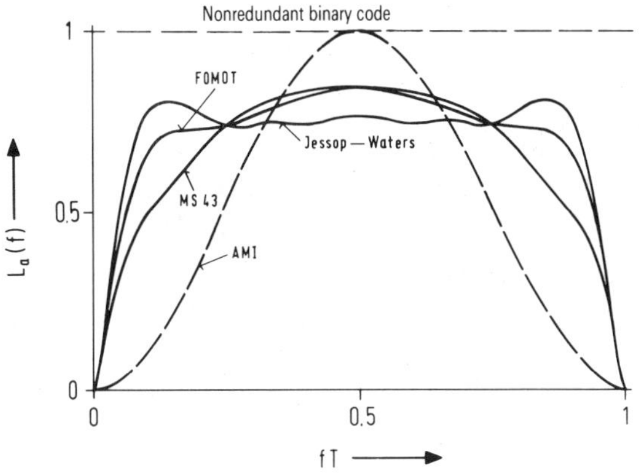

Fig. 4.10 Normalized power spectrum $L_a(f)$ of 4B3T codes [4.6].

The different results for the nonredundant ternary code (Fig. 4.7(b)) and the 4B3T codes (Fig. 4.7(f)) are caused by the fact that other basic detection pulses are produced by the different symbol rates $1/T = 0.633R$ and $1/T = 0.75R$ with the same pulse-shaper cut-off frequency f_I. In this case ($H_I(0) \neq 0$) the individual 4B3T codes differ only marginally.

The most important parameters for the codes described above are compared in Table 4.3. A comparison of the codes with respect to the error probability follows in Sections 8.1.2 and 8.1.3.

Table 4.3 Comparison of the characteristic parameters of various codes

Type	*(1)* T_q/T_c	*(2)* r_c	*(3)* $\overline{a_v^2}$	*(4)* $P(-1)$	$P(0)$	$P(+1)$	*(5)* γ_{FF}	*(6)* ± 1	0	*(7)*
Binary code (nonredundant)	1.000	0.000	1.000	0.500	—	0.500	1.0	∞	—	No
Sequential codes										
First-order bipolar (AMI)	1.000	0.369	0.500	0.250	0.500	0.250	≈1.0	1	∞	Yes
Second-order bipolar	1.000	0.369	0.500	0.250	0.500	0.250	≈1.0	2	∞	Yes
Duo-binary	1.000	0.369	0.500	0.250	0.500	0.250	≈1.0	∞	∞	No
B6ZS code	1.000	0.369	0.532	0.266	0.468	0.266	≈1.0	1	5	Yes
HDB3 code	1.000	0.369	0.551	0.275	0.450	0.275	≈1.0	1	3	Yes
Block codes										
Ternary code (nonredundant)	0.633	0.000	0.666	0.333	0.333	0.333	≈0.6	∞	∞	No
Quaternary code (nonredundant)	0.500	0.000	0.556	0.250	—	0.250	≈0.5	∞	—	No
Jessop–Waters	0.750	0.159	0.687	0.344	0.312	0.344	≈2.0	6	4	Yes
MS43 code	0.750	0.159	0.647	0.324	0.352	0.324	≈1.8	5	4	Yes
FOMOT code	0.750	0.159	0.687	0.344	0.312	0.344	≈1.3	5	4	Yes

(1) symbol duration ratio;
(2) relative code redundancy;
(3) normalized mean transmitter power;
(4) probability of occurrence for values of amplitude −1, 0, +1;
(5) error propagation factor due to coding;
(6) maximum number of consecutive symbols;
(7) direct signal free.

Chapter 5

Decision Feedback Equalization

Contents In Section 5.1 the operation of decision feedback equalization (DFE) is described in terms of the block diagram and the signals. The eye opening for a system with ideal DFE is calculated. On the basis of this some practical examples of DFE are described in Section 5.2, and direct signal recovery is treated as a special case of DFE in Section 5.3. The error propagation effect is explained in Section 5.4, and the increase in mean error probability due to error propagation is specified.

Assumptions Except in the case of detectors with memory the assumptions in Chapters 2 and 3 remain valid. All equations are defined for nonredundant binary or multilevel systems. For coded systems the results should be modified as stated in Chapter 4.

5.1 The principle of decision feedback equalization

In the previous chapters linear methods were assumed for the equalization of the digital signal at the receiver. When these techniques are used the frequency components that are subject to particularly strong attenuation are amplified by means of a linear equalizer frequency response $H_E(f)$. The disadvantage of this type of linear equalization is that, although the useful input signal is strengthened, the noise is also amplified. If the channel frequency response exhibits zeros in the frequency range of interest, e.g. at low frequencies, pure linear equalization is not possible.

Therefore, even from the earliest days, **nonlinear methods** of equalization were developed in which no increase in the detector input noise power is caused by equalization of the useful signal. The best known method of this type is **decision feedback equalization** (DFE) which dates back to the patents of Milnor [5.10] and McColl [5.9] of 1929 and 1936 respectively. This makes use of knowledge of the previously decided symbols. A correction signal which compensates for the interference of the postcursors is fed back from the output of the detector to the input and hence intersymbol interference can be completely (or at least partially) removed.

5.1.1 Eye opening of a system with decision feedback equalization

Figure 5.1 shows the block diagram for a receiver with DFE. As before, the input signal to the decision device is the detector input signal $d(t)$ which is linearly pre-equalized (and thereby noise power limited) by means of the receiver–filter (equalizer) $H_E(f)$. From eqns (2.59) and (2.63)

$$d(t) = \sum_{\nu=-\infty}^{+\infty} a_\nu g_d(t - \nu T) + \check{d}(t) \tag{5.1}$$

Here the base component s_0 of the transmitted signal is set to zero.

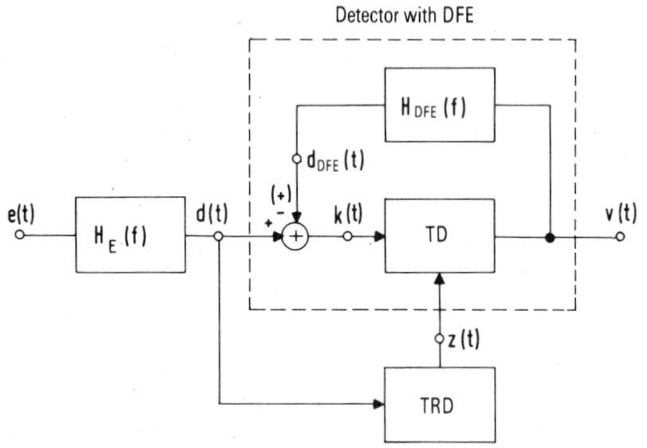

Fig. 5.1 Block diagram of a receiver with DFE. (It should be noted that to describe the direct signal recovery in Section 5.3 it is appropriate to substitute an addition for the minus sign, so that in this case $k(t) = d(t) + d_{DFE}(t)$.)

In a receiver without DFE the detector input signal is fed directly to the threshold detector for the decision process (see Fig. 2.1(a)). In contrast, in a receiver with DFE the **correction signal** $d_{DFE}(t)$ is subtracted prior to the decision process, and so the input signal to the threshold detector is

$$k(t) = d(t) - d_{DFE}(t) \tag{5.2}$$

$k(t)$ is defined as the **corrected detector input signal**.

The correction signal $d_{DFE}(t)$ is derived from the sink signal $v(t)$, which is already regenerated (and thus quasi-noise-free), and is fed back to the decision device input via the linear feedback network with frequency response $H_{DFE}(f) \bullet\!\!-\!\!\circ h_{DFE}(t)$. From the **basic correction pulse**

$$g_{DFE}(t) = g_v(t) * h_{DFE}(t) \tag{5.3}$$

where $g_v(t)$ is the basic sink pulse, and the **receiver-side amplitude coefficients** a_ν' of the decided signal $v(t)$, the correction signal is given by

$$d_{DFE}(t) = \sum_{\nu=-\infty}^{+\infty} a_\nu' g_{DFE}(t - \nu T) \tag{5.4}$$

In this equation it is assumed explicitly that the correction signal has no noise component $\check{d}_{DFE}(t)$ because it is derived from the quasi-noise-free sink signal $v(t)$. The distortions are indicated by erroneous amplitude coefficients ($a_v' \neq a_v$). From eqns (5.1), (5.2) and (5.4) we therefore obtain for the corrected detector input signal of a system with DFE

$$k(t) = \sum_{v=-\infty}^{+\infty} \{a_v g_d(t - vT) - a_v' g_{DFE}(t - vT)\} + \check{d}(t) \tag{5.5}$$

First it is assumed that the detector always makes the correct decision ($a_v' = a_v$ for all v). This is achieved in the case of a reasonably small error probability. (The effect of error propagation is treated in Section 5.4.) With this assumption the corrected detector input signal can be written in the form

$$k(t) = \sum_{v=-\infty}^{+\infty} a_v g_k(t - vT) + \check{d}(t) \tag{5.6}$$

The **corrected basic detection pulse** is therefore the difference

$$g_k(t) = g_d(t) - g_{DFE}(t) \tag{5.7}$$

The basic detection pulse can be improved and an increase in the eye opening can be achieved by suitable sizing of the basic correction pulse $g_{DFE}(t)$. However, the detection noise signal $\check{d}(t)$, and hence also the noise power N_d, are not changed by DFE.

All the results given in Chapters 3 and 4 are equally valid for a receiver with DFE if the detector input signal $d(t)$ is replaced by the corrected signal $k(t)$ and the basic detection pulse $g_d(t)$ is replaced by $g_k(t)$. In particular the vertical eye opening (eqn (3.51)) of a nonredundant M-level bipolar transmission system with DFE is

$$o(T_D) = 2\left\{\underbrace{\frac{g_k(T_D)}{M-1}}_{\text{main value}} - \underbrace{\sum_{v=1}^{v} |g_k(T_D - vT)|}_{\text{precursors}} - \underbrace{\sum_{v=1}^{n} |g_k(T_D + vT)|}_{\text{postcursors}}\right\} \tag{5.8}$$

Here v and n respectively denote the numbers of precursors and postcursors of the corrected basic detector input pulse $g_k(t)$.

5.1.2 Ideal decision feedback equalization

The function of the basic correction pulse $g_{DFE}(t)$ is to reduce the intersymbol interference caused by the sloping trailing edges of the basic detection pulse $g_d(t)$. However, because the amplitude coefficient a_v' is not determined until time $T_D + vT$, for reasons of causality for the vth correction pulse

$$a_v' g_{DFE}(t - vT) = 0 \quad \text{for } t \leqslant T_D + vT \tag{5.9}$$

This yields the following causality conditions:

$$g_{DFE}(t) = 0 \qquad \text{for } t \leqslant T_D \tag{5.10a}$$

$$g_k(t) = g_d(t) \tag{5.10b}$$

This equation makes it clear that only the postcursors can be removed by DFE. For reasons of causality it is not possible to compensate for the precursors.

There is a fundamental difference between ideal and nonideal DFE. With **ideal DFE** the following is valid for the basic correction pulse (Fig. 5.2(a)):

$$g_{\mathrm{DFE}}(t) = \begin{cases} 0 & \text{for } t < T_{\mathrm{D}} + T_{\mathrm{V}} \\ g_d(t) & \text{for } t \geqslant T_{\mathrm{D}} + T_{\mathrm{V}} \end{cases} \tag{5.11}$$

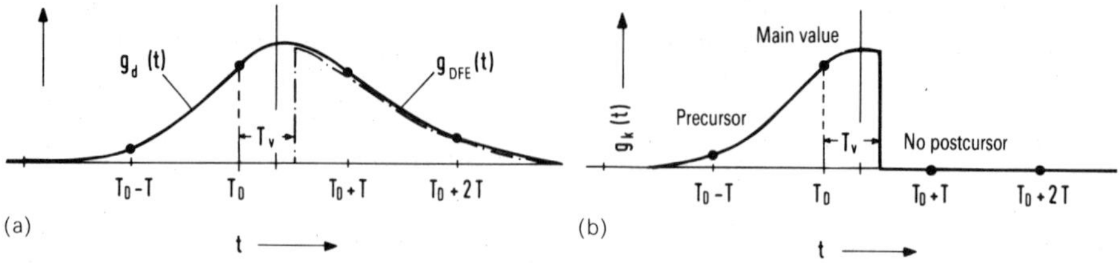

Fig. 5.2 Basic pulses $g_d(t)$, $g_{\mathrm{DFE}}(t)$ and $g_k(t)$ with ideal DFE.

The time delay T_{V} takes account of the decision processing time, for which $0 < T_{\mathrm{V}} < T$ must hold. For decision devices whose processing time T_{V} is greater than the symbol duration T it is not sensible to use DFE because in this case it is not possible to compensate for the first postcursor.

It is clear from eqns (5.7) and (5.11) that with ideal DFE the corrected basic detection pulse $g_k(t)$ is equal to zero for $t \geqslant T_{\mathrm{D}} + T_{\mathrm{V}}$ (Fig. 5.2(b)). In this case from eqn (5.8) we obtain the vertical eye opening of a nonredundant bipolar M-level signal as

$$o(T_{\mathrm{D}}) = 2\left\{\frac{g_d(T_{\mathrm{D}})}{M-1} - \sum_{\nu=1}^{v} |g_d(T_{\mathrm{D}} - \nu T)|\right\} \tag{5.12}$$

Example In the following the effect of DFE will be explained in terms of the signal time histories shown in Fig. 5.3. This figure is valid for a binary bipolar NRZ rectangular transmitter signal and a Gaussian pulse shaper with cut-off frequency $f_{\mathrm{I}} = 0.3R$. In this case the basic detector input pulse $g_d(t)$ with detection at the pulse center exhibits one nonnegligible precursor and one nonnegligible postcursor (Fig. 5.3(a)). Figure 5.3(d) shows the detector input signal $d(t)$ which corresponds to this basic detection pulse and the transmitter signal shown in Fig. 5.3(c). In a receiver without DFE the signal $d(t)$ would be input directly to the threshold device. Noise is not considered in this illustration. It can be seen from the eye pattern in Fig. 5.3(e) that the normalized eye opening here would only amount to about 10%.

In a receiver with ideal DFE the postcursors are completely compensated for, so that they do not contribute to reduction of the eye opening (see Fig. 5.3(b)). The correction signal $d_{\mathrm{DFE}}(t)$ and the corrected detector input signal $k(t)$, for the example considered here, are shown in Figs 5.3(f) and 5.3(g). It is evident from the eye pattern

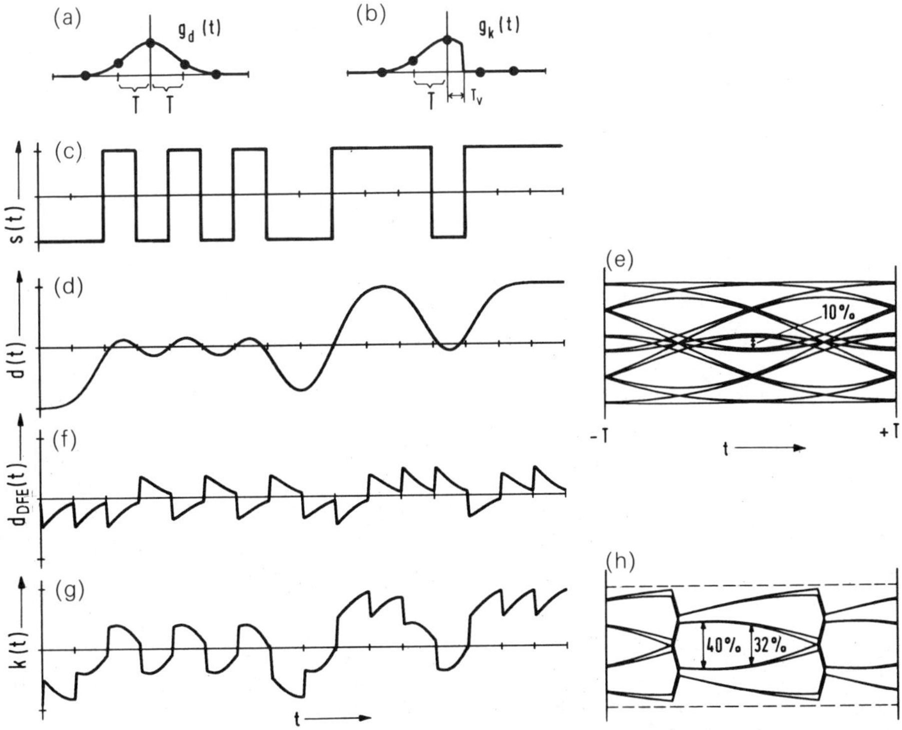

Fig. 5.3 Signal time histories and eye patterns for a system with and without DFE: (a) basic detection pulse; (b) corrected basic detection pulse; (c) transmitted signal; (d) detection signal; (e) eye pattern without DFE; (f) correction signal; (g) corrected detection signal; (h) eye pattern with ideal DFE.

in Fig. 5.3(h) that the (normalized) eye opening is increased to 32% by use of ideal DFE, based on a detection time T_D of zero.

However, because of the DFE the eye pattern is asymmetrical and so the detection time $T_D = 0$ is not optimum. For the optimum detection time, i.e. a shift of T_D to an earlier time ($\mathring{T}_D = -0.4T$), the normalized eye opening is about 40%. As the detection noise power is not changed by DFE, the worst-case signal-to-noise ratio ρ_U is a factor of 4^2 (about 12 dB) greater than that for a system without DFE, i.e. the error probability is significantly reduced.

5.1.3 Nonideal decision feedback equalization

Ideal DFE does not generally occur in practical digital systems. It is more suitable for providing an estimation of the theoretical limitations of DFE.

We obtain a form of DFE more suited to practical application if complete compensation of the basic detection pulse is not required at all times $t \geqslant T_D + T_V$ but only at the equally spaced detection times $T_D + \nu T$. This is then referred to as **complete DFE at the detection instants** (Fig. 5.4(a)). Except at the detection times $T_D + \nu T$ the corrected detection pulse $g_k(t)$ may well be nonzero, which is the case, for example, if the basic correction pulse $g_{DFE}(t)$ has a stepped time history (see Fig. 5.7 later). The result is a smaller horizontal eye opening than that for the basic detection pulse shown in Fig. 5.2(b). However, at the detection time T_D the vertical eye opening is as large as for ideal DFE, and it can be calculated using eqn (5.12) (see Section 5.2.2).

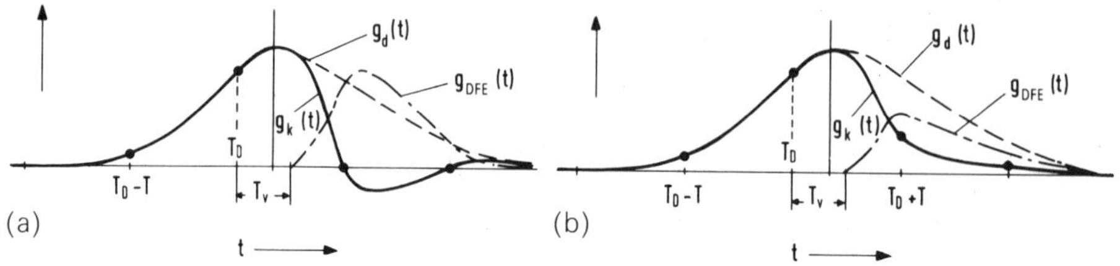

Fig. 5.4 Basic pulses $g_d(t)$, $g_{DFE}(t)$ and $g_k(t)$ with (a) complete DFE and (b) incomplete DFE.

In the general case, however, it has to be assumed that not all postcursors can be completely compensated for, at least not if the effect of tolerances cannot be neglected. We then refer to an **incomplete DFE** (Fig. 5.4(b)). In this case the vertical eye opening has to be calculated using the general equation (5.8).

5.2 Practical examples of decision feedback equalization

In the following two examples we show how the feedback network $H_{DFE}(f)$ should be sized so that the basic correction pulse $g_{DFE}(t)$ best reproduces the trailing edge of $g_d(t)$. Low-pass systems are considered exclusively here. In these the DFE has the task of reconstructing the higher frequency spectral components which are strongly attenuated during the transmission.

In contrast with DFE for direct signal recovery (Section 5.3), the basic detection pulse $g_d(t)$ has only a few marked postcursors which make a significant contribution to the intersymbol interference. It is therefore advantageous with this type of DFE, which is often also referred to as "high frequency DFE", to treat the approximation to the desired correction pulse in the time domain.

5.2.1 Decision feedback equalization for a Gaussian detection pulse

First it is assumed that $g_d(t)$ is Gaussian or can at least be approximated by a Gaussian pulse (Fig. 5.5). In addition it is assumed in the present example that the

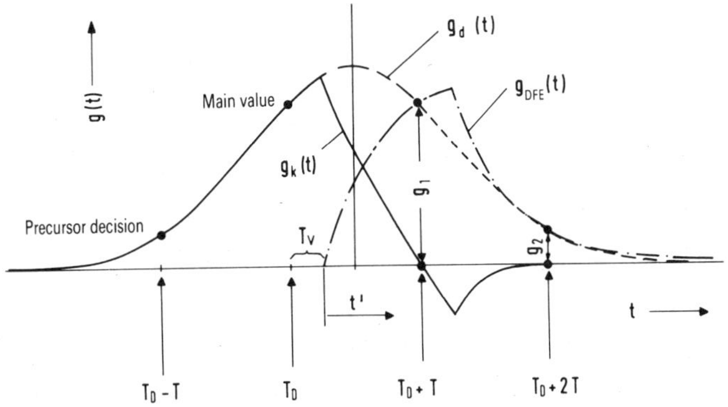

Fig. 5.5 Compensation for the trailing edge of a Gaussian detection pulse ($\Delta t_d = 2T$) by means of the response of a first-order low-pass filter to a rectangular pulse ($T_D = -T/2$; $T_V = T/4$; $f_{DFE} = 0.57/T$; $\hat{g}_{DFE} = 1.12g_1$).

detector outputs a rectangular NRZ signal $v(t)$, and that a first-order low-pass filter with cut-off frequency f_{DFE} is used as the feedback network. Hence the basic correction pulse is the same as the response of the low-pass filter to a rectangular pulse (see the Appendix, Table A3):

$$g_{DFE}(t) = \begin{cases} \hat{g}_{DFE}\{1 - \exp(-4f_{DFE}t')\} & 0 \leqslant t' \leqslant T \\ \hat{g}_{DFE}\{\exp(4f_{DFE}T) - 1\}\exp(-4f_{DFE}t') & t' \geqslant T \end{cases} \tag{5.13}$$

To simplify this illustration a time variable $t' = t - T_D - T_V$ is introduced. The freely selectable parameters of the pulse, i.e. the amplitude $\hat{g}_{DFE}$ and the cut-off frequency f_{DFE}, are also chosen such that both the first two postcursors of $g_d(t)$ are corrected for as well as possible. With the abbreviations $g_1 = g_d(T_D + T)$ and $g_2 = g_d(T_D + 2T)$ we obtain two equations for these two parameters:

$$g_{DFE}(t = T_D + T) = \hat{g}_{DFE}[1 - \exp\{-4f_{DFE}(T - T_V)\}] \overset{!}{=} g_1 \tag{5.14a}$$

$$g_{DFE}(t = T_D + 2T) = \hat{g}_{DFE}\exp\{-4f_{DFE}(T - T_V)\}\{1 - \exp(-4f_{DFE}T)\} \overset{!}{=} g_2 \tag{5.14b}$$

This set of equations cannot in general be solved analytically. Only simple solutions can be obtained in certain special cases. For example, when $T_V = 0$

$$f_{DFE} = \frac{\ln(g_1/g_2)}{4T} \qquad \hat{g}_{DFE} = \frac{{g_1}^2}{g_1 - g_2}$$

For $T_V = T/2$ we obtain a quadratic equation in the variable $\beta = \exp(-2f_{DFE})$

$$\beta^2 + \beta - \frac{g_2}{g_1} = 0 \tag{5.15}$$

from which the second parameter $\hat{g}_{DFE}$ can be determined through eqn (5.14). In most cases a set of transcendental equations results, which can only be solved numerically. For example, for $T_V = T/4$ and $g_2 = 0.2g_1$ we obtain the values shown in Fig. 5.5.

5.2.2 Realization of decision feedback equalization with a transversal filter

The **transversal filter** (TF) offers an implementation method of compensation for the postcursors at detection time. As shown in Fig. 5.6 it consists of a number of delay elements of symbol duration T, the filter coefficients $k_1,\ldots,k_n$ and a summing element.

The input signal to the transversal filter is the sink signal

$$v(t) = \sum_{\nu=-\infty}^{+\infty} a_\nu' g_v(t - \nu T) \tag{5.16}$$

which is assumed to be rectangular in Fig. 5.7(a). Therefore, with the rectangular function rec(x) defined in the Appendix, Table A1, we have for the basic sink pulse

$$g_v(t) = \hat{g}_v \operatorname{rec}\left(\frac{t - T_D - T_V - T/2}{T}\right) = \begin{cases} \hat{g}_v & \text{for } T_D + T_V < t < T_D + T_V + T \\ 0 & \text{otherwise} \end{cases} \tag{5.17}$$

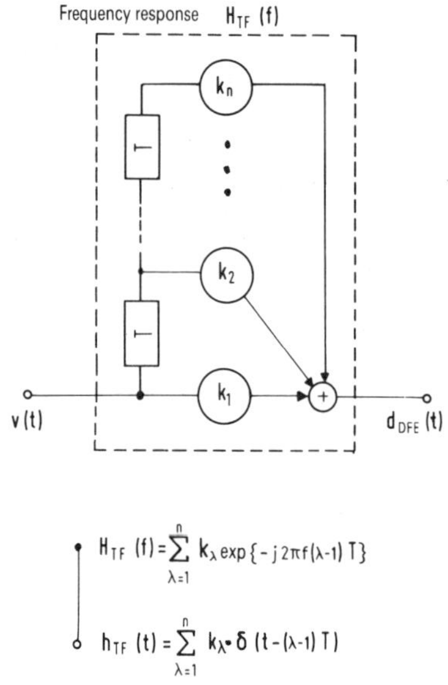

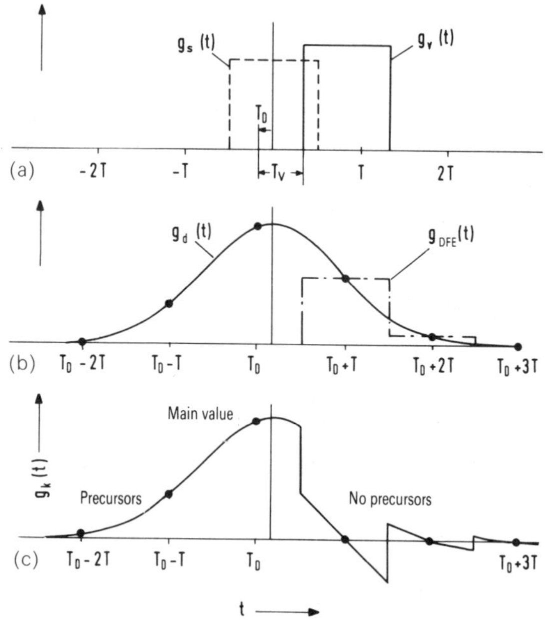

Fig. 5.6 Transversal filter of length n.

Fig. 5.7 Basic pulses of the DFE with a transversal filter: (a) transmitter and sink pulse; (b) detection and correction pulse; (c) corrected detection pulse.

With the pulse response $h_{\text{TF}}(t)$ defined in Fig. 5.6 we therefore obtain the correction signal

$$d_{\text{DFE}}(t) = \sum_{\lambda=1}^{n} k_\lambda v\{t-(\lambda-1)T\}$$
$$= \sum_{\lambda=1}^{n} \sum_{\nu=-\infty}^{+\infty} k_\lambda a_\nu' \hat{g}_v \operatorname{rec}\left(\frac{t-T_{\text{D}}-T_{\text{V}}+T/2-\nu T-\lambda T}{T}\right) \tag{5.18}$$

Comparison with eqn (5.4) shows that the basic correction pulse

$$g_{\text{DFE}}(t) = \sum_{\lambda=1}^{n} k_\lambda \hat{g}_v \operatorname{rec}\left(\frac{t-T_{\text{D}}-T_{\text{V}}+T/2-\lambda T}{T}\right) \tag{5.19}$$

can be represented by the sum of the shifted rectangular pulses weighted by the filter coefficients k_λ (Fig. 5.7). The first n postcursors of the detection pulse $g_d(t)$ can be wholly compensated for if at the times $T_{\text{D}}+T,\ldots, T_{\text{D}}+nT$

$$g_{\text{DFE}}(T_{\text{D}}+\lambda T) = g_d(T_{\text{D}}+\lambda T) \qquad \lambda = 1,\ldots,n \tag{5.20}$$

Assuming further that the delay T_{V} is smaller than the symbol duration T, we obtain the filter coefficients

$$k_\lambda = \frac{g_d(T_{\text{D}}+\lambda T)}{\hat{g}_v} \qquad \lambda = 1,\ldots,n \tag{5.21}$$

Figure 5.7(c) shows the corrected basic detection pulse $g_k(t)$, which is sawtooth like and is zero at the detection times $T_{\text{D}}+\nu T$ ($\nu = 1,\ldots,n$) of the subsequent symbols. Except at these equidistant points in time, compensation for the postcursors is not complete which, in comparison with ideal DFE, results in a smaller horizontal eye opening.

Example Figure 5.8 shows the (corrected) basic detection pulse and the eye pattern for a binary system with rectangular NRZ transmitter pulses and a Gaussian pulse shaper ($f_{\text{I}} = 0.3R$). A receiver without DFE ($\mathring{T}_{\text{D}} = 0$) has a (normalized) eye opening of about 10% (Fig. 5.8(a)). For all the receivers with DFE considered here, the vertical eye opening at the optimum detection time $\mathring{T}_{\text{D}} = -0.4T$ is the same, amounting to a normalized 40% (Figs 5.8(b), 5.8(c) and 5.8(d)). However, the horizontal eye opening differs in each case.

5.3 Direct signal recovery

It has already been stated in the preceding chapters that a nonredundant transmitter signal causes a closed eye if no direct signal can be transmitted, i.e. when $H_{\text{K}}(0)$ or $H_{\text{E}}(0)$ is zero. The low frequency components of the spectrum suppressed during transmission can be reconstructed at the receiver with the aid of DFE which causes the eye to open again.

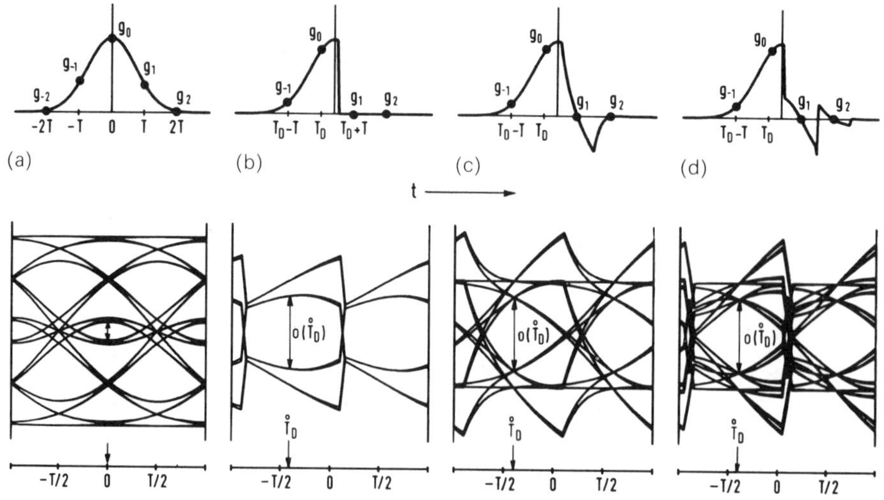

Fig. 5.8 Pulse and eye patterns for systems without and with DFE: (a) without DFE; (b) ideal DFE (see Fig. 5.2); (c) first-order low-pass filter (see Fig. 5.5); (d) transversal filter (see Fig. 5.7).

In contrast with the use of DFE for compensation of high frequency intersymbol interference, we refer in this case to **direct signal recovery**. It serves to reduce the intersymbol interference which is attributable to the lower band limit of the channel. Hence the term "low frequency DFE" is often used.

Note It is advantageous when examining DFE for direct signal recovery to substitute the subtraction of the signals $d(t)$ and $d_{DFE}(t)$ in the block diagram by an addition which represents the idea that missing frequency components are replaced by the correction signal (see Fig. 5.1).

5.3.1 Direct signal recovery in high-pass systems

To describe direct signal recovery we first consider a high-pass system, i.e. the effect of the upper band limit is initially disregarded. The pulse-shaper frequency response is then given by (Fig. 5.9(a))

$$H_I(f) = H_{HP}(f) \tag{5.22}$$

If the basic transmitter pulse $g_s(t)$ shown in Fig. 5.9(c) is applied at the input of this high-pass filter $H_{HP}(f)$, the pulse that results at its output is given by

$$g_d(t) = g_s(t) * h_{HP}(t) \tag{5.23}$$

with a pulse area of zero (Fig. 5.9(d)).

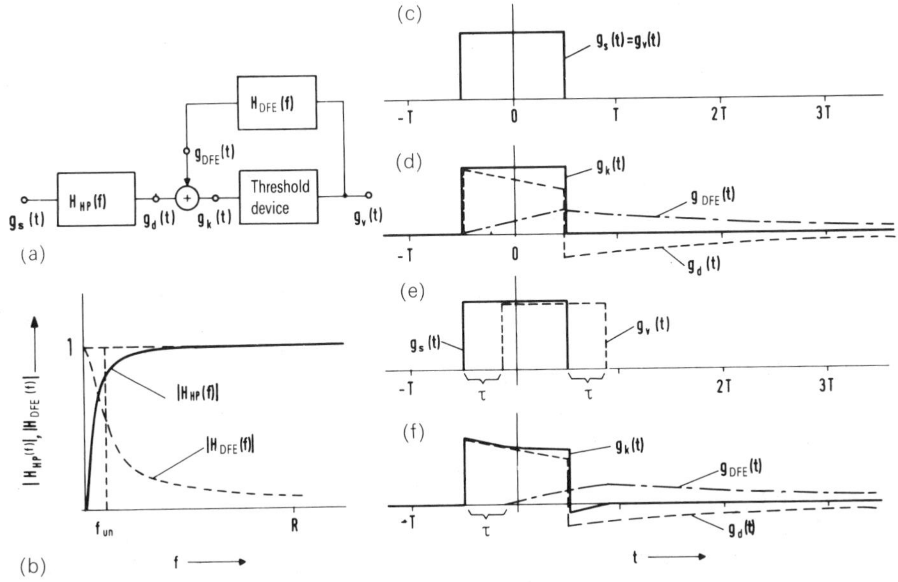

Fig. 5.9 Illustration of the direct signal recovery of a high-pass system: (a) block diagram; (b) frequency responses $|H_{HP}(f)|$ and $|H_{DFE}(f)|$; (c), (e) basic transmitter and sink pulse; (d), (f) basic detection pulse, basic correction pulse and corrected basic detection pulse.

Because of the long trailing edge, the length of which depends on the cut-off frequency f_{lo} of the high-pass filter, a closed eye results in the case of a receiver without DFE. This long trail can be significantly reduced by using a suitably sized low-pass filter in the feedback loop, which is equivalent to restoration of the missing low frequency components of the spectrum and results in an opened eye. With the pulse response $h_{DFE}(t) \circ\!\!-\!\!\bullet H_{DFE}(f)$ of the feedback network the basic correction pulse is

$$g_{DFE}(t) = g_v(t) * h_{DFE}(t) \tag{5.24}$$

and so the corrected basic detection pulse can be described by

$$g_k(t) = g_d(t) + g_{DFE}(t) = g_s(t) * h_{HP}(t) + g_v(t) * h_{DFE}(t) \tag{5.25}$$

First it is assumed that the output pulse $g_v(t)$ of the decision device is identical with the basic transmitter pulse $g_s(t)$ with respect to both shape and timing (Fig. 5.9(c)). For causality reasons this assumption can only be fulfilled if the detection time $T_D = -T/2$ and the delay $T_V = 0$. With this restriction the interference effect of the lower band limit can be completely eliminated, i.e. in this case $g_k(t) = g_s(t)$ is possible. Considering the relationship $g_v(t) = g_s(t)$ we obtain from eqn. (5.25) by means of Fourier transformation

$$G_k(f) = G_s(f)H_{HP}(f) + G_s(f)H_{DFE}(f) \stackrel{!}{=} G_s(f) \tag{5.26}$$

and from this, for the ideal feedback network for direct signal recovery,

$$\begin{aligned} &H_{DFE}(f) = 1 - H_{HP}(f) \\ &h_{DFE}(t) = \delta(t) - h_{HP}(t) \end{aligned} \tag{5.27}$$

This means that, with a missing upper band limit ("high-pass system"), the earliest possible detection time ($T_D = -T/2$) and a detector which produces no delay ($T_V = 0$), direct signal recovery need not produce any reduction in the eye opening even when the lower cut-off frequency is relatively high. The low frequency spectral components, which are suppressed during transmission by the high-pass filter $H_{HP}(f)$, can be added again at the receiver by means of an equivalent low-pass filter $H_{DFE}(f) = 1 - H_{HP}(f)$ (Fig. 5.9(b)). (Note that $H_{DFE}(f) = 1 - H_{HP}(f)$ but $|H_{DFE}(f)| \neq 1 - |H_{HP}(f)|$.)

If the detection occurs later than this earliest possible detection time $T_D = -T/2$ and if the delay T_V induced by the decision device is taken into consideration as well, the basic sink pulse $g_v(t)$ is delayed relative to the basic transmitter pulse $g_s(t)$ by a time difference $\tau = T/2 + T_D + T_V$. Complete direct signal recovery (i.e. $g_k(t) = g_s(t)$) is not possible here for causality reasons (Figs 5.9(e) and 5.9(f)). Also, direct signal recovery shows a significant improvement in this case if a suitably sized low-pass filter is used in the feedback network:

$$H_{DFE}(f) = \alpha_{DFE}\{1 - H_{HP}(f)\} \tag{5.28}$$

Example With rectangular NRZ transmitter pulses and a first-order high-pass filter, i.e.

$$H_{HP}(f) = \frac{1}{1 - j2f_{lo}/\pi f} \tag{5.29}$$

we obtain the following basic detection pulse (see the Appendix, Table A3):

$$g_d(t) = \begin{cases} \hat{g}_s \exp\left\{-4f_{lo}\left(t + \dfrac{T}{2}\right)\right\} & \text{for } -\dfrac{T}{2} \leqslant t \leqslant \dfrac{T}{2} \\ \hat{g}_s\{\exp(-4f_{lo}T) - 1\} \exp\left\{-4f_{lo}\left(t - \dfrac{T}{2}\right)\right\} & \text{for } t > \dfrac{T}{2} \end{cases} \tag{5.30}$$

Here, by analogy with eqn (2.38), f_{lo} is the cut-off frequency of the low-pass filter

$$1 - H_{HP}(f) = \frac{1}{1 + j(\pi f/2f_{lo})^2} \tag{5.31}$$

equivalent to the high-pass filter $H_{HP}(f)$. It is further assumed that the basic sink pulse $g_v(t)$ is shifted relative to the basic transmitter pulse by the time difference τ and that the feedback network is sized according to eqn (5.28). Thus we have for the basic correction pulse (Fig. 5.9(f))

$$g_{\mathrm{DFE}}(t) = \begin{cases} \alpha_{\mathrm{DFE}} \hat{g}_v \left[1 - \exp\left\{ -4f_{\mathrm{lo}}\left(t + \dfrac{T}{2} - \tau \right) \right\} \right] & \text{for } -\dfrac{T}{2} + \tau \leqslant t \leqslant \dfrac{T}{2} + \tau \\ \alpha_{\mathrm{DFE}} \hat{g}_v \{\exp(4f_{\mathrm{lo}}T) - 1\} \exp\left\{ -4f_{\mathrm{lo}}\left(t + \dfrac{T}{2} - \tau \right) \right\} & \text{for } t > \dfrac{T}{2} + \tau \end{cases} \tag{5.32}$$

A comparison of eqn (5.30) with eqn (5.32) shows that for $\hat{g}_v = \hat{g}_s$ and $t > T/2 + \tau$ we have

$$g_{\mathrm{DFE}}(t) = -\alpha_{\mathrm{DFE}} \exp(4f_{\mathrm{lo}}\tau)\, g_d(t) \tag{5.33}$$

If we now select $\alpha_{\mathrm{DFE}} = \exp(-4f_{\mathrm{lo}}\tau)$ the corrected basic detection pulse $g_k(t)$ is identically equal to zero when $t \geqslant T/2 + \tau$ (Fig. 5.9(f)). The normalized eye opening is then about 85%.

5.3.2 Direct signal recovery in band-pass systems

In high-pass systems, complete direct signal recovery for $T_{\mathrm{D}} = -T/2$ and $T_{\mathrm{V}} = 0$ is also possible in principle if the lower cut-off frequency f_{lo} is chosen to be arbitrarily high. The noise power is infinitely large, however, because of the missing upper band limit.

In comparison, the complete reconstruction of the low frequency spectral components is not possible in a band-pass system. If the upper cut-off frequency f_{up} of the whole system is finite, the basic detection pulse $g_d(t)$ does not exhibit a step increase as shown in Fig. 5.9(d) and does not reach a value as high as its original maximum. Consequently the amplitude $g_d(0)$ of the basic detection pulse is smaller the more high and low frequency spectral components are filtered out of the spectrum $G_d(f)$ (because $g_d(0) = \int G_d(f)\,\mathrm{d}f$).

The description is simplified when the frequency response of the band-pass filter is the product of the frequency responses of a high-pass and a low-pass filter:

$$H_{\mathrm{I}}(f) = H_{\mathrm{HP}}(f) H_{\mathrm{LP}}(f) \tag{5.34}$$

By means of this division of the pulse shaper the direct signal recovery can be related to the equations of Section 5.3.1 and the pulse shaping to the results from Chapter 3. However, it has to be assumed here that the low-pass cut-off frequency f_{up} is significantly higher than the high-pass cut-off frequency f_{lo}. In accordance with eqn (5.28) we have for the sizing of the feedback network

$$H_{\mathrm{DFE}}(f) = \alpha_{\mathrm{DFE}}\{1 - H_{\mathrm{HP}}(f)\} H_{\mathrm{LP}}(f) \tag{5.35}$$

Example For illustration we consider a band-pass system (Fig. 5.10(a)) whose frequency response can be constructed from the multiplication of a Gaussian low-pass filter (cut-off frequency f_{up}) and a first-order high-pass filter (cut-off frequency f_{lo}):

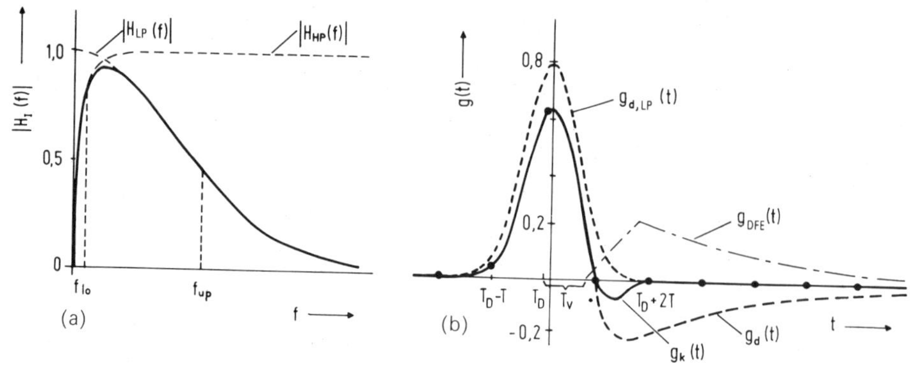

Fig. 5.10 Illustration of the direct signal recovery with a band-pass system: (a) pulse-shaper frequency response from eqn (5.36) ($f_{lo} = 0.1R$; $f_{up} = 0.5R$); (b) basic pulses $g_d(t)$, $g_{DFE}(t)$ and $g_k(t)$ with a first-order low-pass feedback filter (see eqn (5.35)).

$$H_I = \frac{1}{1 - j2f_{lo}/\pi f} \exp\left\{-\pi\left(\frac{f}{2f_{up}}\right)^2\right\} \tag{5.36}$$

The corresponding basic detection pulse $g_d(t)$ is illustrated in Fig. 5.10(b) for parameters $f_{lo} = 0.1R$ and $f_{up} = 0.5R$, where the delay time is not considered. The basic detection pulse $g_{d,LP}(t)$ for the corresponding low-pass system ($f_{lo} = 0$; $f_{up} = 0.5R$) is also shown for comparison. $g_d(t)$ has a long negative trail because of the missing low frequency spectral components. When we add to this pulse a correction pulse consisting of the response $g_{DFE}(t)$ of a first-order low-pass filter with cut-off frequency $f_{lo} = 0.1R$ to a rectangular pulse, the corrected detection pulse $g_k(t)$ has practically no postcursor. The detection time T_D and the amplitude factor α_{DFE} are to be optimized with regard to an eye opening which is as large as possible. For the values $f_{lo} = 0.1R$, $f_{up} = 0.5R$ and $T_V = 0.8T$ considered here, we obtain the optimum values $\mathring{T}_D = -T/10$ and $\mathring{\alpha}_{DFE} = 0.65$. The resultant normalized eye opening in this case is 57% which is about 1% less than that for the comparable low-pass system ($f_{lo} = 0$; $f_{up} = 0.5R$) without DFE.

The loss due to the cut-off frequency f_{lo} becomes greater as the upper cut-off frequency f_{up} becomes lower. This is treated in more detail in system optimization in Section 8.1.

5.4 Error propagation due to decision feedback equalization

Until now we have always assumed in the description of DFE that the detector has identified preceding symbols correctly. If in a system with DFE a transmission error due to noise occurs then a correction pulse of erroneous amplitude is generated. The result is that the postcursors of the incorrectly identified pulse are not eliminated but increased so that the subsequent symbols are in error with higher probability. We refer to this effect as **error propagation due to DFE**.

If the basic detection pulse $g_d(t)$ has n nonnegligible postcursors, on average the error probability of the n subsequent symbols is made decidedly greater by a single incorrect decision. The determination for the following n symbols is affected by each subsequent error. Thus there is an increase in the value of the mean error probability.

5.4.1 Symbol error probability with regard to error propagation

In the following, for the purpose of illustration, a binary system ($M = 2$) with ideal DFE is assumed. The same result is obtained if a binary system with complete DFE at the detection instants is considered (see Fig. 5.4(a)). However, the method of calculation has to be modified slightly for a system with incomplete DFE or a multilevel system.

Example The basic detection pulse shown in Fig. 5.11(a) is used to explain the error propagation effect. If the effect of error propagation is neglected, a system with ideal DFE produces the corrected basic detection pulse of Fig. 5.11(b).

If an amplitude coefficient $a_\nu' = -a_\nu \in \{-1; +1\}$ is wrongly determined, the n subsequent postcursors of the detection pulse $a_\nu g_d(t - \nu T)$ are not reduced but doubled. Therefore in this case the "corrected" basic detection pulse $g_k(t)$ shown in Fig. 5.11(c) is produced.

Now let us consider the symbol error probability p_{S_ν} as defined in eqn (3.1) which, for a receiver with (ideal) DFE and $T_D = 0$, can be determined from the corrected useful sample values $\tilde{k}(\nu T)$ and the detection noise power N_d. In Fig. 5.11(d) it is assumed that all amplitude coefficients were correctly determined ($a_\nu' = a_\nu$ for all ν), so that $\tilde{k}(t)$ can be constructed from the corrected basic detection pulse (Fig. 5.11(b)). In this case and with an assumed constant noise power N_d we obtain the values specified in Fig. 5.11(d) for the symbol error probabilities p_{S_ν}.

However, if an amplitude coefficient is incorrectly determined ($a_\kappa' \neq a_\kappa$), the subsequent sample values $\tilde{k}(\nu T)$, $\nu > \kappa$, deviate from the other values by $\pm 2g_{\nu-\kappa}$. According to the signs of a_ν, a_κ and $g_{\nu-\kappa}$, the error probabilities p_{S_ν} for $\nu > \kappa$ can be significantly increased or reduced. For example, in Fig. 5.11(e) an incorrect decision is assumed at the time $t = 2T$ so that the amplitude coefficients are $a_2 = -1$ and $a_2' = +1$. The symbol error probability $p_{S_3} = 0.02$ for the subsequent symbol is markedly increased (from 10^{-14}) by the erroneous correction of the DFE. However, the amplitude coefficient a_4 is corrupted with the lower error probability $p_{S_4} = 10^{-10}$ by the preceding incorrect compensation. p_{S_5} and p_{S_6} remain almost unchanged because the third and further postcursor samples of the basic detection pulse $g_d(t)$ considered are very small. This example shows that, as a rule, the error propagation effect is limited to only a few symbols.

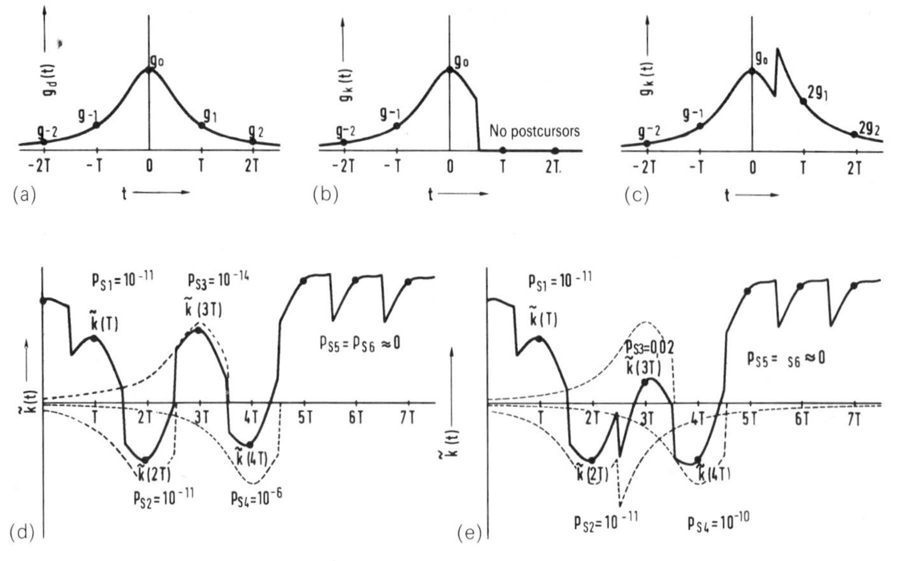

Fig. 5.11 Illustration of the error propagation effect: (a) basic detection pulse $g_d(t)$; (b) corrected basic detection pulse for ideal DFE without consideration of the error propagation effect; (c) corrected basic detection pulse for ideal DFE with consideration of the error propagation effect; (d) corrected useful detection signal for ideal DFE without consideration of the error propagation effect; (e) corrected useful detection signal for ideal DFE with consideration of the error propagation effect.

To calculate the mean error probability of a nonredundant binary system without DFE, averaging has to be done over 2^{v+n+1} symbol sequences where v and n define the number of precursors and postcursors of the basic detection pulse $g_d(t)$. If we neglect the effect of error propagation on the mean symbol error probability, only 2^{v+1} symbol sequences have to be averaged for a system with complete DFE. For this case we obtain, by analogy with eqn (3.37),

$$p_{\mathrm{M}}(\text{neglecting error propagation}) = 2^{-(v+1)} \sum_{i=1}^{2^{v+1}} p_{\mathrm{S}_i} \tag{5.37}$$

The symmetry with regard to the decision value $E = 0$ allows the calculation cost to be halved in both cases.

When the error propagation effect is considered the cost of the calculation of the mean error probability p_{M} increases greatly. The method of Andexser [5.1, 5.14], in which the mean error probability is calculated as an average of all possible error patterns, is presented in the following. For this, the **logical error variables** ξ_v are introduced:

$$\xi_v = \begin{cases} \mathrm{L} & \text{when } a_v' \neq a_v \\ \mathrm{O} & \text{when } a_v' = a_v \end{cases} \tag{5.38}$$

Further, for quantitative description of the statistical dependences of the individual decisions, the **mean follow-up probabilities** are defined as follows:

$$P_F(\xi_0|\xi_{-1}\ldots\xi_{-n}) \tag{5.39}$$

If we assume that the previous and the third previous amplitude coefficients were falsely determined (i.e. $a_{-1}' \neq a_{-1}$, $a_{-3}' \neq a_{-3}$), the follow-up probability P_F(L|LOLO) relates to the conditional probability that the amplitude coefficient $a_0' \neq a_0$ is incorrectly determined. The follow-up probability P_F(L|OOOO...) is therefore equal to the mean error probability neglecting the error propagation (see eqn (5.37)).

In order to clarify that the follow-up probabilities are themselves mean values, we now consider a pulse without precursors ($v = 0$) and assume further that the previous and third previous decisions were incorrect. As a result of error propagation the useful sample value $\tilde{k}(T_D)$ of the corrected detection signal can have one of the eight values given in Table 5.1 depending on the values of the amplitude coefficients a_0, a_{-1} and a_{-3}.

Table 5.1

i	a_0	a_{-1}	a_{-3}	a_{-1}'	a_{-3}'	$\tilde{k}(T_D)$
1	+1	+1	+1	−1	−1	$g_0 + 2g_1 + 2g_3$
2	+1	+1	−1	−1	+1	$g_0 + 2g_1 - 2g_3$
3	+1	−1	+1	+1	−1	$g_0 - 2g_1 + 2g_3$
4	+1	−1	−1	+1	+1	$g_0 - 2g_1 - 2g_3$
5	−1	+1	+1	−1	−1	$-g_0 + 2g_1 + 2g_3$
6	−1	+1	−1	−1	+1	$-g_0 + 2g_1 - 2g_3$
7	−1	−1	+1	+1	−1	$-g_0 - 2g_1 + 2g_3$
8	−1	−1	−1	+1	+1	$-g_0 - 2g_1 - 2g_3$

Correspondingly the mean follow-up probability P_F(L|LOLO...) can be determined as the mean value of the eight error probabilities P_i(L|LOLO...) if possible symmetries around the decision value $E = 0$ are not taken into consideration. $P_i(\mathrm{L}|\xi_{-1}\ldots\xi_{-n})$ is hence the error probability for a particular error pattern $\xi_{-1}\ldots\xi_{-n}$ and for a particular sequence $\langle a_v \rangle_i$ of amplitude coefficients.

As an example, the probability that the symbol $a_0 = +1$ is falsely identified, assuming that the previous coefficients $a_{-1} = -1$ and $a_{-3} = +1$ were falsely identified but the other foregoing coefficients were correctly determined ($\xi_0 = \xi_{-1} = \xi_{-3} = \mathrm{L}$; $\xi_{-2} = \xi_{-4} = \ldots = \xi_{-n} = \mathrm{O}$), is given by

$$P_i(\mathrm{L}|\xi_{-1}\ldots\xi_{-n}) = P_3(\mathrm{L}|\mathrm{LOLO}\ldots) = Q\left(\frac{g_0 - 2g_1 + 2g_3}{N_d^{1/2}}\right) \tag{5.40}$$

Using z to denote the number of wrong decisions in the previous n decisions, when considering v precursors, we obtain

$$P_{\mathrm{F}}(\mathrm{L}|\xi_{-1}\ldots\xi_{-n}) = 2^{-(v+z+1)} \sum_{i=1}^{2^{v+z+1}} P_i(\mathrm{L}|\xi_{-1}\ldots\xi_{-n}) \tag{5.41}$$

To determine the mean error probability the individual error patterns (for which the variable j is used in the following), which were previously assumed to be constant have to be calculated and averaged. Hence for a nonredundant binary system with ideal DFE, we have for the mean symbol error probability when considering error propagation

$$p_{\mathrm{M}} = \sum_{j=1}^{2^n} P_{\mathrm{F}}(\xi_0 = \mathrm{L}|\xi_{-1}\ldots\xi_{-n})P(\xi_{-1}\ldots\xi_{-n}) \tag{5.42}$$

Here $P(\xi_{-1}\ldots\xi_{-n})$ defines the probability of the occurrence of the error pattern $\xi_{-1}\ldots\xi_{-n}$. The problem encountered when evaluating this equation lies in the calculation of these 2^n probabilities of occurrence. The method of calculating the probabilities is outlined briefly below; a detailed description is given in ref. 5.14.

As the occurrence of an error in the decision ξ_0 depends on the n preceding decisions $\xi_{-1}\ldots\xi_{-n}$, this process can be described as an nth-order Markov process with two states O and L. However, to provide simpler start conditions, individual steps are not considered; instead every n steps are considered together. Thus the process of transition from one error pattern to another (i.e. after n steps) is represented as a first-order Markov process with 2^n states. The Markov diagram for the case $n = 2$ is shown in Fig. 5.12. $P_{\mathrm{T}}(\mathrm{LO}|\mathrm{LL})$ in this diagram defines, for example, the conditional probability that the error pattern LO follows the error pattern $\xi_{-1}\xi_{-2} = \mathrm{LL}$ after $n = 2$ steps. This transition probability can be expressed as the product of $n = 2$ follow-up probabilities:

$$P_{\mathrm{T}}(\mathrm{LO}|\mathrm{LL}) = P_{\mathrm{F}}(\mathrm{O}|\mathrm{LL})P_{\mathrm{F}}(\mathrm{L}|\mathrm{OL}) \tag{5.43}$$

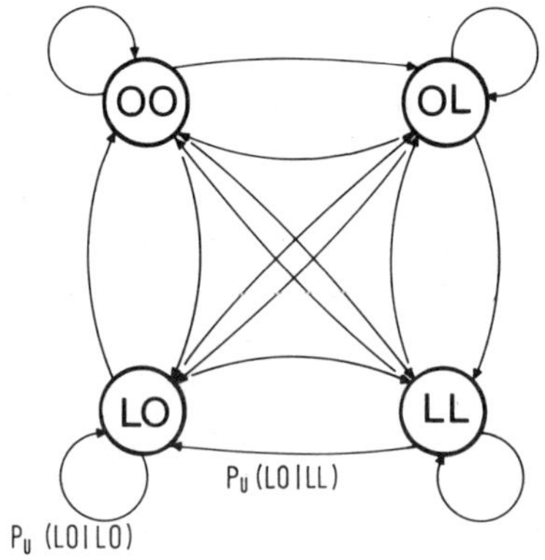

Fig. 5.12 Markov diagram describing the transition from one error pattern to another ($n = 2$).

The follow-up probabilities $P_F(\mathrm{O}|\mathrm{LL})$ and $P_F(\mathrm{L}|\mathrm{OL})$ respectively are again the mean values of all symbol sequences and can be calculated from eqn (5.41).

For an arbitrary value of n the 2^{2n} possible transition probabilities P_T between the individual error patterns can each be defined as the product of n averaged follow-up probabilities P_F. The probabilities of these transitions can be combined in a stochastic $2^n \times 2^n$ matrix. For example, for $n = 2$,

$$\mathbb{P} = \begin{bmatrix} P_T(\mathrm{OO}|\mathrm{OO}) & P_T(\mathrm{LO}|\mathrm{OO}) & P_T(\mathrm{OL}|\mathrm{OO}) & P_T(\mathrm{LL}|\mathrm{OO}) \\ P_T(\mathrm{OO}|\mathrm{LO}) & P_T(\mathrm{LO}|\mathrm{LO}) & P_T(\mathrm{OL}|\mathrm{LO}) & P_T(\mathrm{LL}|\mathrm{LO}) \\ P_T(\mathrm{OO}|\mathrm{OL}) & P_T(\mathrm{LO}|\mathrm{OL}) & P_T(\mathrm{OL}|\mathrm{OL}) & P_T(\mathrm{LL}|\mathrm{OL}) \\ P_T(\mathrm{OO}|\mathrm{LL}) & P_T(\mathrm{LO}|\mathrm{LL}) & P_T(\mathrm{OL}|\mathrm{LL}) & P_T(\mathrm{LL}|\mathrm{LL}) \end{bmatrix} \tag{5.44}$$

The 2^n probabilities of occurrence $P(\xi_{-1} \dots \xi_{-n})$ of the individual error patterns can be determined from this stochastic matrix $\mathbb{P}$. If we define the column vectors of the probabilities of occurrence at the times zero and r of the Markov process as

$$\mathbf{P}_0 = \begin{bmatrix} P_0(\mathrm{OO}) \\ P_0(\mathrm{LO}) \\ P_0(\mathrm{OL}) \\ P_0(\mathrm{LL}) \end{bmatrix} \quad \mathbf{P}_r = \begin{bmatrix} P_r(\mathrm{OO}) \\ P_r(\mathrm{LO}) \\ P_r(\mathrm{OL}) \\ P_r(\mathrm{LL}) \end{bmatrix} \tag{5.45}$$

then

$$\mathbf{P}_r = [\mathbb{P}^{\mathrm{T}}]^r \mathbf{P}_0 \tag{5.46}$$

where $\mathbb{P}^{\mathrm{T}}$ is the transpose of $\mathbb{P}$. Starting from every arbitrary initiation vector $\mathbf{P}_0$ the vector $\mathbf{P}$ of the stationary transition probabilities can be calculated for the limit $r \to \infty$:

$$\mathbf{P} = \lim_{r \to \infty} [\mathbb{P}^{\mathrm{T}}]^r \mathbf{P}_0 \tag{5.47}$$

We obtain the same transition probabilities if the stochastic matrix is multiplied $r \to \infty$ times by itself:

$$\lim_{r \to \infty} \mathbb{P}^r = \begin{bmatrix} P(\mathrm{OO}) & P(\mathrm{LO}) & P(\mathrm{OL}) & P(\mathrm{LL}) \\ P(\mathrm{OO}) & P(\mathrm{LO}) & P(\mathrm{OL}) & P(\mathrm{LL}) \\ P(\mathrm{OO}) & P(\mathrm{LO}) & P(\mathrm{OL}) & P(\mathrm{LL}) \\ P(\mathrm{OO}) & P(\mathrm{LO}) & P(\mathrm{OL}) & P(\mathrm{LL}) \end{bmatrix} \tag{5.48}$$

This stationary matrix is notable in that all the elements in a column are identical. The elements of the first column are equal to the probability of occurrence $P(\mathrm{OO})$ for the error pattern OO, those of the second column are equal to $P(\mathrm{LO})$ etc. Inserting these probabilities and the corresponding follow-up probabilities into eqn (5.42) we obtain the desired mean error probability.

Example To illustrate this calculation process we now consider a basic detection pulse with $n = 2$ postcursors and no precursors ($v = 0$). For simplification the sample values are expressed as multiples of the effective value of the noise signal:

$$g_0 = g_d(T_D) = 4N_d^{1/2} \qquad g_1 = g_d(T_D + T) = 2N_d^{1/2} \qquad g_2 = g_d(T_D + 2T) = N_d^{1/2}$$

If we take into account the symmetry about the decision value E, the following error probabilities for a nonredundant binary system are produced.

(a) *System without DFE* From eqn (3.58)

$$p_U = Q\left(\frac{g_0 - |g_1| - |g_2|}{N_d^{1/2}}\right) = Q(1) = 0.159$$

From eqn (3.37)

$$p_M = \tfrac{1}{4}\{Q(7) + Q(5) + Q(3) + Q(1)\} = 0.040$$

(b) *System with DFE neglecting error propagation* From eqn (5.12)

$$p_U = Q\left(\frac{g_0}{N_d^{1/2}}\right) = Q(4) = 0.32 \times 10^{-4}$$

From eqn (5.37)

$$p_M(\text{neglecting error propagation}) = Q(4) = 0.32 \times 10^{-4}$$

(c) *System with DFE taking account of error propagation*

(1) Calculation of the follow-up probabilities using eqns (5.40) and (5.41):

$P_F(L|OO) = Q(4) = 0.32 \times 10^{-4}$ $\qquad P_F(O|OO) \approx 1$

$P_F(L|LO) = \tfrac{1}{2}\{Q(0) + Q(8)\} = 0.25$ $\qquad P_F(O|LO) = 0.75$

$P_F(L|OL) = \tfrac{1}{2}\{Q(2) + Q(6)\} = 0.011$ $\qquad P_F(O|OL) = 0.989$

$P_F(L|LL) = \tfrac{1}{4}\{Q(-2) + Q(2) + Q(6) + Q(10)\} = 0.25$ $\qquad P_F(O|LL) = 0.75$

(2) Calculation of the transition probabilities using eqn (5.43):

$$P_T(OO|OO) = P_F(O|OO)P_F(O|OO) = 1$$

$$\vdots$$

$$P_T(OL|LO) = P_F(L|LO)P_F(O|LL) = 0.1875$$

$$\vdots$$

$$P_T(LL|LL) = P_F(L|LL)P_F(L|LL) = 0.0625$$

(3) Stochastic matrix according to eqn (5.44):

$$\mathbb{P} = \begin{bmatrix} 1.0000 & 0.32 \times 10^{-4} & 0.24 \times 10^{-4} & 0.79 \times 10^{-5} \\ 0.7415 & 0.85 \times 10^{-2} & 0.1875 & 0.0625 \\ 0.9886 & 0.32 \times 10^{-4} & 0.85 \times 10^{-2} & 0.28 \times 10^{-2} \\ 0.7415 & 0.85 \times 10^{-2} & 0.1875 & 0.0625 \end{bmatrix}$$

(4) Stationary matrix according to eqn (5.48):

$$\lim_{r\to\infty} \mathbb{P}^r \approx \mathbb{P}^3 = \begin{bmatrix} 0.9999 & 0.32\times 10^{-4} & 0.32\times 10^{-4} & 0.11\times 10^{-4} \\ 0.9999 & 0.32\times 10^{-4} & 0.32\times 10^{-4} & 0.11\times 10^{-4} \\ 0.9999 & 0.32\times 10^{-4} & 0.32\times 10^{-4} & 0.11\times 10^{-4} \\ 0.9999 & 0.32\times 10^{-4} & 0.32\times 10^{-4} & 0.11\times 10^{-4} \end{bmatrix}$$

(5) Mean error probability according to eqn (5.42):

$$p_M = (0.32\times 10^{-4})\times 0.9999 + 0.25(0.32\times 10^{-4}) + 0.011(0.32\times 10^{-4}) + 0.25(0.11\times 10^{-4})$$

$$\approx 0.43\times 10^{-4}$$

Comparison of the values of p_M(neglecting error propagation) and p_M shows that in this example the mean error probability increases by a factor of only 1.34 when error propagation is taken into account.

Note To average over all error patterns and all symbol sequences a total of

$$2^{v+n}\sum_{i=0}^{n}\binom{n}{i}2^i$$

cases must be considered, even when symmetry is taken into account. In addition, $2^n \times 2^n$ matrices have to be processed. It is obvious that the computation time and the required accuracy increase very rapidly with increasing n and v.

5.4.2 Error propagation factor

Finally an example will show to what extent the mean error probability is increased by the error propagation effect. For this purpose the **error propagation factor due to DFE** is introduced:

$$\gamma_{EP,\,DFE} = \frac{p_M}{p_M(\text{without error propagation})} \tag{5.49}$$

The error propagation factor is plotted against the mean error probability p_M in Fig. 5.13(b). This is based on the use of rectangular transmitted pulses and of a Gaussian low-pass filter with cut-off frequency $f_I = 0.3R$, and the corresponding optimum detection time $\mathring{T}_D = -0.3T$ (Fig. 5.13(a)). It was shown in Section 5.1.2 (see Fig. 5.3) that for this pulse the eye opening can be increased from 10% to about 40% by the use of an ideal DFE.

Figure 5.13(b) shows that the mean error probability is increased by slightly less than a factor of 2 by consideration of the error propagation effect. Further, this increase is independent of the noise power, i.e. of the actual value of p_M. Therefore, even when the error propagation effect is considered the mean symbol error probability is still smaller than the worst-case error probability p_U, which is

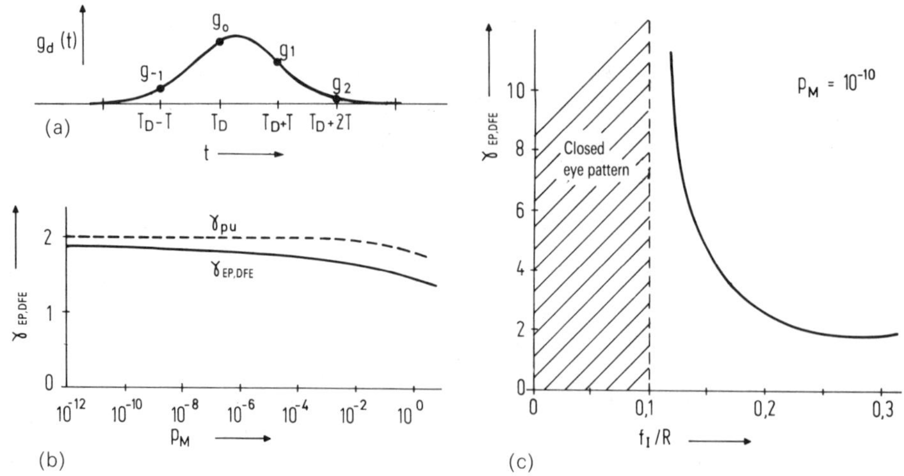

Fig. 5.13 Dependence of the error propagation factor for the basic detection pulse shown in (a) on (b) the mean error probability p_M and (c) the pulse-shaper cut-off frequency f_I.

calculated from the vertical eye opening and the noise power and is used in subsequent chapters as an optimization criterion.

The error propagation factor is very much larger if a large number n of postcursors contribute to the error propagation. In Fig. 5.13(c) the error propagation factor is plotted against the pulse-shaper cut-off frequency f_I. The detection time is assumed to be the optimum corresponding to eqn (8.14). $\gamma_{EP,\,DFE}$ increases hyperbolically as the cut-off frequency becomes smaller and has a value of 10 when $f_I = 0.1R$. However, since even a system with ideal DFE exhibits a closed eye pattern for $f_I < 0.13R$, the effect of error propagation can in general be neglected. With a mean error probability of 10^{-10} a signal-to-noise ratio improvement of slightly more than 0.1 dB is already sufficient to compensate for the error propagation effect.

Note With incomplete DFE the error propagation factor due to DFE is somewhat lower. It should be mentioned in addition that the error propagation factor due to direct signal recovery with a binary band-pass system does not exceed a value of 2 (compare ref. 5.6).

Chapter 6

Optimum Digital Receiver

The authors thank Dr.-Ing. U. Peters for his assistance in the preparation of this chapter. The contents and the description are based on his dissertation [6.20].

Contents Receivers which result in the minimum error probability for a given basic receiver pulse and given noise power spectrum are described in this chapter. First the receiver strategy of an optimum receiver is derived in general terms (Section 6.1), and then some different realizations are described in terms of their block diagrams (Section 6.2). While in these optimum receivers the detection of the whole message is achieved in a single decision process, the decision rules are modified in Section 6.3 so that parts of the message can already be detected optimally before the receiver has received the whole signal. As these receivers use the Viterbi algorithm, either directly or in a modified form, they are referred to as "Viterbi detectors". The mean error probability of an optimum receiver is calculated in Section 6.4, and approximations to this which are valid under certain boundary conditions are derived.

Assumptions Here, in contrast with Chapters 2–5, a digital source which only generates a finite number N of M-level source symbols is considered first. The optimum receiver determines the whole symbol sequence (Section 6.2) or at least several symbols (Section 6.3) simultaneously. For additive disturbances of Gaussian distribution it is assumed that the noise power spectrum $L_n(f)$ has no zeros in the frequency range considered.

6.1 Strategy for the optimum receiver

Figure 6.1 shows the block diagram for the digital transmission system under consideration. The M-level source outputs the source symbols $q_1 \dots q_\nu \dots q_N$. To distinguish it from the infinitely long sequence $\langle q_\nu \rangle$ the source symbol sequence

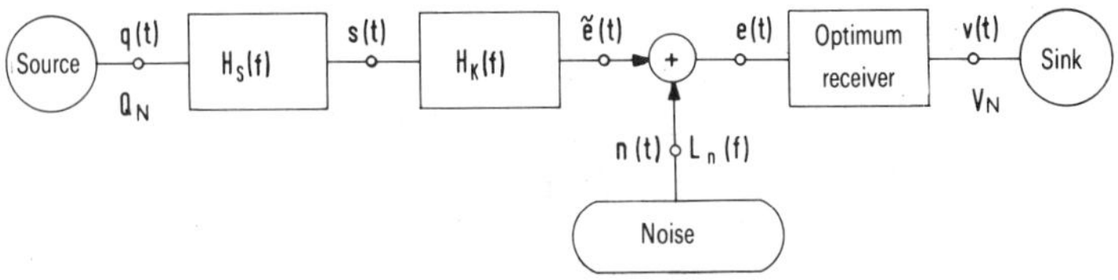

Fig. 6.1 Block diagram of a digital transmission system with an optimum receiver.

consisting of N symbols is denoted Q_N. Similarly the sink symbol sequence with N symbols is denoted V_N. An optimum receiver is defined as one which selects as the received signal $e(t)$ the most probable symbol sequence ${Q_N}^{(j)}$ from all possible source symbol sequences ${Q_N}^{(1)} \ldots {Q_N}^{(i)} \ldots {Q_N}^{(I)}$ and outputs this as the sink symbol sequence

$$V_N = {Q_N}^{(j)} \in \{{Q_N}^{(i)}\} \qquad i = 1 \ldots I \tag{6.1}$$

$I \leqslant M^N$ is the number of permissible source symbol sequences; the equals sign applies for nonredundant sources. The *a priori* probability p_i with which the source symbol sequence ${Q_N}^{(i)}$ is transmitted is defined as follows:

$$p_i = P(Q_N = {Q_N}^{(i)}) \tag{6.2}$$

A receiver which, with a given receiver input signal $e(t)$, selects the most probable symbol sequence ${Q_N}^{(j)}$ and thus is aware of the *a priori* probability p_i is called a **maximum *a posteriori* receiver (MAP receiver).**

With the assumption that the signal $e(t)$ is input to the receiver, the MAP receiver determines all I inference probabilities $P\{{Q_N}^{(i)}|e(t)\}$ which express how likely it is that the corresponding sequence ${Q_N}^{(i)}$ was transmitted. Then it outputs the most probable sequence $V_N = {Q_N}^{(j)}$ for which

$$P\{{Q_N}^{(j)}|e(t)\} > P\{{Q_N}^{(i)}|e(t)\} \qquad \forall \begin{matrix} i = 1 \ldots I \\ i \neq j \end{matrix} \tag{6.3}$$

This decision rule can be expressed according to Bayes' formula as follows:

$$P\{e(t)|{Q_N}^{(j)}\}\frac{P\{{Q_N}^{(j)}\}}{P\{e(t)\}} > P\{e(t)|{Q_N}^{(i)}\}\frac{P\{{Q_N}^{(i)}\}}{P\{e(t)\}} \qquad \forall \begin{matrix} i = 1 \ldots I \\ i \neq j \end{matrix} \tag{6.4}$$

As $e(t)$ is a continuous signal, the probabilities $P\{e(t)\}$ and $P\{e(t)|{Q_N}^{(i)}\}$ tend toward zero. Therefore in the following these parameters will be understood to be the absolute probability and conditional probability respectively that at all times a possible receiver signal differs no more than a certain amount $\pm\varepsilon/2$ from the actual receiver signal $e(t)$. For simplicity, the passage to the limit $\varepsilon \to 0$ is omitted in the following.

The probability $P\{e(t)\}$ is on both sides of the inequality and thus need not be considered further. $P({Q_N}^{(j)})$ and $P({Q_N}^{(i)})$ are the *a priori* probabilities p_j and p_i according to definition (6.2). Thus the two conditional probabilities in eqn (6.4) remain to be determined. With the assumption that the source symbol sequence $Q_N^{(i)}$ was transmitted, the useful received signal is $\tilde{e}_i(t)$. The actual received signal $e(t)$ differs from it by the difference signal

$$\Delta e_i(t) = e(t) - \tilde{e}_i(t) \tag{6.5}$$

The conditional probability that the received signal $e(t)$ occurs with the assumed source symbol sequence $Q_N^{(i)}$ is thus equal to the probability that the noise signal $n(t)$ has the same value as the difference signal $\Delta e_i(t)$:

$$P\{e(t)|Q_N^{(i)}\} = P\{n(t) = \Delta e_i(t)\} \tag{6.6}$$

$\Delta e_i(t)$ is thus the noise signal that must occur to corrupt the ith useful received signal $\tilde{e}_i(t)$ in the actual received signal $e(t)$. If we consider only a fixed time $t = t_1$, this probability can be expressed as a boundary value by means of the probability density function $f_n(n)$:

$$P\{n(t_1) = \Delta e_i(t_1)\} = \lim_{\varepsilon \to 0} \int_{\Delta e_i(t_1) - \varepsilon/2}^{\Delta e_i(t_1) + \varepsilon/2} f_n(n)\, \mathrm{d}n = \lim_{\varepsilon \to 0} [\varepsilon f_n\{\Delta e_i(t_1)\}] \tag{6.7}$$

In a similar manner the joint probability

$$P\{n(t_1) = \Delta e_i(t_1); n(t_2) = \Delta e_i(t_2); \ldots; n(t_k) = \Delta e_i(t_k)\}$$
$$= \lim_{\varepsilon \to 0} [\varepsilon^k f_n\{\Delta e_i(t_1); \Delta e_i(t_2); \ldots; \Delta e_i(t_k)\}] \tag{6.8}$$

for k different instants $t_1, t_2, \ldots, t_k$ can be replaced by the joint probability density function. Finally, if we consider an infinite number of infinitely densely placed adjacent time instants, the following can be used to describe the conditional probability of eqn (6.4) [6.22]:

$$P\{e(t)|Q_N^{(i)}\} = P\{n(t) = \Delta e_i(t)\} = \lim_{k \to \infty} \lim_{\varepsilon \to 0} \varepsilon^k f_n\{\Delta e_i(t)\} \tag{6.9}$$

When this result is inserted into eqn (6.4) the MAP receiver decision rule for arbitrary noise distribution becomes

$$V_N = Q_N^{(j)} \quad \text{if} \quad p_j f_n\{\Delta e_j(t)\} > p_i f_n\{\Delta e_i(t)\} \quad \forall \begin{array}{l} i = 1 \ldots l \\ i \neq j \end{array} \tag{6.10}$$

As the passages to the limits $k \to \infty$ and $\varepsilon \to 0$ appear on both sides of the inequality they can be dropped here.

It is assumed in the following that the noise $n(t)$ is of Gaussian distribution. Hence, according to refs 1.26 and 6.13 the joint probability density function becomes

$$f_n\{\Delta e_i(t)\} = K \exp\left\{ -\frac{1}{2L_0} \int_{-\infty}^{+\infty} \Delta e_i(t) \int_{-\infty}^{+\infty} \Delta e_i(\tau) w_n(t - \tau)\, \mathrm{d}\tau\, \mathrm{d}t \right\} \tag{6.11}$$

in which k is a normalization constant of no further interest. $w_n(\tau)$ is the (normalized) **inverse autocorrelation function** of the noise signal $n(t)$, which is defined as the Fourier integral of the reciprocal value of the normalized noise power spectrum:

$$w_n(\tau) = \int_{-\infty}^{+\infty} \frac{L_0}{L_n(f)} \exp(\mathrm{j}2\pi f\tau)\,\mathrm{d}f \quad (\text{in s}^{-1}) \tag{6.12}$$

L_0 is a normalization constant with the dimensions of a power spectrum. Hence the following relationship exists between the inverse autocorrelation function $w_n(\tau)$ and the ACF $l_n(\tau)$ of the noise signal $n(t)$:

$$\begin{array}{ccccc} \dfrac{L_0}{L_n(f)} & & L_n(f) & = & L_0 \\ \bullet\!\!-\!\!\circ & & \bullet\!\!-\!\!\circ & & \bullet\!\!-\!\!\circ \\ w_n(\tau) & * & l_n(\tau) & = & L_0\delta(\tau) \end{array} \tag{6.13}$$

The inner integral in eqn (6.11) corresponds to the convolution operation $w_n(t)*\Delta e_i(t)$, so that the joint probability density function can also be represented as follows:

$$f_n\{\Delta e_i(t)\} = K \exp\left[-\frac{1}{2L_0}\int_{-\infty}^{+\infty} \Delta e_i(t)\{w_n(t)*\Delta e_i(t)\}\,\mathrm{d}t\right] \tag{6.14}$$

As $\Delta e_i(t)$ represents the difference between $e(t)$ and $\tilde{e}_i(t)$, the exponent is given by

$$\begin{aligned} [\ldots] = &-\frac{1}{2L_0}\int_{-\infty}^{+\infty} e(t)\{w_n(t)*e(t)\}\,\mathrm{d}t + \frac{1}{2L_0}\int_{+\infty}^{-\infty} e(t)\{w_n(t)*\tilde{e}_i(t)\}\,\mathrm{d}t \\ &+\frac{1}{2L_0}\int_{-\infty}^{+\infty} \tilde{e}_i(t)\{w_n(t)*e(t)\}\,\mathrm{d}t - \frac{1}{2L_0}\int_{-\infty}^{+\infty} \tilde{e}_i(t)\{w_n(t)*\tilde{e}_i(t)\}\,\mathrm{d}t \end{aligned} \tag{6.15}$$

The second and third integrals are the same because

$$\int_{-\infty}^{+\infty} e(t)\int_{-\infty}^{+\infty} \tilde{e}_i(\tau)w_n(t-\tau)\,\mathrm{d}\tau\,\mathrm{d}t = \int_{-\infty}^{+\infty} \tilde{e}_i(t)\int_{-\infty}^{+\infty} e(\tau)w_n(t-\tau)\,\mathrm{d}\tau\,\mathrm{d}t \tag{6.16}$$

Inserting the result in eqn (6.10) and taking the logarithm of both sides of the inequality we obtain, after some rearrangement, the following general decision rule for an MAP receiver with Gaussian noise: the MAP receiver chooses from all $i = 1 \ldots I$ possible symbol sequences $Q_N^{(i)}$ the sequence $Q_N^{(j)}$ for which

$$\begin{aligned} &\int_{-\infty}^{+\infty} e(t)\{w_n(t)*\tilde{e}_j(t)\}\,\mathrm{d}t - \frac{1}{2}\int_{-\infty}^{+\infty} \tilde{e}_j(t)\{w_n(t)*\tilde{e}_j(t)\}\,\mathrm{d}t + L_0\ln(p_j) \\ &\quad > \int_{-\infty}^{+\infty} e(t)\{w_n(t)*\tilde{e}_i(t)\}\,\mathrm{d}t - \frac{1}{2}\int_{-\infty}^{+\infty} \tilde{e}_i(t)\{w_n(t)*\tilde{e}_i(t)\}\,\mathrm{d}t + L_0\ln(p_i) \\ &\qquad\qquad \forall \begin{array}{l} i = 1\ldots I \\ i \neq j \end{array} \end{aligned} \tag{6.17}$$

The sink symbol sequence is thus $V_N = Q_N^{(j)}$.

This means that, to realize an MAP receiver, all I possible useful receiver signals $\tilde{e}_i(t)$, the inverse ACF $w_n(\tau)$ of the noise (or the corresponding power spectrum) as well as the *a priori* probabilities p_i of the possible source symbol sequences $Q_N^{(i)}$ must be known.

A simplification of the MAP receiver is the **maximum likelihood receiver.** This has no knowledge of the *a priori* probabilities and therefore assumes for simplification that all I *a priori* probabilities p_i are equal. The maximum likelihood receiver selects the most probable sequence $V_N = Q_N^{(j)}$ using this simplified assumption. Therefore both terms $L_0 \ln(p_j)$ and $L_0 \ln(p_i)$ drop out of the decision rule (eqn (6.17)). For a nonredundant source ($p_i = M^{-N}$ for all i) the maximum likelihood receiver possesses the same characteristics as the MAP receiver.

For the special case of Gaussian-distributed white noise with the noise power density L_0, eqn (6.12) yields the normalized inverse ACF

$$w_n(\tau) = \delta(\tau) \tag{6.18}$$

so that the decision rule of the maximum likelihood receiver given in eqn (6.17) can be further simplified: decide on $V_N = Q_N^{(j)}$ if the following is true for all $i \neq j$:

$$\int_{-\infty}^{+\infty} e(t)\tilde{e}_j(t)\,\mathrm{d}t - \frac{1}{2}\int_{-\infty}^{+\infty} \{\tilde{e}_j(t)\}^2\,\mathrm{d}t > \int_{-\infty}^{+\infty} e(t)\tilde{e}_i(t)\,\mathrm{d}t - \frac{1}{2}\int_{-\infty}^{+\infty} \{\tilde{e}_i(t)\}^2\,\mathrm{d}t \tag{6.19}$$

If, further, the energies

$$E_e^{(i)} = \int_{-\infty}^{+\infty} \{\tilde{e}_i(t)\}^2\,\mathrm{d}t \qquad i = 1 \ldots M^N \tag{6.20}$$

of all useful received signals are equal, then we obtain the simplest version of a maximum likelihood receiver with the decision rule

$$\int_{-\infty}^{+\infty} e(t)\tilde{e}_j(t)\,\mathrm{d}t > \int_{-\infty}^{+\infty} e(t)\tilde{e}_i(t)\,\mathrm{d}t \qquad \forall \begin{matrix} i = 1 \ldots M^N \\ i \neq j \end{matrix} \tag{6.21}$$

6.2 Optimum receiver for detection of the complete message

There are several possible, fundamentally different, ways of implementing the general optimum receiver decision rule given in eqn (6.17). The block diagrams of three alternative practical forms in which the detection of the whole message occurs in a single decision process are presented in this section.

6.2.1 Correlation receiver

The **cross-correlation function** (CCF) of two ergodic processes $x(t)$ and $y(t)$, by analogy with the ACF, is given by the following expression (cf. eqn (2.18)):

$$l_{xy}(\tau) = \lim_{T_0 \to \infty} \frac{1}{2T_0} \int_{-T_0}^{+T_0} x(t)\,y(t+\tau)\,\mathrm{d}t \qquad \text{(in V}^2\text{ for example)} \tag{6.22}$$

If at least one of the two signals is time limited, and hence energy limited, then it is necessary to use the **energy CCF** by analogy with eqn (2.23):

$$l^{\cdot}_{xy}(\tau) = \int_{-\infty}^{+\infty} x(t)\,y(t+\tau)\,\mathrm{d}t \qquad \text{(in V}^2\text{ s for example)} \tag{6.23}$$

In view of the decision rule in eqn (6.17) the first integral can be represented as the energy CCF of the two functions $e(t)$ and $w_n(t) * \tilde{e}_i(t)$ at time $\tau = 0$. The received signal $e(t)$ therefore has to be correlated with all the $I \leqslant M^N$ possible signals $w_n(t) * \tilde{e}_i(t)$, i.e. be multiplied and then integrated (Fig. 6.2(a)). The convolution between the ith useful received signal $\tilde{e}_i(t)$ and the inverse ACF $w_n(t)$ can be realized, for example, using a filter with the frequency response $L_0/L_n(f)$ (see eqn 6.12)).

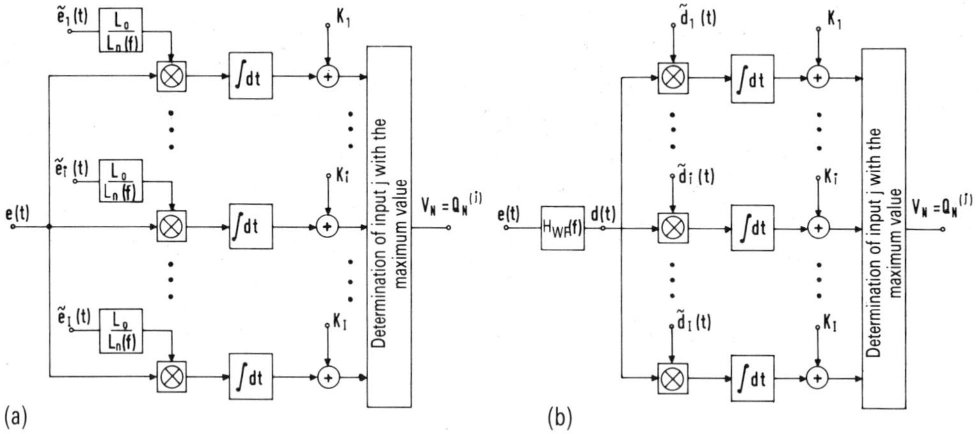

Fig. 6.2 Block diagrams of a correlation receiver in two different configurations.

The second integral in eqn (6.17) corresponds to a fixed energy, independent of the received signal $e(t)$, and is abbreviated to

$$E_d^{(i)} = \int_{-\infty}^{+\infty} \tilde{e}_i(t)\{w_n(t) * \tilde{e}_i(t)\}\,\mathrm{d}t \tag{6.24}$$

For white noise, i.e. when $w_n(t) = \delta(t)$, $E_d^{(i)}$ is identical with $E_e^{(i)}$ from eqn (6.20). From Parseval's theorem

$$\int_{-\infty}^{+\infty} x(t)\,y(t)\,\mathrm{d}t = \int_{-\infty}^{+\infty} X(f)\,Y^*(f)\,\mathrm{d}f \tag{6.25}$$

the following can also be stated for this case:

$$E_d^{(i)} = \int_{-\infty}^{+\infty} \frac{L_0}{L_n(f)} |\tilde{E}_i(f)|^2\,\mathrm{d}f \tag{6.26}$$

The energies $E_d^{(i)}$ corresponding to the individual symbol sequences $Q_N^{(i)}$ must be known to the receiver and, in a similar manner to the probabilities p_i, are accounted for by the constants (Fig. 6.2)

$$K_i = L_0 \ln(p_i) - \tfrac{1}{2}E_d^{(i)} \qquad i = 1 \ldots I \tag{6.27}$$

The correlation receiver therefore determines the quantities

$$\int_{-\infty}^{+\infty} e(t)\{w_n(t) * \tilde{e}_i(t)\}\,\mathrm{d}t + K_i$$

for all I possible sequences $Q_N^{(i)}$ and selects those with the highest value. The source symbol sequence $Q_N^{(j)}$ corresponding to this quantity is then the most probable of all possible received symbol sequences and is output as the sink symbol sequence V_N. If all *a priori* probabilities are equal and all useful received signals have the same energy, the constants K_i do not have to be taken into consideration.

Figure 6.2(b) shows that the correlation receiver can be realized more simply by using a suitable receiver filter $H_E(f) = H_{WF}(f)$ without information loss. Hence $H_{WF}(f)$ is referred to as a **whitening filter** or a decorrelation filter. This filter has the characteristic that the noise component $\check{d}(t)$ of its output signal shows no statistical dependence, i.e. for the noise power spectrum

$$L_{\check{d}}(f) = L_n(f)|H_{WF}(f)|^2 = L_0 = \text{constant} \tag{6.28}$$

Hence the amplitude response of the whitening filter can be derived as follows:

$$|H_{WF}(f)| = \left\{\frac{L_0}{L_n(f)}\right\}^{1/2} \tag{6.29}$$

The phase is arbitrary because it does not affect the detection noise signal.

The detection signal results from the convolution operation $d(t) = e(t) * h_{WF}(t)$. Hence, with the possible useful detection signals

$$\tilde{d}_i(t) = \tilde{e}_i(t) * h_{WF}(t) \qquad i = 1 \ldots I \tag{6.30}$$

we obtain for the first integral in the decision rule (eqn (6.17)) of the MAP receiver by double use of Parseval's theorem (see eqn (6.25))

$$\int_{-\infty}^{+\infty} e(t)\{w_n(t)*\tilde{e}_i(t)\}\mathrm{d}t = \int_{-\infty}^{+\infty} E^*(f)\left\{\frac{L_0}{L_n(f)}\tilde{E}_i(f)\right\}\mathrm{d}f$$
$$= \int_{-\infty}^{+\infty}\left[E(f)\left\{\frac{L_0}{L_n(f)}\right\}^{1/2}\right]^*\left[\left\{\frac{L_0}{L_n(f)}\right\}^{1/2}\tilde{E}_i(f)\right]\mathrm{d}f$$
$$= \int_{-\infty}^{+\infty} D^*(f)\tilde{D}_i(f)\,\mathrm{d}f$$
$$= \int_{-\infty}^{+\infty} d(t)\,\tilde{d}_i(t)\,\mathrm{d}t \qquad (6.31)$$

The second integral, which was already abbreviated in eqn (6.24) to $E_d^{(i)}$, in fact corresponds to the energy of the ith useful detection signal $\tilde{d}_i(t)$ if a whitening filter is used:

$$E_d^{(i)} = \int_{-\infty}^{+\infty} \{\tilde{d}_i(t)\}^2\,\mathrm{d}t \qquad (6.32)$$

Hence the decision rule for the correlation receiver with a whitening filter is

$$V_N = Q_N^{(j)} \quad \text{if} \quad \int_{T_1}^{T_2} d(t)\tilde{d}_j(t)\,\mathrm{d}t + K_j > \int_{T_1}^{T_2} d(t)\tilde{d}_i(t)\,\mathrm{d}t + K_i \qquad \forall\, \begin{matrix} i = 1\ldots I \\ i \neq j \end{matrix} \qquad (6.33)$$

As all useful detection signals $\tilde{d}_i(t)$ before the time T_1 (before the first symbol was transmitted) and from a certain time T_2 are identically zero, it is sufficient to integrate only from time $t = T_1$ to $t = T_2$.

6.2.2 Matched-filter receiver

The integrations needed in eqn (6.17) for the MAP receiver's decision rule can also be implemented by linear filtering and subsequent sampling. Figure 6.3 shows the corresponding arrangement, which is denoted in the following as a **matched-filter receiver.**

The linear filter in the ith branch of the receiver is matched to the useful received signal $\tilde{e}_i(t)$ ○—● $\tilde{E}_i(f)$ and to the noise power spectrum $L_n(f)$. The frequency and pulse response of this filter are given by

$$H_{\mathrm{MF}}^{(i)}(f) = K_{\mathrm{MF}}\frac{L_0}{L_n(f)}\tilde{E}_i^*(f) \qquad (6.34)$$

$$h_{\mathrm{MF}}^{(i)}(t) = K_{\mathrm{MF}}w_n(t)*\tilde{e}_i(-t) \qquad (6.35)$$

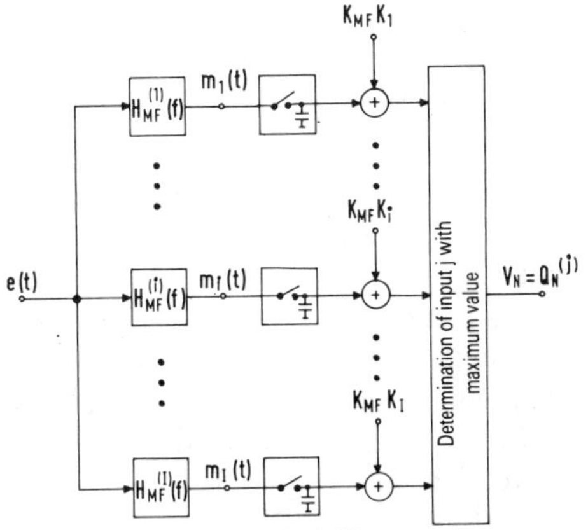

Fig. 6.3 Matched-filter receiver.

The arbitrary constant K_{MF} with units reciprocal volts seconds is needed to ensure a dimensionless frequency response. If the received signal $e(t)$ is applied to the matched-filter input, the output signal obtained is

$$m_i(t) = e(t) * h_{\mathrm{MF}}{}^{(i)}(t) = K_{\mathrm{MF}} \int_{-\infty}^{+\infty} e(\tau) h_{\mathrm{MF}}{}^{(i)}(t-\tau)\,\mathrm{d}\tau \tag{6.36}$$

It follows from consideration of eqn (6.35) that at time $t = 0$

$$m_i(0) = K_{\mathrm{MF}} \int_{-\infty}^{+\infty} e(\tau)\{w_n(-\tau) * \tilde{e}_i(\tau)\}\,\mathrm{d}\tau \tag{6.37}$$

Thus, with the symmetrical property $w_n(-\tau) = w_n(\tau)$ and renaming τ as t, we have

$$\int_{-\infty}^{+\infty} e(t)\{w_n(t) * \tilde{e}_i(t)\}\,\mathrm{d}t = \frac{m_i(0)}{K_{\mathrm{MF}}} \tag{6.38}$$

Inserting this result in the general MAP receiver rule given by eqn (6.17) we obtain the following decision rule for the matched-filter receiver signal with constants K_j and K_i as given by eqn (6.27):

$$V_N = Q_N{}^{(j)} \quad \text{if} \quad m_j(0) + K_{\mathrm{MF}}K_j > m_i(0) + K_{\mathrm{MF}}K_i \qquad \forall \begin{array}{l} i = 1 \ldots I \\ i \neq j \end{array} \tag{6.39}$$

The output signals $m_i(t)$ of the total of I different matched filters have to be sampled at time $T_{\mathrm{D}} = 0$ and the constant values $K_{\mathrm{MF}}K_i$ have to be added to them. The largest of these I values must then be sought, and the symbol sequence $Q_N{}^{(j)}$ which corresponds to this value is output as the most likely message (Fig. 6.3).

Note The filters $H_{\mathrm{MF}}{}^{(i)}(f)$ are assumed to be noncausal here. For practical applications they can be substituted by causal filters. The additional delay time needed for them can be taken into consideration by a suitable choice of sampling time $T_{\mathrm{D}} > 0$.

6.2.3 Receiver with noise-energy detector

In both the optimum receivers introduced so far, the implementation was based directly on the decision rule given by eqn (6.17). In this case the maximum similarity between the receiver input signal and the possible useful receiver input signal is the basis of the detection. These similarities were determined by cross-correlation or, rather, by a matched filter (equivalent to a correlation filter).

A further practical realization results if we consider the differences between the (possibly filtered) receiver input signal $e(t)$ and the possible useful signals $\tilde{e}_i(t)$. Figure 6.4 shows the block diagram of a decision device which works according to this principle. In the following this type is referred to as a receiver with a **noise-energy detector** [6.20]. First the received input signal $e(t)$ is passed through a whitening filter $H_{\mathrm{WF}}(f)$ as defined by eqn (6.29), so that the noise component $\check{d}(t)$ of the detection signal $d(t)$ exhibits no statistical dependences, i.e. $L_{\check{d}}(f) = L_0 = \text{constant}$.

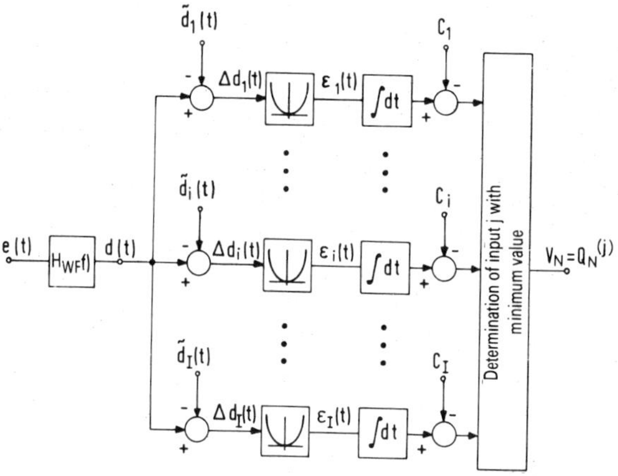

Fig. 6.4 Receiver with noise-energy detector.

We derive from eqns (6.10) and (6.11) the decision rule of the MAP receiver:

$$\int_{-\infty}^{+\infty} \Delta e_j(t)\{w_n(t)*\Delta e_j(t)\}\,dt - 2L_0\ln(p_j)$$

$$< \int_{-\infty}^{+\infty} \Delta e_i(t)\{w_n(t)*\Delta e_i(t)\}\,dt - 2L_0\ln(p_i) \qquad \forall\, \begin{matrix} i=1\ldots I \\ i\neq j \end{matrix} \tag{6.40}$$

By analogy with eqn (6.31) we can also say that, when a whitening filter is used,

$$\int_{-\infty}^{+\infty} \{\Delta d_j(t)\}^2\,\mathrm{d}t - 2L_0\ln(p_j) < \int_{-\infty}^{+\infty} \{\Delta d_i(t)\}^2\,\mathrm{d}t - 2L_0\ln(p_i) \tag{6.41}$$

Then, by analogy with definition (6.5), the difference between the actual detection signal $d(t)$ and the ith possible useful detection signal $\tilde{d}_i(t)$ is abbreviated to

$$\Delta d_i(t) = d(t) - \tilde{d}_i(t) \tag{6.42}$$

This is referred to later as the **linear error quantity.** The **squared error quantity** can be calculated from it as the instantaneous squared deviation between the detector input signal $d(t)$ and the useful signal $\tilde{d}_i(t)$ corresponding to the sequence $Q_N^{(i)}$:

$$\varepsilon_i(t) = \{\Delta d_i(t)\}^2 = \{d(t) - \tilde{d}_i(t)\}^2 \tag{6.43}$$

Integration of the squared error quantity $\varepsilon_i(t)$ specifies the energy of the difference signal $\Delta d_i(t)$, which is denoted in the following as the **noise energy:**

$$E_{\Delta d}{}^{(i)} = \int_{T_1}^{T_2} \varepsilon_i(t)\,\mathrm{d}t = \int_{T_1}^{T_2} \{d(t) - \tilde{d}_i(t)\}^2\,\mathrm{d}t \tag{6.44}$$

T_1 and T_2 determine the limits of the time interval during which the individual useful detection signals are different. For $t < T_1$ and $t > T_2$ it is assumed that all the useful detection signals $\tilde{d}_i(t)$ are identical.

The noise energy $E_{\Delta d}{}^{(i)}$ defined here denotes the energy that a noise signal in the time interval $T_1 \ldots T_2$ would have to possess in order that the useful detection signal $\tilde{d}_i(t)$ corresponding to the sequence $Q_N^{(i)}$ considered would be corrupted in the actual detection signal $d(t)$. If the sequence $Q_N = Q_N^{(k)}$ were transmitted, the noise energy $E_{\Delta d}{}^{(k)}$ represents the real noise energy of the detection signal.

Figure 6.4 shows the block diagram of a receiver with a noise-energy detector. The receiver forms the squared error quantities $\varepsilon_i(t)$ for all I possible symbol sequences $Q_N^{(i)}$ and from them the noise energy by integration. The different *a priori* probabilities are accounted for by the constants

$$C_i = -2L_0 \ln(p_i) \qquad i = 1 \ldots I \tag{6.45}$$

With the noise-energy detector the effect of the various energies of the individual useful signals is already implicitly accounted for and hence does not need to be incorporated into the constants as in the correlation receiver and the matched-filter receiver.

Thus for the receiver with a noise-energy detector the decision rule is written

$$V_N = Q_N^{(j)} \quad \text{if} \quad E_{\Delta d}{}^{(j)} + C_j < E_{\Delta d}{}^{(i)} + C_i \qquad \forall \begin{array}{l} i = 1 \ldots I \\ i \neq j \end{array} \tag{6.46}$$

If we divide the noise energy $E_{\Delta d}{}^{(i)}$ by the integration time $T_2 - T_1$, the mean squared deviation (error) of the detection signal $d(t)$ from the useful signal $\tilde{d}_i(t)$ results. With equiprobable source symbol sequences, therefore, the noise-energy detector selects that sequence which exhibits the smallest mean-squared error.

6.3 Optimum receivers for the detection of partial messages (Viterbi receiver)

It has been assumed previously that the detector determines the sink symbol sequence in a single decision process. The cost of implementing such a detector increases exponentially with the length of the sequence to be detected. In addition it has been assumed that, after N transmitted symbols, no more occur until the detection process ends. This means that after N symbols a pause is necessary before a further N symbols can be transmitted.

In this section the decision rules are modified so that optimum detection of part of the received message can be carried out before the receiver has obtained the whole signal. As these receivers use the Viterbi algorithm, either directly or in a modified form, they are referred to as Viterbi receivers.

This concept has the advantage that it can be implemented at significantly lower cost. The optimum detection of an infinitely long symbol sequence is then possible although, as is the case in every practical realization, a more or less marked deviation from the optimum detection must be tolerated.

As individually determined symbols will already be output before the whole message has been transmitted, it is in general not possible to include the *a priori* probabilities p_i in the detection process. Therefore only the maximum likelihood receiver is considered in the following.

6.3.1 Optimum maximum likelihood detection with periodic sampling

Figure 6.5 shows the block diagram of an optimum Viterbi receiver. Firstly, suitable sample values

$$d_\nu = d(T_\mathrm{D} + \nu T) = \tilde{d}_\nu + \overset{\times}{d}_\nu \tag{6.47}$$

which the Viterbi detector needs to determine the most probable symbol sequence are extracted from the continuous input signal $e(t)$. This is achieved by the use of a matched filter, an ideal sampler and a transversal filter. It will be shown in the following that the quantities required for the optimum receiver can similarly be determined from these sample values d_ν, and so the implementation cost is significantly lower than that of the maximum likelihood receivers described in Section 6.2.

With an optimum maximum likelihood receiver a total of M^N different integrals of the form

$$I_i = \int_{-\infty}^{+\infty} e(t)\{w_n(t) * \tilde{e}_i(t)\}\,\mathrm{d}t \tag{6.48}$$

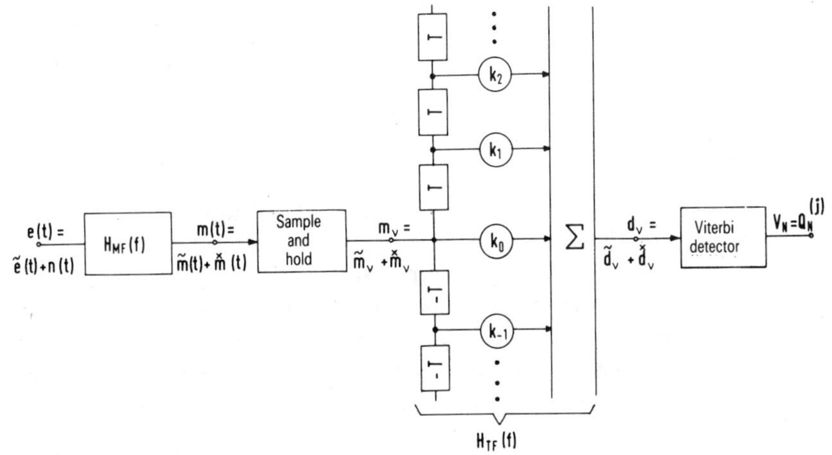

Fig. 6.5 Optimum receiver with matched filter, ideal sampler, discrete whitening filter and Viterbi detector.

have to be calculated according to the general decision rule in eqn (6.17). By analogy with eqn (2.43) the following is true for the ith useful received signal:

$$\tilde{e}_i(t) = \sum_{\nu=1}^{N} a_\nu^{(i)} g_e(t - \nu T) \tag{6.49}$$

Inserting this expression into eqn (6.48) after interchanging the summation and integration and writing the convolution integral in full we obtain

$$I_i = \sum_{\nu=1}^{N} a_\nu^{(i)} \int_{-\infty}^{+\infty} e(\tau) \int_{-\infty}^{+\infty} w_n(\tau') g_e(\tau - \tau' - \nu T)\, \mathrm{d}\tau'\, \mathrm{d}\tau \tag{6.50}$$

For illustration purposes the integration variable t is redesignated as τ.

The double integral in the summation is equal, up to a constant value, to the output signal $m(t)$ of a matched filter at time $t = \nu T$, which is matched to the basic receiver pulse $g_e(t)$ and the noise power spectrum $L_n(f)$. By analogy with eqn (6.35) we have for the frequency response and pulse response of the matched filter

$$\begin{aligned} H_{\mathrm{MF}}(f) &= K_{\mathrm{MF}} \frac{L_0}{L_n(f)} G_e^*(f) \\ &\ \ \updownarrow \\ h_{\mathrm{MF}}(t) &= K_{\mathrm{MF}} \{ w_n(t) * g_e(-t) \} \end{aligned} \tag{6.51}$$

When the received signal $e(t)$ is applied at the input of the matched filter, the following output signal is obtained:

$$m(t) = \int_{-\infty}^{+\infty} e(\tau) h_{\mathrm{MF}}(t - \tau)\, \mathrm{d}\tau = K_{\mathrm{MF}} \int_{-\infty}^{+\infty} e(\tau) \int_{-\infty}^{+\infty} w_n(\tau') g_e(\tau - \tau' - t)\, \mathrm{d}\tau'\, \mathrm{d}\tau \tag{6.52}$$

Here we assume that the inverse ACF $w_n(\tau)$ as defined by eqn (6.12) is symmetrical. A comparison of eqn (6.50) with eqn (6.52) makes it clear that the integrals required for optimum detection can be written as follows:

$$I_i = \frac{1}{K_{\mathrm{MF}}} \sum_{\nu=1}^{N} a_\nu^{(i)} m_\nu \tag{6.53}$$

where $m_\nu = m(\nu T)$ is the sample value of the matched-filter output signal at times νT (Fig. 6.5). The sum of the sample values m_ν rated with the possible amplitude coefficients $a_\nu^{(i)}$ is identical with I_i up to a dimensional constant.

In the following it will be shown further that the energies $E_d^{(i)}$ required for the general decision rule of eqn (6.17) can be determined as terms dependent on the possible amplitude coefficients. To do this we need the basic matched-filter pulse $g_m(t)$ as the response of the matched filter to a basic received pulse $g_e(t)$ at its input. For this, eqn (6.51) gives

$$g_m(t) = g_e(t) * h_{\mathrm{MF}}(t) = K_{\mathrm{MF}}\{w_n(t) * g_e(t) * g_e(-t)\} \tag{6.54}$$

$g_e(t) * g_e(-t)$ is the energy ACF $l^{\bullet}_{ge}(t)$ of the basic received pulse (see eqn (2.23)). Hence the basic matched-filter pulse can also be written

$$g_m(t) = K_{\mathrm{MF}}\{w_n(t) * l^{\bullet}_{ge}(t)\} \tag{6.55}$$

As $w_n(t)$ and $l^{\bullet}_{ge}(t)$ are symmetrical functions, $g_m(t)$ is also.

The energy $E_d^{(i)}$ defined in eqn (6.24) can be written, using eqn (6.49), as

$$E_d^{(i)} = \int_{-\infty}^{+\infty} \sum_{\nu=1}^{N} a_\nu^{(i)} g_e(t-\nu T)\left\{w_n(t) * \sum_{\lambda=1}^{N} a_\lambda^{(i)} g_e(t-\lambda T)\right\} \mathrm{d}t \tag{6.56}$$

from which using eqn (6.54) and after some rearrangement we obtain

$$E_d^{(i)} = \frac{1}{K_{\mathrm{MF}}} \sum_{\nu=1}^{N} a_\nu^{(i)} \sum_{\lambda=1}^{N} a_\lambda^{(i)} g_m\{(\nu-\lambda)T\} \tag{6.57}$$

Then, on substituting $\kappa = \nu - \lambda$ we obtain

$$E_d^{(i)} = \frac{1}{K_{\mathrm{MF}}} \sum_{\nu=1}^{N} a_\nu^{(i)} \sum_{\kappa=\nu-N}^{\nu-1} a_{\nu-\kappa}^{(i)} g_m(\kappa T) \tag{6.58}$$

The integrals needed for maximum likelihood detection according to eqn (6.17) are now determined. Since in addition the constant K_{MF} appears in both eqn (6.53) and eqn (6.58), we can write the decision rule of the maximum likelihood receiver using a matched filter and an ideal sampler:

$$\sum_{\nu=1}^{N} a_\nu^{(j)}\left\{2m_\nu^{(j)} - \sum_{\kappa=\nu-N}^{\nu-1} a_{\nu-\kappa}^{(j)} g_m(\kappa T)\right\}$$

$$> \sum_{\nu=1}^{N} a_\nu^{(i)}\left\{2m_\nu^{(i)} - \sum_{\kappa=\nu-N}^{\nu-1} a_{\nu-\kappa}^{(i)} g_m(\kappa T)\right\} \quad \forall \begin{matrix} i = 1 \ldots M^N \\ i \neq j \end{matrix} \tag{6.59}$$

Thus the N sample values $m_\nu = \tilde{m}_\nu + \overset{\times}{m}_\nu$ of the matched-filter output signal contain all the information that is needed for optimum detection, and so the practical implementation cost of the optimum receiver can be reduced.

By using an additional filter, the decision rule can be simplified still further. The noise components $\overset{\times}{m}_\nu$ of the signals sampled at times νT exhibit statistical dependences which are described by the discrete ACF

$$l_{\overset{\times}{m}}(\lambda T) = \overline{\overset{\times}{m}_\nu \overset{\times}{m}_{\nu+\lambda}} \tag{6.60}$$

By using eqns (2.19), (2.68) and (6.51) this ACF can be written

$$\begin{aligned} l_{\overset{\times}{m}}(\lambda T) &= \int_{-\infty}^{+\infty} L_n(f)\,|H_{\mathrm{MF}}(f)|^2 \exp(\mathrm{j}2\pi f\lambda T)\,\mathrm{d}f \\ &= K_{\mathrm{MF}}{}^2 \int_{-\infty}^{+\infty} \frac{L_0{}^2}{L_n(f)} |G_e(f)|^2 \exp(\mathrm{j}2\pi f\lambda T)\,\mathrm{d}f \end{aligned} \tag{6.61}$$

With a suitably dimensioned filter we can now obtain uncorrelated noise-sample values $\overset{\times}{d}_\nu = \overset{\times}{d}(\nu t)$. This means that the ACF $l_{\overset{\times}{d}}(\tau)$ corresponding to the times $\tau = \nu T$ has null points. Away from the equidistant sampling times νT, $l_{\overset{\times}{d}}(\tau)$ may have an arbitrary value so that the use of a transversal filter is possible (Fig. 6.5). This filter is referred to below as a **discrete whitening filter**. By analogy with Fig. 5.6 the frequency response and the pulse response are given by

$$H_{\mathrm{TF}}(f) = \sum_{\lambda=-\infty}^{+\infty} k_\lambda \exp(-\mathrm{j}2\pi f\lambda T) \tag{6.62}$$

$$h_{\mathrm{TF}}(t) = \sum_{\lambda=-\infty}^{+\infty} k_\lambda \delta(t - \lambda T) \tag{6.63}$$

The coefficients k_λ of this filter can be determined from

$$l_{\overset{\times}{d}}(\lambda T) = l_{\overset{\times}{m}}(\lambda T) * \{h_{\mathrm{TF}}(\lambda T) * h_{\mathrm{TF}}(-\lambda T)\} \overset{!}{=} 0 \quad \text{for } \lambda \neq 0 \tag{6.64}$$

Here we observe that the ACF contains no statement about the phase response of the filter $H_{\mathrm{TF}}(f)$, and hence an arbitrary number of solutions is possible for the coefficients of the transversal filter [6.20].

If the noise power spectrum $L_n(f)$ contains no null points in the frequency range of interest, an optimum decision can be derived from the detection sample values d_ν. As the noise sample values $\overset{\times}{d}_\nu$ are uncorrelated, by analogy with eqn (6.41) the following decision rule holds for the maximum likelihood receiver ($p_i = p_j$ for all i, j):

$$V_N = Q_N{}^{(j)} \quad \text{if} \quad \sum_{\nu=-\infty}^{+\infty} \{d_\nu - \tilde{d}_\nu{}^{(j)}\}^2 < \sum_{\nu=-\infty}^{+\infty} \{d_\nu - \tilde{d}_\nu{}^{(i)}\}^2 \quad \forall\, i \neq j \tag{6.65}$$

The detection sample value d_ν can be computed from the sampled values of the matched-filter output signal $m(t)$ and the coefficients k_λ of the discrete whitening filter:

$$d_\nu = d(\nu T) = m(t) * h_{\mathrm{TF}}(t)\ |_{t=\nu T} = \sum_{\lambda=-\infty}^{+\infty} k_{\nu-\lambda} m_\lambda \tag{6.66}$$

If the symbol sequence $Q_N^{(i)}$ has been transmitted, in accordance with eqn (3.10) the following defines the possible useful detection sample values:

$$\tilde{d}_\nu^{(i)} = \sum_{\kappa=-\infty}^{+\infty} a_{\nu-\kappa}{}^{(i)} g_\kappa \tag{6.67}$$

Here $g_\kappa = g_d(T_\mathrm{D} + \kappa T)$ denotes the basic detection pulse values from definition (3.11) which can be calculated from the basic matched-filter pulse $g_m(t)$ and the filter coefficients k_λ.

For uniformity and to present the simplest possible description of the detection process, we assume for the remainder of this chapter that the basic detection pulse $g_d(t)$ can be described totally by the $\upsilon + 1$ values $g_{-\upsilon} \cdots g_0$:

$$g_\kappa = g_d(T_\mathrm{D} + \kappa T) = 0 \quad \text{for} \quad \kappa < -\upsilon \quad \text{and} \quad \kappa > 0 \tag{6.68}$$

This is not a critical limitation, as every basic detection pulse $g_d(t)$ can fulfil this condition with a suitable choice of the detection time T_D. As in preceding sections, υ denotes the number of precursors.

With this precondition, when $\nu < 1 - \upsilon$ and $\nu > N$ all possible useful detection sample values $\tilde{d}_\nu^{(i)}$ are zero, and so these sample values do not have to be taken into consideration for the decision. Thus we obtain the following decision rule for the maximum likelihood receiver with matched filter and discrete whitening filter (see eqn (6.65)):

$$V_N = Q_N^{(j)} \quad \text{if} \quad \sum_{\nu=1-\upsilon}^{N} \left(d_\nu - \sum_{\kappa=-\upsilon}^{0} a_{\nu-\kappa}{}^{(j)} g_\kappa \right)^2$$
$$< \sum_{\nu=1-\upsilon}^{N} \left(d_\nu - \sum_{\kappa=-\upsilon}^{0} a_{\nu-\kappa}{}^{(i)} g_\kappa \right)^2 \qquad \forall\, i \neq j \tag{6.69}$$

This result states that for optimum detection of a sequence of N symbols only $N + \upsilon$ sample values are needed, where υ defines the number of precursors of the basic detection pulse considered.

6.3.2 Iterative calculation of the error quantities

In the following we define the νth **error quantity** of the symbol sequence $Q_N^{(i)}$ as the squared error between the actual sample value d_ν and the corresponding useful sample value $\tilde{d}_\nu^{(i)}$ of the sequence $Q_N^{(i)}$ (see eqn (6.43)):

$$\varepsilon_\nu^{(i)} = (d_\nu - \tilde{d}_\nu^{(i)})^2 \qquad \nu = 1 - v, \ldots, N \tag{6.70}$$

The sum of all the error quantities $\varepsilon_\nu^{(i)}$ up to the time νT is defined as the νth **total error quantity**:

$$\gamma_\nu^{(i)} = \sum_{k=1-v}^{\nu} \varepsilon_k^{(i)} \qquad \nu = 1 - v, \ldots, N \tag{6.71}$$

For $\nu < -v$, let $\gamma_\nu^{(i)} = 0$. With this definition and eqn (6.65) we obtain the decision rule of the optimum receiver:

$$V_N = Q_N^{(j)} \quad \text{if} \quad \gamma_N^{(j)} < \gamma_N^{(i)} \qquad \forall \begin{matrix} i = 1 \ldots M^N \\ i \neq j \end{matrix} \tag{6.72}$$

As can be seen from eqn (6.71), the total error quantity can be calculated iteratively,

$$\gamma_\nu^{(i)} = \gamma_{\nu-1}^{(i)} + \varepsilon_\nu^{(i)} \qquad \nu = 1 - v, \ldots, N \tag{6.73}$$

where according to eqn (6.70) we have for the νth error quantity

$$\varepsilon_\nu^{(i)} = \left(d_\nu - \sum_{\kappa=-v}^{0} a_{\nu-\kappa}^{(i)} g_\kappa \right)^2 \tag{6.74}$$

It is clear from this that the νth error quantity only depends on the amplitude coefficients $a_\nu^{(i)}, \ldots, a_{\nu+v}^{(i)}$ corresponding to the source symbols $q_\nu^{(i)}, \ldots, q_{\nu+v}^{(i)}$. However, the remaining symbols $q_1, \ldots, q_{\nu-1}$ and $q_{\nu+v+1}, \ldots, q_N$ of the ith sequence do not have to be taken into consideration for determination of the error quantity $\varepsilon_\nu^{(i)}$.

To make this plain, the nomenclature below is used in the following for the νth error quantity:

$$\varepsilon_\nu^{(i)} = \varepsilon_\nu(q_\nu^{(i)} \ldots q_{\nu+v}^{(i)}) \tag{6.75}$$

In Table 6.1 those symbols of the sequence $Q_N^{(i)}$ which affect the error values $\varepsilon_\nu^{(i)}$ at various times are indicated by a cross. As the total error quantity $\gamma_\nu^{(i)}$ can be calculated from the error quantities $\varepsilon_{1-v}^{(i)}$ to $\varepsilon_\nu^{(i)}$, it is dependent on the symbols $q_1^{(i)}$ to $q_{\nu+v}^{(i)}$; this will be expressed using the nomenclature

$$\gamma_\nu^{(i)} = \gamma_\nu(q_1^{(i)} \ldots q_{\nu+v}^{(i)}) \tag{6.76}$$

Using this expression in eqn (6.73) we obtain the following rule for iterative calculation of the νth total error quantity:

$$\gamma_\nu(q_1^{(i)} \ldots q_{\nu+v}^{(i)}) = \gamma_{\nu-1}(q_1^{(i)} \ldots q_{\nu+v-1}^{(i)}) + \varepsilon_\nu(q_\nu^{(i)} \ldots q_{\nu+v}^{(i)}) \tag{6.77}$$

With a nonredundant M-level source there are therefore up to $M^{\nu+v}$ different values for the total error quantity γ_ν. With each step of the iteration the number of total error quantities to be considered increases by a factor M. The number of error quantities ε_ν to be calculated at each step of the iteration amounts to M^{v+1} and thus is independent of ν.

Table 6.1 Dependence of the error quantities ε_ν and γ_ν on the source symbols q_ν

	$q_1^{(i)}$	$q_2^{(i)}$	…	$q_{1+v}^{(i)}$	$q_{2+v}^{(i)}$	…	$q_{\nu-1}^{(i)}$	$q_\nu^{(i)}$	$q_{\nu+1}^{(i)}$	…	$q_{\nu+v-1}^{(i)}$	$q_{\nu+v}^{(i)}$	$q_{\nu+v+1}^{(i)}$	…	$q_N^{(i)}$
$\varepsilon_{1-v}^{(i)}$	×	—	—	—	—	—	—	—	—	—	—	—	—	—	—
$\varepsilon_{2-v}^{(i)}$	×	×	—	—	—	—	—	—	—	—	—	—	—	—	—
⋮															
$\varepsilon_1^{(i)}$	×	×	×	×	—	—	—	—	—	—	—	—	—	—	—
$\varepsilon_2^{(i)}$	—	×	×	×	×	—	—	—	—	—	—	—	—	—	—
⋮															
$\varepsilon_{\nu-1}^{(i)}$	—	—	—	—	—	—	×	×	×	×	×	—	—	—	—
$\varepsilon_\nu^{(i)}$	—	—	—	—	—	—	—	×	×	×	×	×	—	—	—
$\varepsilon_{\nu+1}^{(i)}$	—	—	—	—	—	—	—	—	×	×	×	×	×	—	—
⋮															
$\gamma_{\nu-1}^{(i)}$	×	×	×	×	×	×	×	×	×	×	×	—	—	—	—
$\gamma_\nu^{(i)}$	×	×	×	×	×	×	×	×	×	×	×	×	—	—	—
⋮															

Calculation of the total error quantities can be illustrated graphically using a tree structure. Figure 6.6 is for a binary system ($M = 2$) and a basic detection pulse with one precursor ($v = 1$). In this tree the total error quantities $\gamma_\nu^{(i)}$ are placed at nodes; the branches correspond to error quantities $\varepsilon_\nu^{(i)}$. It is assumed that at the starting node the symbol O was transmitted at least v times before the actual message to be transmitted:

$$q_{1-v} = \mathrm{O} \quad \ldots \quad q_0 = \mathrm{O} \tag{6.78}$$

According to eqn (6.73) the total error quantity $\gamma_\nu^{(i)}$ is calculated as the sum of the preceding total error quantity $\gamma_{\nu-1}^{(i)}$ and the error quantity $\varepsilon_\nu^{(i)}$. This means that the value at a node is equal to the sum of the value at the preceding node and the value on the intervening branch. The symbol sequences corresponding to the total error quantities emerge when we follow the path from the starting node to the node being considered. In Fig. 6.6 upward pointing branches correspond to the symbol L and downward pointing branches correspond to the symbol O.

Example A message of $N = 3$ unipolar coded binary symbols ($a_\nu = 0$ or $a_\nu = 1$) is transmitted via a channel. The basic detection pulse $g_d(t)$ after the matched filter and the discrete whitening filter has only two sample values differing from zero,

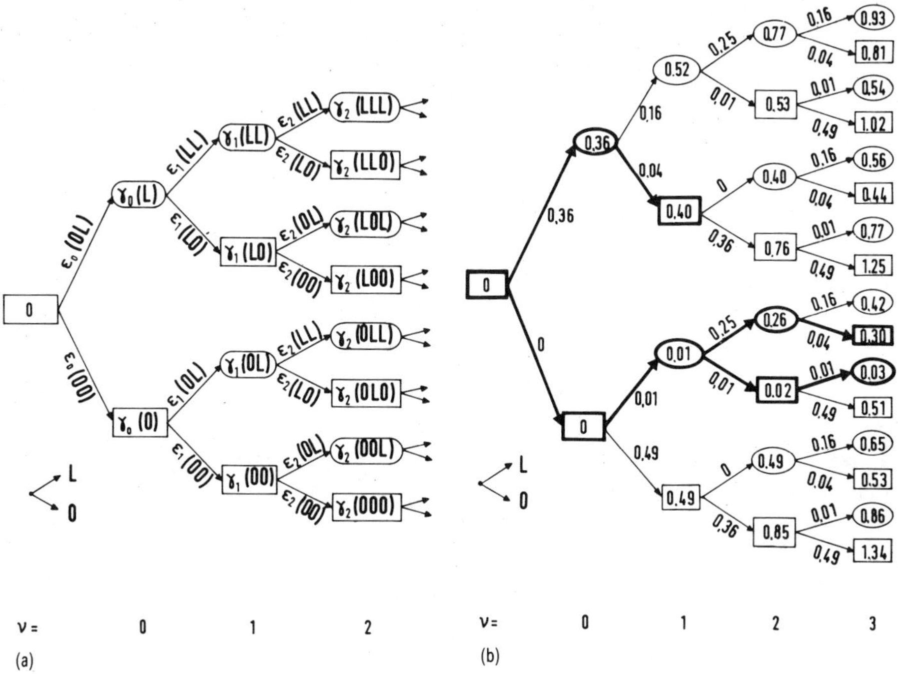

Fig. 6.6 Illustration of the error quantities ε_ν and the total error quantities γ_ν in a tree structure ($M = 2$, $v = 1$): (a) general case; (b) the example quoted in the text.

which are normalized for simplicity: $g_{-1} = 0.6$ and $g_0 = 0.5$. At the iteration steps $v = 0$, $v = 1$, $v = 2$ and $v = 3$ the following are the sample values at the detector input: $d_0 = 0$, $d_1 = 0.7$, $d_2 = 0.6$ and $d_3 = 0.4$.

For the iteration step $v = 0$ there are only two error quantities, which can be calculated from eqn (6.74):

$$\varepsilon_0(\mathrm{OO}) = \{0 - (0 \times 0.6 + 0 \times 0.5)\}^2 = 0 \quad \varepsilon_0(\mathrm{OL}) = \{0 - (1 \times 0.6 + 0 \times 0.5)\}^2 = 0.36$$

The total error quantities $\gamma_0(\mathrm{OO})$ and $\gamma_0(\mathrm{OL})$ are identical with the corresponding error quantities $\varepsilon_0(\mathrm{OO})$ and $\varepsilon_0(\mathrm{OL})$ (see Fig. 6.6(b)). For the iteration step $v = 1$ at time T the following four error quantities have to be calculated:

$$\varepsilon_1(\mathrm{OO}) = \{0.7 - (0 \times 0.6 + 0 \times 0.5)\}^2 = 0.49 \qquad \varepsilon_1(\mathrm{OL}) = \{0.7 - (1 \times 0.6 + 0 \times 0.5)\}^2 = 0.01$$

$$\varepsilon_1(\mathrm{LO}) = \{0.7 - (0 \times 0.6 + 1 \times 0.5)\}^2 = 0.04 \qquad \varepsilon_1(\mathrm{LL}) = \{0.7 - (1 \times 0.6 + 1 \times 0.5)\}^2 = 0.16$$

The total error quantities $\gamma_1(\mathrm{OO})$, $\gamma_1(\mathrm{OL})$, $\gamma_1(\mathrm{LO})$ and $\gamma_1(\mathrm{LL})$ are each obtained from the sum of the value at the preceding node and the value on the intervening branch. There are also four different error quantities at each of the subsequent iteration steps $v > 1$, which are again calculated from eqn (6.74). Their values for this example can be read from the tree structure shown in Fig. 6.6(b).

In the following, two total error quantities $\gamma_v^{(j)}$ and $\gamma_v^{(i)}$ are assumed whose corresponding symbol sequences differ in the first v symbols and are equal in the subsequent υ symbols (υ is again the number of precursors of the basic detection pulse). For both these total error quantities we assume that

$$\gamma_v(q_1^{(j)} \dots q_v^{(j)} q_{v+1}^{(l)} \dots q_{v+\upsilon}^{(l)}) < \gamma_v(q_1^{(i)} \dots q_v^{(i)} q_{v+1}^{(l)} \dots q_{v+\upsilon}^{(l)}) \tag{6.79}$$

The table below shows the dependence of the total error quantities $\gamma_v^{(j)}$ and $\gamma_v^{(i)}$ on the individual symbols of the sequence, where an identical dependence is denoted by a cross:

	q_1	q_2	$\cdots$	q_v	q_{v+1}	$\cdots$	$q_{v+\upsilon}$	$q_{v+\upsilon+1}$
$\gamma_v^{(j)}$	0	0	0	0	×	×	×	...
$\gamma_v^{(i)}$	□	□	□	□	×	×	×	...

For each of these two total error quantities, M new total error quantities are determined at the subsequent iteration step $v + 1$. Therefore for $\mu = 1, \dots, M$ the following are obtained from eqn (6.77) when the assumption of eqn (6.79) is considered:

$$\begin{aligned}\gamma_{v+1}^{(j,\mu)} &= \gamma_{v+1}(q_1^{(j)} \dots q_v^{(j)}, q_{v+1}^{(l)} \dots q_{v+\upsilon}^{(l)}, q_{v+\upsilon+1}^{(\mu)}) \\ &= \gamma_v^{(j)} + \varepsilon_{v+1}(q_{v+1}^{(l)} \dots q_{v+\upsilon}^{(l)}, q_{v+\upsilon+1}^{(\mu)})\end{aligned} \tag{6.80a}$$

$$\begin{aligned}\gamma_{v+1}^{(i,\mu)} &= \gamma_{v+1}(q_1^{(i)} \dots q_v^{(i)}, q_{v+1}^{(l)} \dots q_{v+\upsilon}^{(l)}, q_{v+\upsilon+1}^{(\mu)}) \\ &= \gamma_v^{(i)} + \varepsilon_{v+1}(q_{v+1}^{(l)} \dots q_{v+\upsilon}^{(l)}, q_{v+\upsilon+1}^{(\mu)})\end{aligned} \tag{6.80b}$$

Thus for all $\mu = 1, \ldots, M$ the same error quantity is added to both the total error quantities $\gamma_v^{(j)}$ and $\gamma_v^{(i)}$, and we obtain for the difference between the total error quantities for all $\mu = 1, \ldots, M$

$$\gamma_{v+1}^{(i,\mu)} - \gamma_{v+1}^{(j,\mu)} = \gamma_v^{(i)} - \gamma_v^{(j)} \tag{6.81}$$

This means that the difference between the two total error quantities of the differing partial symbol sequences $q_1^{(j)}, \ldots, q_v^{(j)}$ and $q_1^{(i)}, \ldots, q_v^{(i)}$ are not affected. At further iteration steps also, these symbols have no effect on the difference between the two total error quantities.

The decision rule for the maximum likelihood receiver (eqn (6.72)) states that from all possible symbol sequences $Q_N^{(i)}$ the sequence $Q_N^{(j)}$ selected is that which exhibits the smallest total error quantity $\gamma_N^{(j)}$. In fact, the M^N error quantities would have to be evaluated and compared at the time NT. However, if the quantity relationship in eqn (6.79) is fulfilled at time vT, it is possible to predict that the following is true at later times NT:

$$\gamma_N(q_1^{(j)} \ldots q_v^{(j)} q_{v+1}^{(l)} \ldots q_N^{(l)}) < \gamma_N(q_1^{(i)} \ldots q_v^{(i)} q_{v+1}^{(l)} \ldots q_N^{(l)}) \tag{6.82}$$

However, this also means that at time vT it is already sure that there is no (most probable) sequence whose first v symbols are equal to $q_1^{(i)}, \ldots, q_v^{(i)}$. Therefore the total error quantity $\gamma_v^{(i)}$ corresponding to this sequence need be considered no further.

The following important result for practical application emerges from this: for a basic detection pulse with υ precursors only M^υ of the total $M^{v+\upsilon}$ different total error quantities need be considered because only these belong to symbol sequences which are possibly part of the most probable symbol sequence $V_N = Q_N^{(j)}$. The number of error quantities to be considered is therefore independent of the length N of the message to be detected. Hence it is sufficient for optimum detection to calculate and store the relevant M^υ different total error quantities. In the following these quantities to be considered are defined as the **minimum total error quantities**:

$$\mathring{\gamma}_v^{(l)} = \mathring{\gamma}_v(q_{v+1}^{(l)} \ldots q_{v+\upsilon}^{(l)}) = \min_{i=1 \ldots M^v} \{\gamma_v(q_1^{(i)} \ldots q_v^{(i)} q_{v+1}^{(l)} \ldots q_{v+\upsilon}^{(l)})\} \tag{6.83}$$

The variable l runs from 1 to M^υ, whereas $i = 1, \ldots, M^v$.

Determination of the minimum total error quantities $\mathring{\gamma}_v^{(l)}$ can be very clearly illustrated using a **trellis** [6.8]. Figure 6.7 shows sections from the trellis for a binary system ($M = 2$) with $\upsilon = 1$ and $\upsilon = 2$ precursors. In a similar way to the tree structure in Fig. 6.6 branches are introduced which correspond to the error quantities $\varepsilon_v(q_v^{(i)} \ldots q_{v+\upsilon}^{(i)})$. Nodes relate to the total of M^υ different minimum total error quantities. The advantage of the trellis compared with the tree structure lies in the fact that in the former the number of branches and nodes is not increased by a factor M at every iteration step. By selection of the minimum total error quantities only those symbol sequences that still actually come into consideration as part of the most probable sequence have to be taken into account.

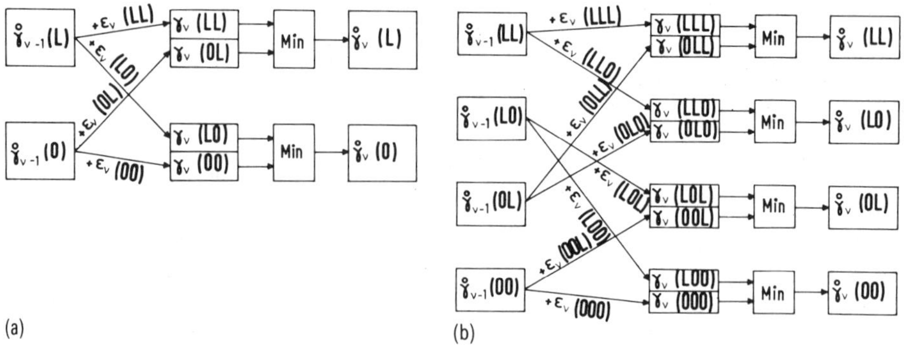

Fig. 6.7 Trellis for a binary system with v precursors: (a) $v = 1$; (b) $v = 2$.

Example When these results are applied to the numerical example discussed above ($M = 2$, $v = 1$), only two total error quantities $\gamma_\nu(q_1 \dots q_{\nu+1})$ have to be considered further at every time νT. These are enclosed in bold frames in the tree structure of Fig. 6.6(b). As an example, at the iteration step $\nu = 3$ we obtain

$$\mathring{\gamma}_3(\mathrm{O}) = \min_{i=1\dots8} \{\gamma_3(q_0{}^{(i)}, q_1{}^{(i)}, q_2{}^{(i)}, q_3{}^{(i)} = \mathrm{O})\} = \gamma_3(\mathrm{OLLO}) = 0.30$$

$$\mathring{\gamma}_3(\mathrm{L}) = \min_{i=1\dots8} \{\gamma_3(q_0{}^{(i)}, q_1{}^{(i)}, q_2{}^{(i)}, q_3{}^{(i)} = \mathrm{L})\} = \gamma_3(\mathrm{OLOL}) = 0.03$$

Only these two minimum total error quantities are needed for the evaluation of further error quantities at the next iteration step $v = 4$. Similarly only the corresponding sequences OLLO and OLOL can be part of the most probable symbol sequence.

Figure 6.8 shows that the trellis results from the tree structure if only the minimum total error quantities $\mathring{\gamma}_\nu{}^{(l)}$ shown in bold in Fig. 6.6(b) are considered and if the values to be compared are written next to one another. As only one precursor is present in this example ($v = 1$), only two decisions need to be made at each iteration step ν, i.e. whether $\gamma_\nu(\mathrm{OO})$ is greater or smaller than $\gamma_\nu(\mathrm{LO})$ and whether $\gamma_\nu(\mathrm{OL})$ is greater or smaller than $\gamma_\nu(\mathrm{LL})$.

6.3.3 Determination of the tentatively detected partial symbol sequences

Previously the sink symbol sequence was determined in a single decision process (at time NT), and this caused a considerable delay in the received signal. In this section the decision rules are changed so that optimum detection of parts of the received message can be conducted before the receiver has received the whole signal.

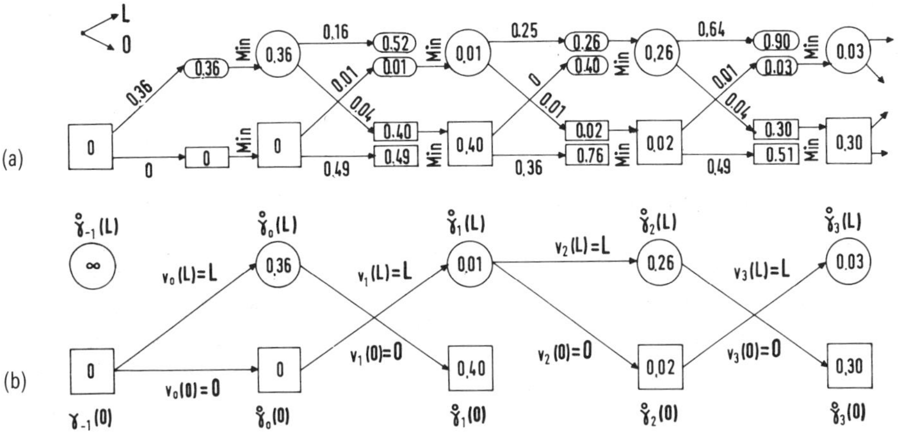

Fig. 6.8 Trellis for the numerical example in Fig. 6.6(b) in two different forms ($M = 2, v = 1$).

Firstly the **tentatively detected partial symbol sequences** $V_\nu^{(l)}$ ($l = 1 \ldots M^v$) are defined. $V_\nu^{(l)}$ is the most probable symbol sequence $q_1^{(j)} \ldots q_\nu^{(j)}$ corresponding to the minimum total error quantity $\mathring{\gamma}_\nu^{(l)}$, which was selected with the assumption that the next v symbols represent the sequence $q_{\nu+1}^{(l)} \ldots q_{\nu+v}^{(l)}$. The tentatively detected partial symbol sequence $V_\nu^{(l)}$ consists of the symbols $q_1^{(j)}$ to $q_\nu^{(j)}$ if, for the νth minimum total error quantity (see eqn (6.83)),

$$\mathring{\gamma}_\nu^{(l)} = \mathring{\gamma}_\nu(q_1^{(j)} \ldots q_\nu^{(j)} q_{\nu+1}^{(l)} \ldots q_{\nu+v}^{(l)}) = \min_{i=1 \ldots M^\nu} \{\gamma_\nu(q_1^{(i)} \ldots q_\nu^{(i)} q_{\nu+1}^{(l)} \ldots q_{\nu+v}^{(l)})\} \quad (6.84)$$

In a similar manner the **tentatively detected symbol** $v_\nu^{(l)}$ is that symbol $q_\nu^{(j)}$ which fulfils the following condition:

$$\mathring{\gamma}_\nu^{(l)} = \mathring{\gamma}_\nu(q_\nu^{(j)} q_{\nu+1}^{(l)} \ldots q_{\nu+v}^{(l)}) = \min_{\mu=1 \ldots M} \{\gamma_\nu(q_\nu^{(\mu)} q_{\nu+1}^{(l)} \ldots q_{\nu+v}^{(l)})\} \quad (6.85)$$

The tentatively detected symbol $v_\nu^{(l)}$ is therefore also selected under the supposition that the next v symbols represent the sequence $q_{\nu+1}^{(l)} \ldots q_{\nu+v}^{(l)}$. It can be calculated easily from the trellis (see Fig. 6.8). At each node that corresponds to a minimal total error quantity $\mathring{\gamma}_\nu^{(l)}$, there are M branches which lead from M different nodes $\mathring{\gamma}_{\nu-1}^{(l)}$. The following holds true:

$$\mathring{\gamma}_\nu(q_{\nu+1}^{(l)} \ldots q_{\nu+v}^{(l)}) = \min_{\mu=1 \ldots M} \{\gamma_{\nu-1}(q_\nu^{(\mu)} q_{\nu+1}^{(l)} \ldots q_{\nu+v-1}^{(l)}) + \varepsilon_\nu(q_\nu^{(\mu)} q_{\nu+1}^{(l)} \ldots q_{\nu+v}^{(l)})\}$$

$$= \gamma_{\nu-1}(v_\nu^{(l)} q_{\nu+1}^{(l)} \ldots q_{\nu+v-1}^{(l)}) + \varepsilon_\nu(v_\nu^{(l)} q_{\nu+1}^{(l)} \ldots q_{\nu+v}^{(l)}) \quad (6.86)$$

One of the preceding nodes is selected. The branch leading from this node to the node being considered is defined as the selected branch, to which the tentatively detected symbol $v_\nu^{(l)}$ is assigned. The continuous path from the first node to the νth node corresponds to the tentatively detected partial symbol sequence $V_\nu^{(l)}$.

Example Figure 6.9 shows the trellis for a binary system ($M = 2$) and a basic detection pulse with $v = 2$ precursors. As a consequence of these two assumptions there are exactly $M^v = 4$ nodes at each iteration step ν which correspond to the minimum total error quantities $\mathring{\gamma}_\nu(\mathrm{OO})$, $\mathring{\gamma}_\nu(\mathrm{OL})$, $\mathring{\gamma}_\nu(\mathrm{LO})$ and $\mathring{\gamma}_\nu(\mathrm{LL})$. Since for detection it is only necessary to consider the selected branches, only these are drawn in the trellis. The symbols on each individual branch are the tentatively detected symbols $v_\nu(\mathrm{OO})$, $v_\nu(\mathrm{OL})$, $v_\nu(\mathrm{LO})$ and $v_\nu(\mathrm{LL})$. It can be seen from this trellis that, with certainty, the first symbol in the sink symbol sequence V_N is $v_1 = \mathrm{L}$, as here all tentatively detected symbols $v_1(\mathrm{OO})$, $v_1(\mathrm{OL})$, $v_1(\mathrm{LO})$ and $v_1(\mathrm{LL})$ are equal to L. Therefore the symbol v_1 can already be output at the iteration step $v = 1$ and does not need to be considered further in the detection of the subsequent symbols v_2, v_3, v_4 etc.

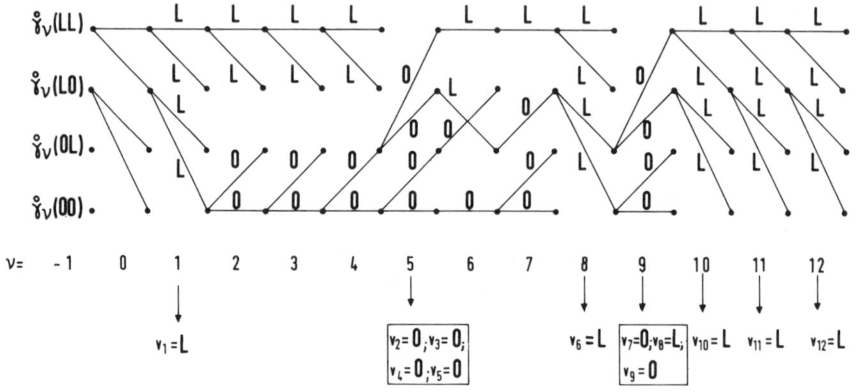

Fig. 6.9 Example of a trellis for a binary system ($M = 2$) and a basic detection pulse with $v = 2$ precursors.

The partial symbol sequences $V_\nu(\mathrm{OO})$, $V_\nu(\mathrm{OL})$, $V_\nu(\mathrm{LO})$ and $V_\nu(\mathrm{LL})$ corresponding to the four minimum total error quantities are summarized in Table 6.2. If, for example, we consider the iteration step $\nu = 4$, we see that only the two sequences LOOO and LLLL are suitable as (possible) most probable sequences; LOOO is the most probable sequence under the assumption that the next pair of symbols $v_5 v_6$ is OO or OL, and LLLL is the most probable sequence if it is assumed that the fifth and sixth symbols are LO or LL.

At the iteration step $\nu = 5$ all four tentatively detected partial symbol sequences are identical and equal to LOOOO (see Fig. 6.9 and Table 6.2). This means that, independent of which symbols follow, this sequence is the most probable, so that at this iteration step the symbols $v_2 = \mathrm{O}$, $v_3 = \mathrm{O}$, $v_4 = \mathrm{O}$ and $v_5 = \mathrm{O}$ can already be outputted.

However, at the iteration steps $\nu = 6$ and $\nu = 7$ the tentatively detected partial symbol sequences differ so that no definite decision is possible. The four tentatively

Table 6.2 Tentatively detected partial symbol sequences for the example in Fig. 6.9 ($M = 2, v = 2$)

	Tentatively detected partial symbol sequences						
	$V_v(OO)$	$V_v(OL)$	$V_v(LO)$	$V_v(LL)$	Z_v	L_v	*Outputted sink symbols*
1	L	L	L	L	1	0	$v_1 = L$
2	LO	LO	LL	LL	0	1	
3	LOO	LOO	LLL	LLL	0	2	
4	LOOO	LOOO	LLLL	LLLL	0	3	
5	LOOOO	LOOOO	LOOOO	LOOOO	4	0	$v_2 = v_3 = v_4 = v_5 = O$
6	LOOOOO	LOOOOL	LOOOOO	LOOOOL	0	1	
7	LOOOOOO	LOOOOOO	LOOOOLO	LOOOOLL	0	2	
8	LOOOOLOL	LOOOOLOL	LOOOOLLL	LOOOOLLL	1	2	$v_6 = L$
9	LOOOOLOLO	LOOOOLOLO	LOOOOLOLO	LOOOOLOLO	3	0	$v_7 = v_9 = O; v_8 = L$
10	LOOOOLOLOL	LOOOOLOLOL	LOOOOLOLOL	LOOOOLOLOL	1	0	$v_{10} = L$
11	LOOOOLOLOLL	LOOOOLOLOLL	LOOOOLOLOLL	LOOOOLOLOLL	1	0	$v_{11} = L$
12	LOOOOLOLOLLL	LOOOOLOLOLLL	LOOOOLOLOLLL	LOOOOLOLOLLL	1	0	$v_{12} = L$

detected partial symbol sequences also remain different at $\nu = 8$. However, it can be seen that the sixth symbol in all four partial symbol sequences is the same, so that at the iteration step $\nu = 8$ the symbol $v_6 = \mathrm{L}$ can be outputted. Likewise the symbols v_7, v_8 and v_9 can be detected at the following step ($\nu = 9$) in this example.

The maximum likelihood receiver decision rule illustrated by this example can be summarized by the following theorem.

Theorem If all M^v tentatively detected partial symbol sequences $V_\nu^{(l)}$ agree in the first Z_ν symbols at time νT, these Z_ν symbols can be outputted as part of the most probable sequence V_N. These Z_ν symbols do not have to be considered any more in the further detection process, so that the part of the tentatively detected partial symbol sequence to be considered reduces by Z_ν symbols.

In the following the tentatively detected partial symbol sequence $V_\nu^{(l)}$ contains only that part of the sequence which contains the symbols which have not yet been definitively detected. This portion is underlined in Table 6.2. L_ν defines the length (equal to the number of symbols still to be considered at the time νT) of the tentatively detected partial symbol sequence, where at iteration step ν

$$L_\nu = L_{\nu-1} + 1 - Z_\nu \tag{6.87}$$

The length L_ν and the number Z_ν of the symbols to be outputted are given in Table 6.2 for the example above.

The detected symbols are determined in blocks of various lengths and with various delays. The times at which the detected symbols are outputted and the length of the blocks of symbols are distributed statistically. However, it can be shown that the probability that the length L_ν of the tentatively detected partial sequence is greater than a maximum length $L_{\max}$ tends to zero if the selected value of $L_{\max}$ is sufficiently large [6.20].

6.3.4 Flow diagram for a Viterbi receiver

In 1967 Viterbi [6.29] introduced a new algorithm for decoding sequential codes. Later, Forney [6.7, 6.8] showed that this algorithm is also suitable for the detection of digital signals, and that the Viterbi algorithm leads to a maximum likelihood decision.

The Viterbi detector is based on a combination of iterative error quantity calculation, derived in previous sections, and iterative detection. Figure 6.10 shows a flow diagram with the individual steps of the Viterbi decision. These steps, which are identified alphabetically, are described below.

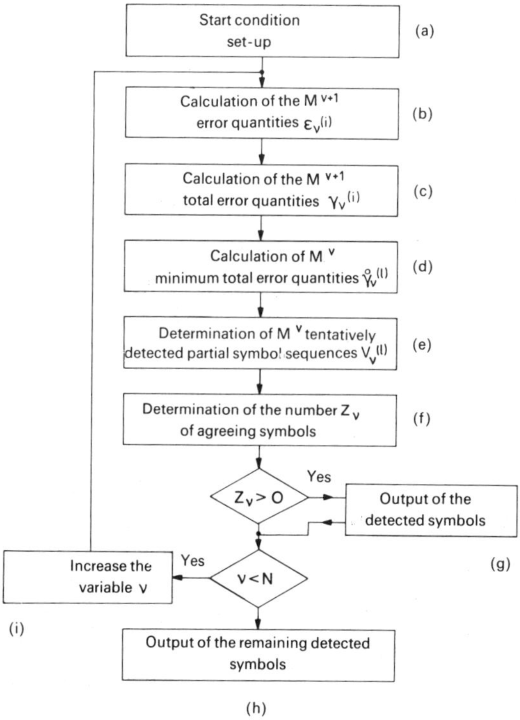

Fig. 6.10 Flow diagram for an optimum Viterbi receiver.

(a) For the first iteration step of the Viterbi detection those values which would normally be calculated in the preceding detection step are suitably pre-assigned. Hence it is assumed that a sequence of v equal symbols (e.g. O ... O) has been transmitted prior to the first transmission symbol. The minimum total error value $\mathring{\gamma}_{-v}(\mathrm{O} \ldots \mathrm{O})$ corresponding to this sequence is then set to zero and the others are set to $+\infty$. The parameter ν, which defines the topical time, is only required if a finite number N of symbols are to be detected. In this case ν is pre-assigned as $1-v$ as the vth precursor of the first transmission symbol q_1 occurs at time $(1-v)T$. It must be further guaranteed that the symbol v_1 is detected as the first symbol. This can be achieved, for example, by assigning the value $-v$ to $L_{\nu-1}$, which defines the length of the tentatively detected partial symbol sequence at the preceding decision point (see eqn (6.87)).

(b) The M^{v+1} error quantities $\varepsilon_\nu^{(i)}$ are determined from eqn (6.70). Figure 6.11(a) shows a possible circuit arrangement of a Viterbi receiver for a binary system ($M = 2$) and a basic detection pulse with only one precursor ($v = 1$). The detection sample value d_ν is obtained from the received signal $e(t)$ using a matched filter, a sample-and-hold device and a discrete whitening filter $H_{\mathrm{WF}}(f)$ (see Section 6.3.1).

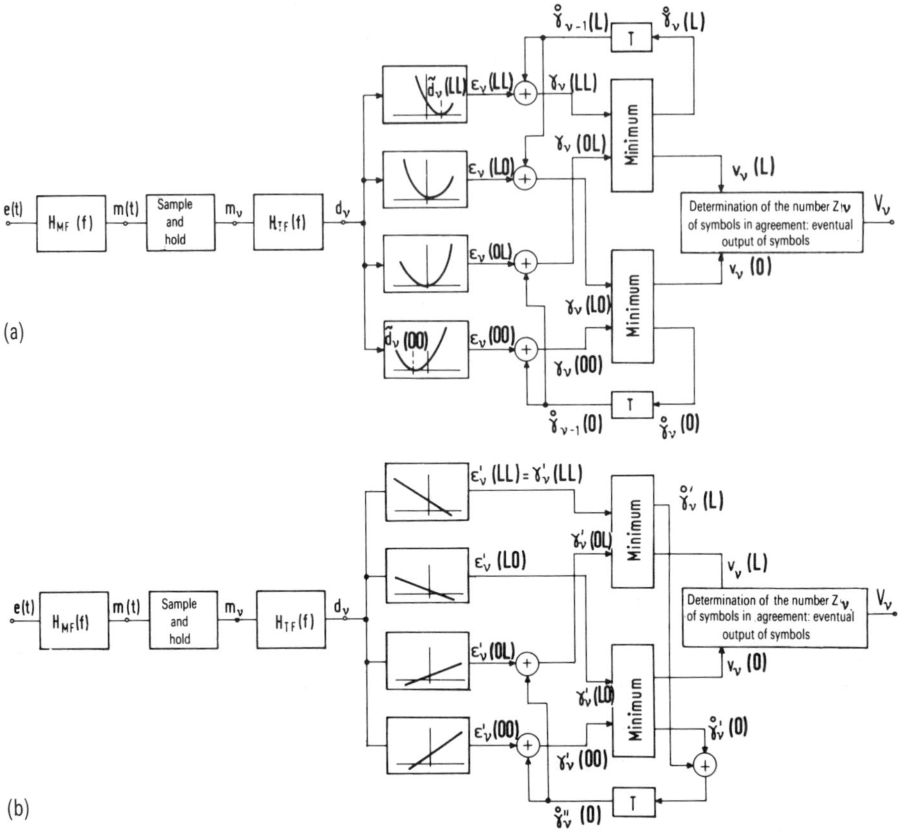

Fig. 6.11 Block diagram of a Viterbi receiver ($M = 2$; $v = 1$): (a) corresponding to the flow diagram in Fig. 6.10; (b) modified as described in Section 6.3.5.

The error quantities $\varepsilon_\nu^{(i)}$ are determined on a quadratic characteristic, the mean values of which are the useful detection sample values $\tilde{d}_\nu^{(i)}$ defined by eqn (6.67).

(c) The M^{v+1} total error quantities $\gamma_\nu^{(i)}$ are determined using eqn (6.73) from the error quantities $\varepsilon_\nu^{(i)}$ calculated in (b) and the M^v minimum total error quantities $\mathring{\gamma}_{\nu-1}^{(l)}$ determined in the preceding detection step. The correct matching between the parameters i and l must be observed here (see Fig. 6.11(a)).

(d) M^v minimum total error quantities $\mathring{\gamma}_\nu^{(l)}$ are selected from the M^{v+1} total error quantities using eqn (6.83). For example, for $M = 2$ and $v = 1$ (see Fig. 6.11(a)), we have $\mathring{\gamma}_\nu(\mathrm{O}) = \min\{\gamma_\nu(\mathrm{OO}), \gamma_\nu(\mathrm{LO})\}$ and $\mathring{\gamma}_\nu(\mathrm{L}) = \min\{\gamma_\nu(\mathrm{OL}), \gamma_\nu(\mathrm{LL})\}$.

(e) The corresponding tentatively detected symbols $v_\nu^{(l)}$ and the tentatively detected partial symbol sequences $V_\nu^{(l)}$ can be calculated as described in Section 6.3.3.

(f) Next a check is made whether the first Z_ν symbols are equal for all M^v tentatively detected partial symbol sequences. The new length L_ν of the tentatively detected partial symbol sequence is calculated from eqn (6.87).

(g) If a definite detection of the individual symbols is possible ($Z_\nu > 0$), these are outputted.

(h) When the index ν has reached the value N, the subseqent detection sample values contain no information about the message to be detected, and the remaining symbols v_ν, not yet determined, can be outputted. In this case it is assumed that after N transmitted symbols the symbol O was transmitted at least v times. The tentatively detected partial symbol sequence $V_\nu(\mathrm{O}\ldots\mathrm{O})$ that corresponds to this sequence can then be outputted as the remainder of the sink symbol sequence V_N.

(i) In preparation for the next detection step the value of ν is increased by unity.

6.3.5 Modified Viterbi receiver

In the following we describe some modifications to the Viterbi receiver by means of which the implementation cost can be reduced. In each case only the fundamental principle is explained. A detailed description of the modified Viterbi receiver can be found in ref. 6.20.

(a) Calculation of the linear error quantity In selecting the tentatively detected symbol sequence the only important question is which of the M values to be compared possesses the minimum value. Hence nothing is changed in the operation of the Viterbi detector if an arbitrary value is added to all M^{v+1} error quantities. The maximum likelihood receiver decision rule is also preserved if the square of the detection sample value d_ν is subtracted from the error quantity $\varepsilon_\nu^{(l)}$ in eqn (6.70). The following defines the modified error quantity ε_ν':

$$\varepsilon_\nu'^{(l)} = \varepsilon_\nu^{(l)} - d_\nu^2 = -(2\tilde{d}_\nu^{(l)})d_\nu + (\tilde{d}_\nu^{(l)})^2 \tag{6.88}$$

Therefore, to calculate this modified error quantity the detection sample value d_ν has to be multiplied by the constant $-2\tilde{d}_\nu^{(l)}$ and be increased by the value $(\tilde{d}_\nu^{(l)})^2$. This linear operation is easier to implement than the squaring required by eqn (6.70). Therefore the quadratic characteristics in the block diagram can be replaced by linear characteristics (see Fig. 6.11(b)).

(b) Reduction of the number of feedback connections by one The total error quantities $\gamma_\nu^{(l)}$ are a measure of the noise energies occurring up to the time νT. This has the consequence that the total error quantities increase monotonically with increasing ν. Now, we subtract an arbitrary quantity $\mathring{\gamma}_\nu^{(k)}$ from all minimum total error quantities $\mathring{\gamma}_\nu^{(l)}$, so that the original relations of the subsequent minimum development are maintained. However, the subtraction reduces the total error quantities so that they remain within a restricted range which simplifies the implementation. In addition, one of the total of M^v minimum total error quantities (i.e. $\mathring{\gamma}_v^{(k)}$) is always zero and does not need to be considered further. Hence the number of feedback connections can be reduced by one.

The Viterbi receiver in Fig. 6.11(b) with modifications (a) and (b) has the same error probability as the receiver in Fig. 6.11(a).

(c) Nonoptimum abbreviated basic detection pulse In the main the cost of implementing an optimum digital receiver depends on the number $v + 1$ of basic detection pulse values which differ from zero. For optimum detection, noise components $\overset{\times}{d}_v$ of the detection sample values are required which are (statistically) independent of each other. The duration $(v + 1)T$ of the basic detection pulse is thus determined by the given channel $H_K(f)$ and given noise power spectrum $L_n(f)$. To determine the most probable message, M^{v+1} error quantities have to be calculated. Even in the case of binary transmission ($M = 2$) this number is often so large that a practical application at higher bit rates is impossible. However, if the requirement for statistically independent noise components $\overset{\times}{d}_v$ of the detection signal is removed, another transversal filter $H_{TF}(f)$ can be used instead of the discrete whitening filter for a given cost M^{v+1} of Viterbi detector. In $h_{TF}(t) \circ\!\!-\!\!\bullet H_{TF}(f)$ the number of basic detection pulse values corresponds to the given number $v + 1$ and at the same time the ACF $l_{\overset{\times}{d}}(\lambda T)$ differs from the optimum ACF $L_0\delta(\lambda T)$ by as little as possible. Falconer and Magee [6.5] and Qureshi and Newhall [6.21] were the first to point out this means of simplification. Cantoni and Kwong [6.3] have shown that the noise power at the detector input falls monotonically with increasing cost of the Viterbi detector. Beare [6.1] compares several possible optimization strategies for the choice of the transversal filter $H_{TF}(f)$ and hence the system response.

(d) Receiver with a constrained decision According to Section 6.3.3, a sink symbol can only be outputted when all M^{v+1} tentatively detected partial symbol sequences $V_v^{(l)}$ agree in the first Z_v symbols. Therefore, the times at which the symbols are outputted are statistically distributed. In general, however, the requirement exists for the detected symbols to be outputted with a constant delay $L_{max}T$.

If, at time vT, the length L_v of the tentatively detected partial symbol sequence has reached the allowable length L_{max}, a **constrained decision** has to be made. In this case the symbol which represents the first symbol in the tentatively detected partial symbol sequence $V_v^{(l)}$ is outputted as the sink symbol $v_{v-L_{max}}$. For choice of the sequences $V_v^{(l)}$ it is checked, for example, which of the total of M^v minimum total error quantities $\mathring{\gamma}_v^{(l)}$ has the smallest value at that time.

It is shown in ref. 6.20 that the mean error probability p_M is increased an insignificant amount by a constrained decision if the maximum length L_{max} of the tentatively detected partial symbol sequence is given by

$$L_{max} \geqslant M^{2v+1} 2^{2v} \tag{6.89}$$

6.4 Error probability of the maximum likelihood receiver

To close this chapter it will be shown how the error probability of an optimum digital receiver can be calculated. For purposes of illustration, a nonredundant M-level source which outputs a sequence of N symbols is assumed. Therefore the probability of occurrence for all M^N possible source symbol sequences is equal to $1/M^N$.

6.4.1 Exact calculation of the mean error probability

We define by

$$p_{j|k} = P(V_N = Q_N{}^{(j)} \mid Q_N = Q_N{}^{(k)}) \qquad \begin{array}{l} j = 1 \dots M^N \\ k = 1 \dots M^N \end{array} \tag{6.90}$$

the conditional probability that the detector selects the sequence $Q_N{}^{(j)}$ under the assumption that the kth symbol sequence was transmitted. Hence the mean symbol error probability according to definition (3.35) can be written

$$p_{\mathrm{M,E}} = \frac{1}{NM^N} \sum_{k=1}^{M^N} \sum_{j=1}^{M^N} n_{kj} p_{j|k} \tag{6.91}$$

where the subscript E indicates that the error probability of an optimum receiver is being treated. $1/M^N$ is the probability of occurrence of the transmitted sequence $Q_N{}^{(k)}$. n_{kj} defines the number of symbols in which the sequence $Q_N{}^{(j)}$ chosen by the detector differs from the transmitted sequence $Q_N{}^{(k)}$. For $j = k$, $n_{kj} = 0$, so that the case $j = k$ does not have to be excluded explicitly in the summation of the error probabilities. The division by N is needed because the conditional error probability $p_{j|k}$ refers to the whole sequence, while $p_{\mathrm{M,E}}$ is the mean error probability of a symbol.

In the following the conditional error probabilities $p_{j|k}$ for an optimum maximum likelihood receiver with whitening filter and noise-energy detector are derived (see Fig. 6.4). The other optimum maximum likelihood receivers mentioned in this chapter have the same error probability because they are based on the same decision rule.

From eqns (6.44) and (6.46) we obtain the (conditional) probability that the transmitted kth sequence is corrupted into the sequence $Q_N{}^{(j)}$:

$$p_{j|k} = P\left[\int_{T_1}^{T_2} \{d(t) - \tilde{d}_j(t)\}^2 \,\mathrm{d}t < \int_{T_1}^{T_2} \{d(t) - \tilde{d}_i(t)\}^2 \,\mathrm{d}t \quad \text{for all} \quad i \neq j \mid Q_N = Q_N{}^{(k)} \right] \tag{6.92}$$

where T_1 and T_2 define the time interval in which the possible useful detection signals $\tilde{d}_i(t)$ are different. It is assumed for $t < T_1$ and $t > T_2$ that for all $i = 1, \ldots, M^N$ the useful detection signals $\tilde{d}_i(t)$ are zero, so that the integration can also be carried out from $-\infty$ to $+\infty$.

If it is assumed that the sequence $Q_N{}^{(k)}$ has been transmitted, the detection signal applied at the receiver (see eqn (2.81)) is given by

$$d(t) = \tilde{d}(t) + \overset{\times}{d}(t) = \tilde{d}_k(t) + \overset{\times}{d}(t) \tag{6.93}$$

Thus $p_{j|k}$ is the probability that a noise signal $\overset{\times}{d}(t)$ occurs that causes the detector to decide, according to the decision rule of eqn (6.46), on the jth symbol sequence even though the kth sequence has been transmitted.

The deviation of the kth useful detection signal from the jth useful detection signal is defined as the **useful difference signal**:

$$\Delta\tilde{d}_{kj}(t) = \tilde{d}_k(t) - \tilde{d}_j(t) \qquad \begin{matrix} j = 1 \ldots M^N \\ k = 1 \ldots M^N \end{matrix} \tag{6.94}$$

Thus we obtain the following equation for the decision $V_N = Q_N{}^{(j)}$ if it is assumed that $Q_N = Q_N{}^{(k)}$ has been transmitted:

$$\int_{-\infty}^{+\infty} \{\Delta\tilde{d}_{kj}(t) + \overset{\times}{d}(t)\}^2 \, \mathrm{d}t < \int_{-\infty}^{+\infty} \{\tilde{d}_{ki}(t) + \overset{\times}{d}(t)\}^2 \, \mathrm{d}t \quad \forall \begin{matrix} i = 1 \ldots M^N \\ i \neq j \end{matrix} \tag{6.95}$$

Hence after some algebraic rearrangement we obtain

$$\int_{-\infty}^{+\infty} \Delta\tilde{d}_{kj}(t)\overset{\times}{d}(t)\, \mathrm{d}t + \frac{1}{2}\int_{-\infty}^{+\infty} \{\Delta\tilde{d}_{kj}(t)\}^2 \, \mathrm{d}t$$
$$< \int_{-\infty}^{+\infty} \Delta\tilde{d}_{ki}(t)\overset{\times}{d}(t)\, \mathrm{d}t + \frac{1}{2}\int_{-\infty}^{+\infty} \{\Delta\tilde{d}_{ki}(t)\}^2 \, \mathrm{d}t \tag{6.96}$$

The first integral represents the cross-energy between the useful difference signal $\Delta\tilde{d}_{kj}(t)$ and the detection noise signal $\overset{\times}{d}(t)$. It corresponds to the energy CCF at the point $\tau = 0$ (see eqn (6.23)). The second integral defines the energy of the useful difference signal $\Delta\tilde{d}_{kj}(t)$, which is referred to in the following as the **energy distance** (between the kth and jth useful detection signals):

$$\Delta E_{kj} = \int_{-\infty}^{+\infty} \{\Delta\tilde{d}_{kj}(t)\}^2 \, \mathrm{d}t = \int_{-\infty}^{+\infty} \{\tilde{d}_k(t) - \tilde{d}_j(t)\}^2 \, \mathrm{d}t \tag{6.97}$$

Hence, according to eqn (6.90) we obtain for the conditional error probability $p_{j|k}$

$$p_{j|k} = P\left[\bigcap_{\substack{i = 1 \ldots M^N \\ i \neq j}} \left\{ \int_{-\infty}^{+\infty} \Delta\tilde{d}_{kj}(t)\overset{\times}{d}(t)\, \mathrm{d}t + \tfrac{1}{2}\Delta E_{kj} < \int_{-\infty}^{+\infty} \Delta\tilde{d}_{ki}(t)\overset{\times}{d}(t)\, \mathrm{d}t + \tfrac{1}{2}\Delta E_{ki} \right\}\right] \tag{6.98}$$

This means that $p_{j|k}$ is to be calculated as the probability of an intersection of $M^N - 1$ event sets. As the individual events $\{\ldots\}$ are statistically dependent on one another, the $p_{j|k}$ can in general only be calculated with difficulty. Since in the calculation the mean error probability corresponding to eqn (6.91) has to be averaged over M^{2N} such conditional probabilities, it is very important to develop bounds or approximations for the error probability of an optimum receiver.

6.4.2 Approximations to the mean error probability

The factor n_{kj} is eqn (6.91) defines the number of symbols in which the detector-selected sequence $Q_N^{(j)}$ differs from the transmitted sequence $Q_N^{(k)}$. A first approximation for the mean error probability is obtained by substituting a mean value n_M for this factor so that it can be extracted from the summation:

$$n_{kj} = \begin{cases} n_\mathrm{M} & \text{for } j \neq k \\ 0 & \text{for } j = k \end{cases} \qquad 1 \leqslant n_\mathrm{M} \leqslant N \tag{6.99}$$

n_M does not necessarily have an integer value. With the conditional probability

$$P(V_N \neq Q_N^{(k)} \mid Q_N = Q_N^{(k)}) = \sum_{\substack{j=1 \\ j \neq k}}^{M^N} p_{j|k} \tag{6.100}$$

that the transmitted sequence $Q_N^{(k)}$ is erroneously detected, we thus obtain an approximation for the mean symbol error probability from eqn (6.91):

$$p_\mathrm{M,E} \approx \frac{n_\mathrm{M}}{NM^N} \sum_{k=1}^{M^N} P(V_N \neq Q_N^{(k)} \mid Q_N = Q_N^{(k)}) \tag{6.101}$$

A further simplification comes with the supposition that all M^N possible source symbol sequences are falsified with approximately equal probability

$$P(V_N \neq Q_N^{(k)} \mid Q_N = Q_N^{(k)}) = P(V_N \neq Q_N) \quad \forall\, k = 1 \ldots M^N \tag{6.102}$$

Hence averaging over k can be neglected so that we obtain as a further approximation

$$p_\mathrm{M,E} \approx \frac{n_\mathrm{M}}{N} P(V_N \neq Q_N) \tag{6.103}$$

To calculate the conditional probability $P(V_N \neq Q_N)$ it is assumed as before that the sequence $Q_N^{(k)}$ was transmitted. In view of the decision rule in eqn (6.96) it is realized that the receiver can only decide for the symbol sequence $Q_N^{(j)}$ if the following condition is fulfilled:

$$\int_{-\infty}^{+\infty} \Delta\tilde{d}_{kj}(t)\overset{\times}{d}(t)\,\mathrm{d}t + \tfrac{1}{2}\Delta E_{kj} < 0 \tag{6.104}$$

Here it is taken into account that $\Delta\tilde{d}_{kk}(t) = 0$ and also, consequently, that $\Delta E_{kk} = 0$. The probability that the receiver actually *can* decide for the jth sequence, under the condition that the kth symbol sequence was transmitted, is as follows:

$$P_{j|k} = P\left\{ \int_{-\infty}^{+\infty} \Delta d_{kj}(t)\overset{\times}{d}(t)\,\mathrm{d}t + \tfrac{1}{2}\Delta E_{kj} < 0 \mid Q_N = Q_N^{(k)} \right\} \tag{6.105}$$

This probability $P_{j|k}$ differs fundamentally from the probability $p_{j|k}$ in eqn (6.90) which defines that the receiver *in fact does* decide for the symbol sequence $Q_N^{(j)}$. It is always the case that $p_{j|k} \leqslant P_{j|k}$.

If any $j \neq k$ fulfils the condition in eqn (6.104), then the receiver certainly makes a wrong decision, i.e. in this case $V_N \neq Q_N^{(k)}$. Therefore the conditional probability in eqn (6.102) can be written

$$P(V_N \neq Q_N^{(k)} \mid Q_N = Q_N^{(k)}) = P\left[\bigcup_{\substack{j=1\ldots M^N \\ j\neq k}} \{\Delta \tilde{d}_{kj}(t)\overset{\times}{d}(t)\,\mathrm{d}t + \tfrac{1}{2}\Delta E_{kj} < 0\} \mid Q_N = Q_N^{(k)} \right] \tag{6.106}$$

The individual events $\{\ldots\}$ of the union of sets are statistically dependent on each other, but in general no statement is possible about the statistical dependence. However, if we assume as a coarse simplification that the individual events are incompatible with each other, we obtain the following upper bound from eqn (6.105):

$$P(V_N \neq Q_N^{(k)} \mid Q_N = Q_N^{(k)}) \leqslant \sum_{\substack{j=1 \\ j\neq k}}^{M^N} P_{j|k} \tag{6.107}$$

The conditional probabilities $P_{j|k}$ can at least be specified analytically in some special cases. For a noise signal $\overset{\times}{d}(t)$ of Gaussian distribution with frequency-independent noise power spectrum $L_{\overset{\times}{d}}(f) = L_0$, which is always produced by an optimum receiver with a whitening filter, the following is a valid example:

$$P_{j|k} = Q\left\{ \left(\frac{\Delta E_{kj}}{4L_0} \right)^{1/2} \right\} \tag{6.108}$$

Here $Q(x)$ is the complementary Gaussian error integral (see the Appendix, Tables A1 and A2) and ΔE_{kj} is the energy distance between the kth and jth useful detection signals. The proof of this important relation is treated in refs. 6.7 and 6.20.

If we insert eqns (6.107) and (6.108) into eqn (6.103), we obtain the following upper bound for the mean error probability:

$$p_{\mathrm{M,E}} = \frac{n_{\mathrm{M}}}{N} P(V_N \neq Q_N^{(k)} \mid Q_N = Q_N^{(k)}) \leqslant \frac{n_{\mathrm{M}}}{N} \sum_{j\neq k} Q\left\{ \left(\frac{\Delta E_{kj}}{4L_0} \right)^{1/2} \right\} \tag{6.109}$$

Example This result will be made clear by a simple illustration in which the binary sequence with the coefficients a_1, a_2, a_3, a_4 is considered; the amplitude coefficients a_v are either $+1$ or -1. The basic detection pulse $g_d(t)$ will be an NRZ rectangular pulse with amplitude $\hat{g}_d$. It is assumed that the sequence $Q_N^{(k)} = \mathrm{OLOL}$

has been transmitted; the corresponding useful detection signal $\tilde{d}_k(t)$ is illustrated in Fig. 6.12(a). Figure 6.12(b) shows the useful detection signal $\tilde{d}_j(t)$ and the useful difference signal $\Delta\tilde{d}_{kj}(t)$ for the possible symbol sequence LLOL (see eqn (6.94)). The energy distance ΔE_{kj} between these two sequences is $\Delta E_1 = 4\hat{g}_d^2 T$. By using eqn (6.108) the probability $P_{j|k}$ that the sequence OLOL can be falsified as LLOL can be calculated. However, the same energy distance ΔE_{kj} is obtained for the sequences OOOL, OLLL and OLOO, and hence also the same probability $P_{j|k}$ (Fig. 6.12(c)). In contrast, the energy distances of the sequences LOOL, OOLL, LLOO, LLLL, OOOO and OLLO are twice this size. This means that the $\binom{4}{2} = 6$ symbol sequences, which differ from the transmitted sequence $Q_N^{(k)}$ in the case of two symbols, all possess the energy distance $\Delta E_2 = 2\Delta E_1 = 8\hat{g}_d^2 T$ (Figs 6.12(d) and 6.12(e)). In addition there are four symbol sequences (OOLO, LLLO, LOOO, LOLL) with the energy distance $\Delta E_3 = 3\Delta E_1$ and a sequence (LOLO) with $\Delta E_4 = 4\Delta E_1$. From this the following emerges as an upper bound for the probability of corruption of the sequence $Q_N^{(k)}$ (see eqn (6.107)):

$$P(V_N \neq Q_N^{(k)} \mid Q_N = Q_N^{(k)}) \leq 4Q\left\{\left(\frac{\hat{g}_d^2 T}{L_0}\right)^{1/2}\right\} + 6Q\left\{\left(\frac{2\hat{g}_d^2 T}{L_0}\right)^{1/2}\right\} + 4Q\left\{\left(\frac{3\hat{g}_d^2 T}{L_0}\right)^{1/2}\right\} + Q\left\{\left(\frac{4\hat{g}_d^2 T}{L_0}\right)^{1/2}\right\} \tag{6.110}$$

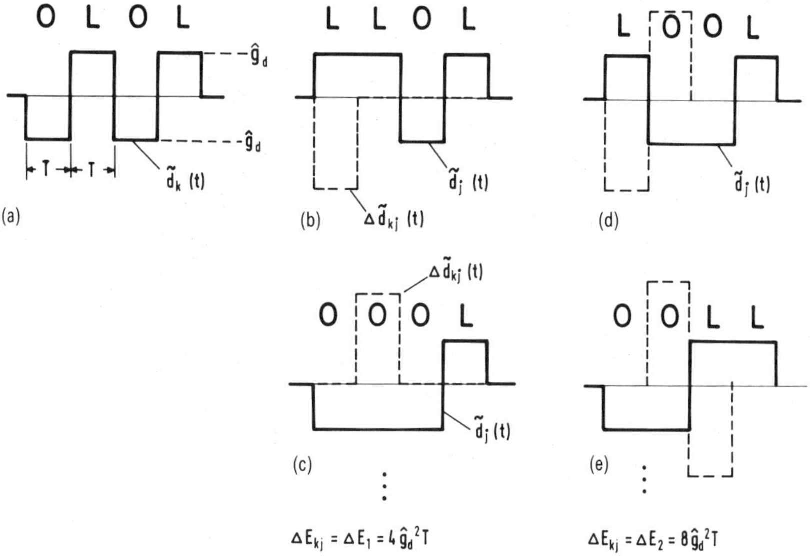

Fig. 6.12 Calculation of the energy distances ΔE_{kj} between the actual useful detection signal $\tilde{d}_k(t)$ and the possible useful detection signals $\tilde{d}_j(t)$.

If $\hat{g}_d^2 T$ is very much larger (by a factor of 10 or more) than the noise power density L_0, the terms with the doubled, trebled and quadrupled energy distance can be neglected, and we obtain from eqn (6.109) as an approximation

$$p_{\mathrm{M,E}} \approx Q\left\{\left(\frac{\hat{g}_d^2 T}{L_0}\right)^{1/2}\right\} \tag{6.111}$$

By generalization an equation for the mean error probability can be derived from this simple example. If we define the **minimum energy distance** between two useful detection signals $\tilde{d}_k(t)$ and $\tilde{d}_j(t)$ as

$$\Delta E_{\min} = \min_{k} \min_{j \neq k} \int_{-\infty}^{+\infty} \{\tilde{d}_k(t) - \tilde{d}_j(t)\}^2 \,\mathrm{d}t \qquad \begin{aligned} j &= 1 \ldots M^N \\ k &= 1 \ldots M^N \end{aligned} \tag{6.112}$$

we have

$$p_{\mathrm{M,E}} = K_{\mathrm{M,E}} Q\left\{\left(\frac{\Delta E_{\min}}{4L_0}\right)^{1/2}\right\} \tag{6.113}$$

The complementary Gaussian error integral $Q(x)$ is a strongly monotonically reducing function and so, in general, the symbol sequences with the minimum energy distance $\Delta E_{\min}$ exert the strongest influence on the mean symbol error probability, a fact which is also manifest from eqn (6.113). The effects which stem from the symbol sequences with nonminimal energy distance and all approximations will be taken into account by the correction term $K_{\mathrm{M,E}}$. If two sequences with minimum energy distance differ in one symbol only, then a good approximation is to put $K_{\mathrm{M,E}}$ equal to unity.

The mean bit error probability p_{B} defined in eqn (4.19) is used in the following chapters as the definitive optimization and comparison criterion. In an M-level transmission the relationship between the symbol and bit error probabilities can also be expressed by a correction term (the error propagation factor due to coding) which, according to Section 4.3.1, lies between $1/\log_2 M$ and 1 for a nonredundant source.

Combining the two correction parameters as $K_{\mathrm{B,E}}$ we obtain for the mean bit error probability of an M-level optimum maximum likelihood receiver with Gaussian distributed disturbances

$$p_{\mathrm{B,E}} = K_{\mathrm{B,E}} Q\left\{\left(\frac{\Delta E_{\min}}{4L_0}\right)^{1/2}\right\} \tag{6.114}$$

For a given transmitter and a given channel no other receiver gives a lower bit error probability.

6.4.3 Calculation of the minimum energy distance

The calculation of the minimum energy distance $\Delta E_{\min}$ in eqn (6.112) is significantly simplified if the amplitude coefficients $\alpha_\nu^{(k,j)}$ of the useful difference signal are introduced:

$$\alpha_\nu^{(k,j)} = a_\nu^{(k)} - a_\nu^{(j)} \qquad \begin{array}{c} k,j = 1 \dots M^N \\ \nu = 1 \dots N \end{array} \tag{6.115}$$

In the following these parameters are referred to as the **difference coefficients**. For binary bipolar amplitude coefficients ($a_\nu = \pm 1$) the difference coefficients α_ν are $[-2; 0; +2]$. For M-level signals with equidistant bipolar amplitude coefficients as in eqn (2.10) we obtain correspondingly ($m = 1, \dots, 2M - 1$)

$$\alpha_\nu \in \{\alpha_m\} = \left\{2\frac{2m - 2M - 1}{2M - 1}\right\} \tag{6.116}$$

For unipolar signals the difference coefficients are half as large.

Therefore if eqns (2.63) and (6.115) are used, the useful difference signal can be written as follows:

$$\Delta \tilde{d}_{kj}(t) = \tilde{d}_k(t) - \tilde{d}_j(t) = \sum_{\nu=1}^{N} \alpha_\nu^{(k,j)} g_d(t - \nu T) \tag{6.117}$$

Inserting this result into eqn (6.97), we obtain for the energy distance between the kth and jth useful detection signals

$$\Delta E_{kj} = \int_{-\infty}^{+\infty} \sum_{\nu=1}^{N} \sum_{\kappa=1}^{N} \alpha_\nu^{(k,j)} \alpha_\kappa^{(k,j)} g_d(t - \nu T) g_d(t - \kappa T)\,\mathrm{d}t \tag{6.118}$$

If we now define the **energy ACF of the basic detection pulse** by analogy with eqn (2.23)

$$l^{\bullet}_{gd}(\tau) = \int_{-\infty}^{+\infty} g_d(t) g_d(t + \tau)\,\mathrm{d}t \tag{6.119}$$

we obtain, after substituting $\lambda = \kappa - \nu$,

$$\Delta E_{kj} = \sum_{\nu=1}^{N} \sum_{\lambda=1-\nu}^{N-\nu} \alpha_\nu^{(k,j)} \alpha_{\nu+\lambda}^{(k,j)}\, l^{\bullet}_{gd}(\lambda T) \tag{6.120}$$

With the discrete ACF of the difference coefficients

$$l_\alpha^{(k,j)}(\lambda) = \sum_{\nu=1}^{N} \alpha_\nu^{(k,j)} \alpha_{\nu+\lambda}^{(k,j)} \tag{6.121}$$

we finally obtain the energy distance between the kth and jth useful detection signals:

$$\Delta E_{kj} = \sum_{\lambda=-\infty}^{+\infty} l_\alpha^{(k,j)}(\lambda)\, l^{\bullet}_{gd}(\lambda T) \tag{6.122}$$

where we assume $N \gg 1$. This equation simplifies the numerical evaluation of the (in total) M^{2N} energy distances ΔE_{kj} which are required for determination of the minimum energy distance $\Delta E_{\min}$. In contrast with the definition of eqn (6.112), here the integration is replaced by a summation which significantly reduces the calculation time required.

To determine the minimum energy distance those difference coefficients $\alpha_v^{(k,j)}$ have to be found which make ΔE_{kj} a minimum. This minimization is particularly easy if the minimum energy distance occurs for single errors. An adequate, not absolutely necessary, condition for this is derived in ref. 6.20. For an optimum receiver the occurrence of single errors is then more likely than a block error if

$$l^{\bullet}_{gd}(0) > 2 \sum_{\lambda=1}^{\infty} l^{\bullet}_{gd}(\lambda T) \tag{6.123}$$

i.e. if the basic detection pulse $g_d(t)$ decays relatively rapidly.

In the following we consider two M-level sequences $Q_N^{(k)}$ and $Q_N^{(j)}$ which differ by only one symbol, e.g. the nth symbol. All other symbols ($v \neq n$) tally. Hence, from eqn (6.121) we obtain the discrete ACF of the difference coefficients

$$l_\alpha^{(k,j)}(\lambda) = \begin{cases} (\alpha_n^{(k,j)})^2 & \text{for } \lambda = 0 \\ 0 & \text{for } \lambda \neq 0 \end{cases} \tag{6.124}$$

and thus the minimum energy distance

$$\Delta E_{\min} = \min_{j \neq k} (\alpha_n^{(k,j)})^2 \, l^{\bullet}_{gd}(0) \tag{6.125}$$

$\Delta E_{\min}$ is a minimum if $a_n^{(k)} = a_\mu$ and $a_n^{(j)} = a_{\mu+1}$ represent neighboring symbols so that the nth difference coefficient $\alpha_n^{(k,j)}$ defined in Section 6.4.3 has the smallest possible value. It thus follows from eqns (2.10) and (6.122) that, for equidistant amplitude coefficients ($a_\mu - a_{\mu-1} = \text{constant}$),

$$\Delta E_{\min} = \frac{4}{(M-1)^2} l^{\bullet}_{gd}(0) = \frac{4}{(M-1)^2} \int_{-\infty}^{+\infty} \{g_d(t)\}^2 \, dt \tag{6.126}$$

For unipolar signals $\Delta E_{\min}$ is a factor of 4 lower.

Inserting this result into eqn (6.114) we obtain the following approximation to the mean bit error probability of a maximum likelihood receiver:

$$p_{\mathrm{B,E}} = K_{\mathrm{B,E}} Q\left[\frac{1}{M-1}\left\{\frac{l^{\bullet}_{gd}(0)}{L_0}\right\}^{1/2}\right] \tag{6.127}$$

This approximation is valid for a nonredundant bipolar M-level transmitter signal and for disturbances of Gaussian distribution. It is further assumed here that individual errors are more likely than block errors. The smaller the error probability itself is, the more exact is the approximation. At very small error probabilities the correction parameter $K_{\mathrm{B,E}}$ can be put equal to unity.

With Parseval's theorem (eqn (6.25)) and eqn (6.30) we have

$$l_{gd}^{\bullet}(0) = \int_{-\infty}^{+\infty} \{g_d(t)\}^2\,\mathrm{d}t = \int_{-\infty}^{+\infty} |G_d(f)|^2\,\mathrm{d}f = \int_{-\infty}^{+\infty} \frac{|G_e(f)|^2}{L_n(f)/L_0}\,\mathrm{d}f \tag{6.128}$$

Here $G_e(f)$ is the spectrum of the basic receiver pulse $g_e(t)$ and $L_n(f)$ is the power spectrum of the noise signal $n(t)$. Hence we obtain the mean bit error probability

$$p_{\mathrm{B,E}} \approx Q\left[\frac{1}{M-1}\left\{\int_{-\infty}^{+\infty} \frac{|G_e(f)|^2}{L_n(f)}\,\mathrm{d}f\right\}^{1/2}\right] \tag{6.129}$$

6.4.4 Minimum energy distance for a coaxial cable system

Equation (6.129) is only valid if single errors are more likely than block errors, i.e. if the condition in eqn (6.123) is fulfilled. In all other cases the minimum energy distance ΔE_{min} has to be calculated from eqn (6.122), and this is illustrated below by the example of a binary coaxial cable system.

For typical cable lengths and bit rates the frequency response of a coaxial cable can be approximated by eqn (2.55):

$$|H_{\mathrm{K}}(f)| = \exp\left\{-a_*\left(\frac{2|f|}{R}\right)^{1/2}\right\} \tag{6.130}$$

Here a_* is the characteristic cable attenuation value defined in eqn (2.54). The noise power spectrum $L_n(f) = L_0$ is frequency independent, so that it can be neglected for the whitening filter ($H_{\mathrm{WF}}(f) = 1$).

If the basic transmitter pulse is an NRZ rectangular pulse ($T_s = T$) with amplitude $\hat{g}_s$, we have for the energy ACF of the basic detection pulse (see eqn (6.119))

$$l_{gd}^{\bullet}(\tau) = \hat{g}_s^{\,2}\,T^2 \int_{-\infty}^{+\infty} \mathrm{si}^2(\pi f T)\exp\left\{-2a_*\left(\frac{2|f|}{R}\right)^{1/2}\right\}\cos(2\pi f\tau)\,\mathrm{d}f \tag{6.131}$$

For a nonredunant binary system $R = 1/T$.

Figure 6.13 shows the energy ACF for characteristic attenuation values a_* of 50 dB and 100 dB. It can be seen that $l_{gd}^{\bullet}(\tau)$ only decays very slowly for these cable attenuation values, so that the condition defined in eqn (6.123) is not fulfilled. A further consequence is that, with these attenuation values, the minimum energy distance does not occur between two symbol sequences which differ by only one symbol. Here two symbol sequences with minimum energy distance differ by several symbols. This is understandable if we consider the following: with the long response of the coaxial cable the useful difference signal, which results from superposition of many pulse responses, can be smaller if the sequences considered differ by several symbols.

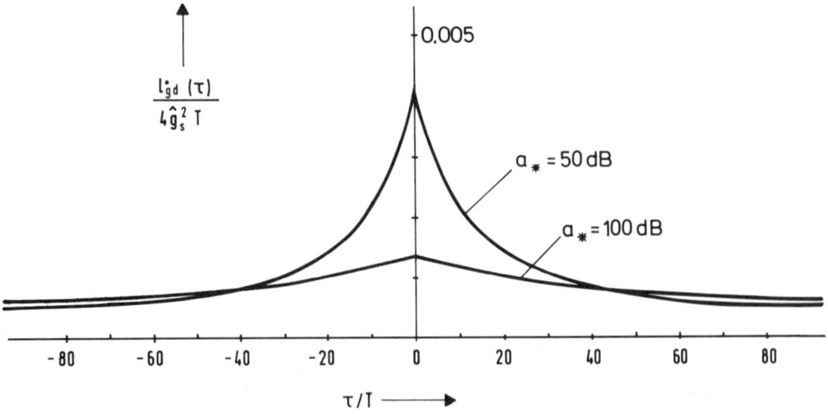

Fig. 6.13 Normalized energy ACF of a coaxial cable with characteristic cable attenuation value a_*.

Figure 6.14 shows the energy distance $\Delta E_{\min}$ normalized by $4\hat{g}_s{}^2 T$ as a function of the characteristic attenuation value a_*. Curve (a) is valid for two symbol sequences which differ by only one symbol, so that $\alpha_n{}^{(k,j)} = \pm 2$ and all other difference coefficients $\alpha_{\nu \neq n}{}^{(k,j)} = 0$. The energy distance ΔE_{kj} of both these symbol sequences is given by eqn (6.126):

$$\Delta E_{kj}{}^{(a)} = 4 l^{\bullet}_{gd}(0) \tag{6.132}$$

This energy distance is the minimum energy distance only if a_* is very small ($a_* < 10\,\text{dB}$). However, if we consider two symbol sequences which differ by two neighboring symbols, the difference coefficient is given by

$$\alpha_\nu{}^{(k,j)} = \begin{cases} \pm 2 & \text{for } \nu = n \\ \mp 2 & \text{for } \nu = n+1 \\ 0 & \text{otherwise} \end{cases} \tag{6.133}$$

The energy distance of the two symbol sequences is curve (b) in Fig. 6.14 and is calculated from eqns (6.121) and (6.122):

$$\Delta E_{kj}{}^{(b)} = 8\{l^{\bullet}_{gd}(0) - l^{\bullet}_{gd}(T)\} \tag{6.134}$$

If the characteristic attenuation value a_* lies between 10 dB and 70 dB then $\Delta E_{kj}{}^{(b)}$ is equal to the minimum energy distance $\Delta E_{\min}$. However, for $70\,\text{dB} < a_* < 130\,\text{dB}$ we have

$$\Delta E_{\min} = \Delta E_{kj}{}^{(c)} = 16 l^{\bullet}_{gd}(0) - 8 l^{\bullet}_{gd}(T) - 16 l^{\bullet}_{gd}(2T) + 8 l^{\bullet}_{gd}(3T) \tag{6.135}$$

Curve (c) is for two symbol sequences which, as shown in Fig. 6.14, differ by four neighboring symbols. With even greater attenuation values, new worst-case symbol sequences occur which differ by more symbols. This means that the minimum energy

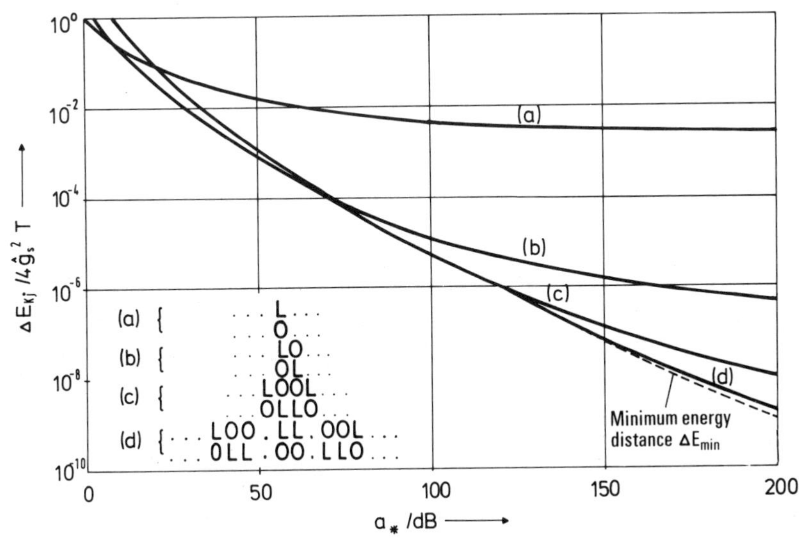

Fig. 6.14 Determination of the minimum energy distance as a function of the characteristic attenuation value a_* of a binary coaxial cable system.

distance ΔE_{min} shown as a broken curve in Fig. 6.14 is formed for different worst-case symbol sequences in different sections of the curve.

It is noteworthy that, when plotted on a logarithmic scale, the function of the minimum energy distance with a characteristic cable attenuation value of $a_* > 80$ dB closely approximates a straight line. The exact calculation of ΔE_{min} for multilevel systems is even more costly. However, with high attenuation values a linear relationship emerges as a good approximation in this case also [6.20].

Chapter 7

Performance Characteristics and Limits of Digital Transmission Systems

Contents The fundamental theory of system optimization and comparison is derived in this chapter. An overview of the most important optimization and comparison criteria is presented in Section 7.1. A very general comparison criterion, the system efficiency, is introduced in Section 7.2. This represents an equivalent measure for the mean bit error probability that relates the achievable signal-to-noise ratio to that of an optimum system consisting of an optimum transmitter, an ideal channel and an optimum receiver. The maximum regenerator-section length and the maximum transmittable bit rate are then determined in Section 7.3. The channel capacity, which is an information-theory upper bound for the error-free transmittable bit rate, is calculated in Section 7.4.

Assumptions The results of Chapter 7 are valid for M-level digital systems in which the information is contained in the amplitude values (pulse amplitude modulation); the possible amplitude coefficients $a_1, \dots a_M$ are equidistant (see eqn (2.10)). Noise is assumed to be stationary and additive and to follow a Gaussian distribution. As well as threshold detector receivers, without and with decision feedback equalization, the optimum digital receiver of Chapter 6 is considered. The receiver's threshold value and the clock signal are generally assumed to be ideal. The effect of tolerances is examined in Section 8.7.

7.1 Optimization criteria and system parameters

Since the advent of digital transmission technology, there have been many studies of system optimization. A comparison of the optimization results shows that the various workers arrived at optimum systems which showed some major differences. However, this is not surprising as the objectives and assumptions underlying the optimizations vary from case to case.

A comparison of the multiplicity of optimization reports allows the following basic optimization targets to be identified, each of which leads to different optimization results, i.e. to different optimum values for the system parameters:

(a) maximization of the regenerator-section length for a given bit rate or maximization of the bit rate for a given regenerator-section length, under the condition that the error probability must not exceed a predefined boundary value;
(b) minimization of the error probability for given bit rate and regenerator-section length;
(c) maximization of permissible tolerances for given bit rate, regenerator-section length and maximum error probability (see Section 8.7);
(d) minimization of the cost of realization for given boundary conditions.

Maximizing the regenerator-section length (or the bit rate) is suitable for general system design. It provides theoretical boundary values which can be approached more or less closely depending on the cost of the practical systems (see Section 7.3). The channel capacity which can be derived from information theory is an example of such a boundary, and is determined in Section 7.4 with various boundary conditions. Other optimum values are generally obtained for the individual system parameters if the aim of the system optimization is the minimization of the bit error probability b rather than the maximization of the regenerator-section length a. However, it is shown in Section 7.3 that each of the two optimization criteria can be converted into the other. As the error probability is easier to calculate than the maximum bridgeable regenerator-section length (or the maximum transmittable bit rate) it is used as the basis of the optimization criterion for the system optimization described in Chapters 8 and 9. For that reason in what follows **system optimization** will be understood to mean the determination of parameters and frequency responses describing the digital system so that the (mean) error probability is a minimum.

Not all the system parameters introduced in Chapter 2 are suitable for an optimization process. For example, the error probability becomes smaller as the transmitted pulse amplitude $\hat{g}_s$ becomes larger or the bit rate R becomes smaller. These quantities are referred to in the following as **nonoptimizable system parameters**. They must be assumed to be constant in the system optimization. In contrast, the **optimizable system parameters** are those system parameters and functions for which—depending on the nonoptimizable system parameters—there are optimum values or optimum functions which result in the minimum error probability.

In addition, certain other boundary conditions must be specified for the optimization. For example, optimization of the system parameters is critically affected by whether power or peak-value limitation of the transmitted signal is required as a co-condition. **Power limitation** means that the mean transmitter power may not exceed a defined maximum value S_s. In this case the following must be true for the mean-squared value of the transmitter signal:

$$\overline{s^2(t)} = \int_{-\infty}^{+\infty} s^2 f_s(s)\, \mathrm{d}s \leqslant S_s \tag{7.1}$$

where $f_s(s)$ is the probability density function of the transmitted signal. **Peak-value limitation** means that the modulation range $\Delta s = s_{max} - s_{min}$ of the transmitted signal is limited (see Fig. 2.3). For a symmetric transmitted signal $s_{min} = -s_{max}$ and therefore

$$|s(t)| \leqslant s_{max} \quad \text{for all } t \tag{7.2}$$

or

$$f_s(s) = 0 \quad \text{for } |s| > s_{max} \tag{7.3}$$

The question of whether power limitation or peak-value limitation is required as the co-condition of the optimization depends on the practical boundary conditions and has to be decided for each case. If, for example, the dimensions and the power loss of the transmitter are critically affected by the mean transmitter power S_s, the optimization must be derived from the power limitation. However, if the modulation range Δs of the transmitter is limited, because of the power loss or the linearity of the components for example, peak-value limitation is a sensible and necessary co-condition. Peak-value limitation is also required if disturbance of other users by cross-talk is not to exceed a given value.

The optimizable and nonoptimizable system parameters of a digital system with a threshold detector are summarized in Table 7.1 (Section 7.2.5) where possible coding (Chapter 4) and decision feedback equalization (DFE) (Chapter 5) are also considered. If the effect of a parameter can also be expressed by other quantities the corresponding rows are marked a, b, ... For example, the transmitted signal amplitude can be described by the three (mutually dependent) values s_{max}, S_s and $\hat{g}_s$.

7.2 Basics of system theory for system comparison

In contrast with system optimization, the effect of the nonoptimizable system parameters on the error probability must also be discussed when systems are compared. Stricter requirements are placed on the comparison criterion than on an optimization criterion. Whereas in the case of optimization it is sufficient that the criterion leads to the true optimum, in the case of system comparison the *value* of the optimum must also be reproduced correctly.

Basically two different philosophies can be considered for the comparison of communication systems:

(a) system comparison at equal transmission channel bandwidth;

(b) system comparison at equal information rate per unit time (bit rate) from the communication source (see eqns (2.8) and (2.9)).

The first type of system comparison has the disadvantage that channels with different bandwidths must be expressed in terms of a comparison channel of constant bandwidth, which can lead to substantial falsification of the results. For digital systems comparison at constant information rate is more appropriate and forms the basis of the approach described below.

Figure 7.1 shows the block diagram of the transmission system considered. To ensure a fair comparison of different systems the same source, i.e. a nonredundant binary source, is always assumed, so that the information rate ϕ and the bit rate R are in agreement. The (mean) bit error probability p_B defined according to eqn (4.19) is used for the optimization criterion so that the effect of transmission codes and the number of levels is taken into consideration.

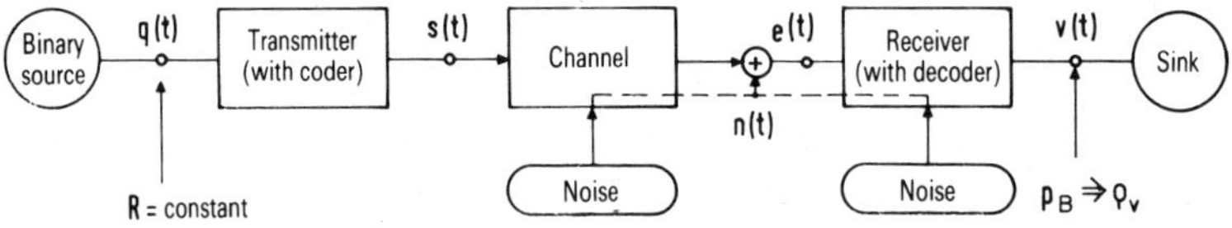

Fig. 7.1 General block diagram of a communication system.

The system efficiency is introduced in the following section as a very general optimization and comparison criterion, which in the case of digital systems provides an equivalent measure for the mean bit error probability p_B. By means of a very slight modification it is also possible to analyze modulated-carrier digital systems and arbitrary analog systems using the system efficiency. Introduction of the system efficiency can make clear how far the performance capability of a communication system is from the theoretically achievable optimum and what improvements in the transmitter, channel and receiver are still possible.

7.2.1 Definition of the system efficiencies

In analog communication systems the signal-to-noise ratio $\rho_v = S_v/N_v$ at the sink is generally used as a measure of the transmission quality. In this case S_v is the useful power and N_v is the noise power of the sink signal $v(t)$. To provide a consistent comparison criterion for analog and digital communication systems the mean bit error probability p_B of a digital system, as the critical criterion of quality, is converted into an equivalent **sink signal-to-noise ratio** ρ_v. For Gaussian-distributed noise

$$p_B = Q(\rho_v^{1/2}) \quad \text{and} \quad \rho_v = \{Q^{-1}(p_B)\}^2 \tag{7.4}$$

where $Q(x)$ is the complementary Gaussian error integral given in Table A2 and $Q^{-1}(x)$ is its inverse function. As $Q(x)$ represents a monotonically decreasing function, the sink signal-to-noise ratio is a measure of the transmission quality which is equivalent to the mean bit error probability. For example $p_B = 10^{-10}$ corresponds to the sink signal-to-noise ratio $10 \lg \rho_v = 16.1$ dB.

The sink signal-to-noise ratio ρ_v depends on the characteristics of the transmitter, the channel and the receiver. By **optimum receiver** we mean a noise-free receiver, which for a given transmitter and a given channel results in the maximum sink signal-to-noise ratio $\rho_{v,E}$:

$$\rho_{v,E} = \rho_v\,(\text{optimum receiver}) \tag{7.5}$$

From this we obtain the **receiver efficiency**

$$\eta_E = \frac{\rho_v}{\rho_{v,E}} = \frac{\rho_v(\text{given receiver})}{\rho_v(\text{optimum receiver})} \qquad 0 \leqslant \eta_E \leqslant 1 \tag{7.6}$$

η_E defines how well the given receiver converts the applied receiver signal $e(t)$ into the largest possible sink signal-to-noise ratio ρ_v; it is therefore a measure of the receiver's performance capability. By definition η_E is equal to unity for a transmission system with an optimum receiver.

The channel efficiency η_K is used as a measure of the quality of the transmission channel compared with an ideal channel, which is defined as a channel over which the transmitter signal is carried unaltered ($\tilde{e}(t) = s(t)$) and which exhibits the minimum thermal noise power density L_{th} (eqn (2.48)). It follows that the frequency response and the noise power density of an **ideal channel** are given by

$$H_K(f) = 1 \tag{7.7a}$$

$$L_n(f) = L_{th} \tag{7.7b}$$

The sink signal-to-noise ratio for an ideal channel, a given transmitter and an optimum receiver is given by

$$\rho_{v,KE} = \rho_v(\text{ideal channel; optimum receiver}) \tag{7.8}$$

Using eqn (7.5) we obtain the **channel efficiency**:

$$\eta_K = \frac{\rho_{v,E}}{\rho_{v,KE}} = \left.\frac{\rho_v(\text{given channel})}{\rho_v(\text{ideal channel})}\right|_{\text{optimum receiver}} \qquad 0 \leqslant \eta_K \leqslant 1 \tag{7.9}$$

An **optimum transmitter** is a transmitter which, with an ideal channel and optimum receiver, provides the largest sink signal-to-noise ratio $\rho_{v,SKE}$:

$$\rho_{v,SKE} = \rho_v(\text{optimum transmitter; ideal channel; optimum receiver}) \tag{7.10}$$

The **transmitter efficiency** relates to the optimum transmitter. Hence,

$$\eta_S = \frac{\rho_{v,\mathrm{KE}}}{\rho_{v,\mathrm{SKE}}} = \frac{\rho_v(\text{given transmitter})}{\rho_v(\text{optimum transmitter})}\bigg|_{\substack{\text{ideal channel}\\ \text{optimum receiver}}} \qquad 0 \leqslant \eta_S \leqslant 1 \tag{7.11}$$

However, the sink signal-to-noise ratio $\rho_{v,\mathrm{SKE}}$ depends significantly on whether the mean power S_s or the peak value $s_{\max}$ of the transmitter signal is limited (see eqns (7.1) and (7.2)). Using the abbreviation

$$\rho_L = \rho_{v,\mathrm{SKE}}\big|_{\text{power limitation}} \tag{7.12}$$

we obtain for the **transmitter efficiency under power limitation**

$$\eta_{S,L} = \frac{\rho_{v,\mathrm{KE}}}{\rho_L} = \frac{\rho_v(\text{given receiver})}{\rho_v(\text{optimum receiver})}\bigg|_{\substack{\text{power limitation}\\ \text{ideal channel}\\ \text{optimum receiver}}} \tag{7.13}$$

The **transmitter efficiency under peak-value limitation** is given by

$$\eta_{S,A} = \frac{\rho_{v,\mathrm{KE}}}{\rho_A} \frac{\rho_v(\text{given transmitter})}{\rho_v(\text{optimum transmitter})}\bigg|_{\substack{\text{peak-value limitation}\\ \text{ideal channel}\\ \text{optimum receiver}}} \tag{7.14}$$

By analogy with eqn (7.12) ρ_A is the maximum sink signal-to-noise ratio with peak-value limitation:

$$\rho_A = \rho_{v,\mathrm{SKE}}\big|_{\text{peak-value limitation}} \tag{7.15}$$

With a nonideal channel or a nonoptimum receiver the maximization of the transmitter efficiency η_S defined here does not necessarily lead to the largest sink signal-to-noise ratio. Rather, the efficiencies describing the total system in Fig. 7.1 must be maximized, such that the **system efficiency under power limitation** is given by

$$\eta_L = \eta_{S,L}\eta_K\eta_E \qquad 0 \leqslant \eta_L \leqslant 1 \tag{7.16}$$

and the **system efficiency under peak-value limitation** is given by

$$\eta_A = \eta_{S,A}\eta_K\eta_E \qquad 0 \leqslant \eta_A \leqslant 1 \tag{7.17}$$

The system efficiency η_L (or η_A) is an equivalent measure to the sink signal-to-noise ratio ρ_v that implicitly takes into account the co-condition of the power limitation (or peak-value limitation) which otherwise can only be formulated with difficulty. An extensive description of these system efficiencies is given in Section 7.2.5.

The system efficiencies defined here can be applied to both digital and analog communication systems. These parameters are calculated below for digital systems.

7.2.2 Receiver efficiency

A knowledge of the contents of Sections 7.2.2–7.2.4 is not necessary to understand Chapters 8 and 9. A reader who has no interest in the study of the theoretically optimum system can omit these sections.

Figure 7.2(a) shows a digital transmission system with symbol-by-symbol threshold detection. The bit rate of the binary nonredundant source signal is R and the symbol rate of the (generally coded) transmitter signal is $1/T$. With nonredundant coding $RT = \log_2 M$ where M is the number of levels. The transmitter pulse shape is characterized by the transmitter frequency response $H_S(f)$; the transmitter pulse amplitude is $\hat{g}_s$. The transmission characteristics of the channel are described by its frequency response $H_K(f)$; the noise $n(t)$ is assumed to be additive and to have a Gaussian distribution, with power spectrum $L_n(f) = F(f)L_{th}$. The spectral noise factor $F(f) \geqslant 1$ (eqn (2.50)) defines the increase in noise power density relative to the unavoidable thermal noise and depends on the characteristics of the channel and the receiver.

The receiver shown in Fig. 7.2(a) consists of an equalizer $H_E(f)$, a threshold detector (TD) and a timing recovery device (TRD). The mean bit error probability p_B of this transmission system can be calculated exactly from eqns (3.53), (3.55) and (4.20). With the restriction that the timing recovery device performs ideally, we obtain from eqn (3.22)

$$p_B = K_B Q(\rho_U{}^{1/2}) \tag{7.18}$$

ρ_U is the worst-case signal-to-noise ratio given by eqn (3.23) and can be calculated from the vertical eye opening $o(T_D)$ and the detector input noise power N_d. With the

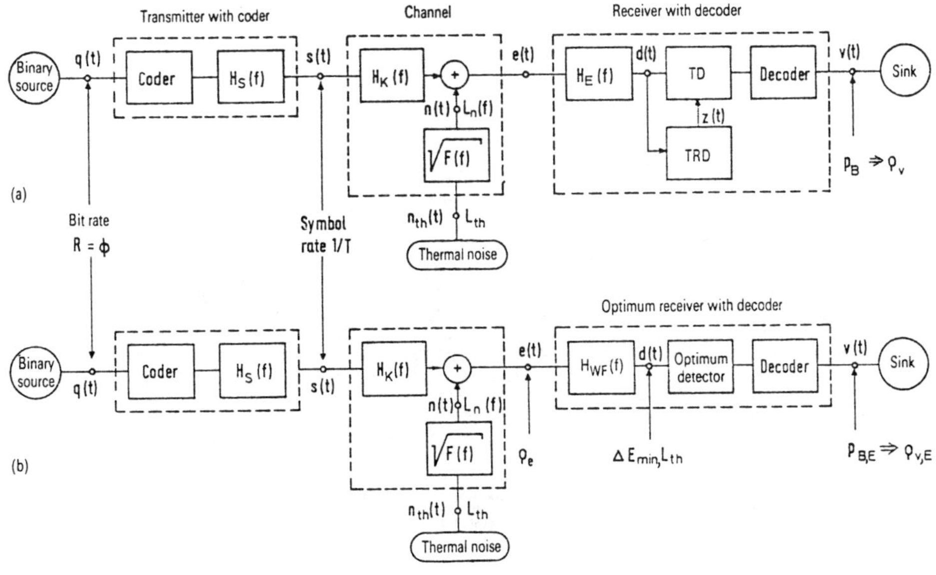

Fig. 7.2 Block diagram of the digital system under consideration with (a) a threshold detector or (b) an optimum receiver.

optimum decision value we have from eqns (2.69) and (3.24)

$$\rho_{\mathrm{U}} = \frac{\{o(T_{\mathrm{D}})/2\}^2}{N_d} \tag{7.19}$$

The correction term K_{B} takes account of, among other things, the error-propagation effect of the coding and is dependent on the number M of levels and the magnitude of the intersymbol interference. K_{B} is approximately constant for low error probabilities.

As the mean bit error probability p_{B} is also related to the sink signal-to-noise ratio ρ_v via the complementary Gaussian error integral $Q(x)$, it follows from eqns (7.4) and (7.18) that

$$\rho_v = [Q^{-1}\{K_{\mathrm{B}}Q(\rho_{\mathrm{U}}^{1/2})\}]^2 = K_\rho \rho_{\mathrm{U}} \tag{7.20}$$

where

$$K_\rho = \frac{1}{\rho_{\mathrm{U}}}[Q^{-1}\{K_{\mathrm{B}}Q(\rho_{\mathrm{U}}^{1/2})\}]^2 \tag{7.21}$$

If the correction term $K_{\mathrm{B}} = 1$, then the correction term K_ρ of the signal-to-noise ratio also has a value of unity. From the characteristics of the complementary Gaussian error integral, however, the correction term K_ρ also approaches the value of unity reasonably closely when $K_{\mathrm{B}} \neq 1$ if ρ_{U} is sufficiently large and hence the error probability is sufficiently small. Figure 7.3 shows the relationship between K_{B} and K_ρ.

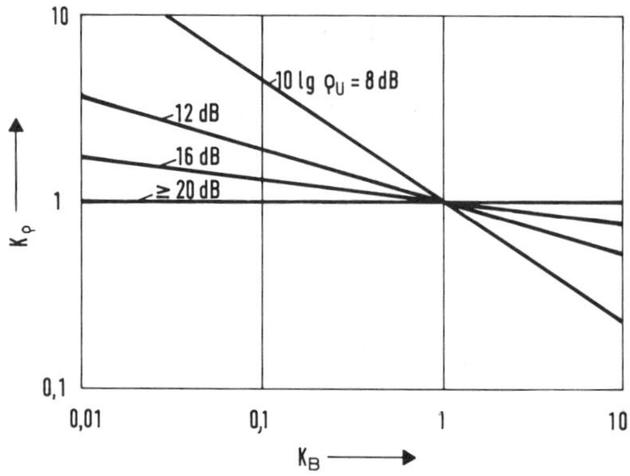

Fig. 7.3 Correction term K_ρ as a function of the correction term K_{B} and the worst-case signal-to-noise ratio $10\lg\rho_{\mathrm{U}}$.

To derive the receiver efficiency $\eta_{\mathrm{E}} = \rho_v/\rho_{v,\mathrm{E}}$ from eqn (7.6) we now consider a digital system with an optimum receiver (see Fig. 7.2(b)). As in Chapter 6 this contains an ideal whitening filter $H_{WF}(f)$ and an optimum detector, e.g. a Viterbi receiver.

For a given transmitter and a given channel this digital system produces the minimum mean bit error probability $p_{B,E}$. No other receiver can produce a smaller bit error probability, i.e. $p_B \geqslant p_{B,E}$ in all cases. Therefore, with Gaussian distributed noise (see eqn (6.114)),

$$p_{B,E} = K_{B,E} Q\left\{\left(\frac{\Delta E_{min}}{4L_{th}}\right)^{1/2}\right\} = K_{B,E} Q(\rho_e{}^{1/2}) \tag{7.22}$$

where ΔE_{min} is the minimum energy distance between the two useful detector input signals $\tilde{d}_k(t) \circ\!\!-\!\!\bullet \tilde{D}_k(f)$ and $\tilde{d}_j(t) \circ\!\!-\!\!\bullet \tilde{D}_j(f)$ according to eqn (6.112). The frequency-independent noise power density after the ideal whitening filter is the same as the thermal noise power density L_{th} (see eqn. (6.28)):

$$H_{WF}(f) = \left\{\frac{L_{th}}{L_n(f)}\right\}^{1/2} = \frac{1}{\{F(f)\}^{1/2}} \tag{7.23}$$

We define the **minimum receiver signal-to-noise ratio** as follows:

$$\rho_e = \frac{\Delta E_{min}}{4L_{th}} \tag{7.24}$$

ρ_e is the smallest energy distance which two permissible useful detection signals can possess normalized with respect to the thermal noise power density. It is a measure of the certainty of detection of the worst-case receiver signals.

By analogy with eqn (7.20) the sink signal-to-noise ratio $\rho_{v,E}$ of the optimum receiver can be written as follows (see eqn (7.5)):

$$\rho_{v,E} = K_{\rho,E}\,\rho_e \tag{7.25}$$

By analogy with eqn (7.21) the constant $K_{\rho,E}$ can be calculated from $K_{B,E}$ and ρ_e. If the minimum receiver signal-to-noise ratio ρ_e is sufficiently large, a good approximation to the correction term $K_{\rho,E}$ is again unity and $\rho_{v,E} \approx \rho_e$.

Using Parseval's theorem (eqn (6.25)) we obtain the following for the minimum receiver signal-to-noise ratio (see eqn (6.112)):

$$\rho_e = \min_k \min_{j \neq k} \int_{-\infty}^{+\infty} \frac{|\tilde{D}_k(f) - \tilde{D}_j(f)|^2}{4L_{th}}\,df = \min_k \min_{j \neq k} \int_{-\infty}^{+\infty} \frac{|\tilde{E}_k(f) - \tilde{E}_j(f)|^2}{4L_{th}F(f)}\,df \tag{7.26}$$

Here $\tilde{E}_k(f)$ and $\tilde{E}_j(f)$ are the spectra of the useful receiver signals $\tilde{e}_k(t)$ and $\tilde{e}_j(t)$ (see Section 6.4). It follows from eqns (7.6), (7.20) and (7.25) that the efficiency of the threshold receiver shown in Fig. 7.2(a) is given by

$$\eta_E = \frac{\rho_v}{\rho_{v,E}} = \frac{K_\rho \rho_U}{K_{\rho,E}\rho_e} \approx \frac{\rho_U}{\rho_e} \tag{7.27}$$

As $\rho_U \leqslant \rho_e$ in every case, the efficiency η_E of every receiver is less than or equal to unity. Therefore $\eta_E = 1$ is the limiting value for an optimum receiver as shown in

Fig. 7.2(b) and is in fact independent of the characteristics of the transmitter and the channel. Hence the receiver efficiency η_E is a measure of how well the receiver manages to convert the receiver signal applied at its input into as large a sink signal-to-noise ratio ρ_v as possible. The efficiency of a given receiver encompasses losses due to nonideal equalization, losses due to nonexistent or less than ideal compensation for intersymbol interference and losses caused by a nonoptimum detector (threshold drift, clock jitter). Finally, η_E is also reduced by the additional noise produced by the receiver.

7.2.3 Channel efficiency

The transmission channel has three different types of unwanted effect on the transmitted signal: it weakens it, distorts it and superimposes noise upon it. These transmission channel characteristics can be sufficiently accurately described by the frequency response $H_K(f)$ and the spectral noise factor $F(f)$.

The channel efficiency $\eta_K = \rho_{v,E}/\rho_{v,KE}$ (eqn (7.9)) is a measure of how these undesirable effects modify the sink signal-to-noise ratio relative to that of an ideal channel when the optimum receiver of Fig. 7.2(b) is assumed. $\rho_{v,E}$ is given by eqns (7.25) and (7.26). Hence the sink signal-to-noise ratio $\rho_{v,KE}$ of a digital system with an ideal channel and an optimum receiver remains to be calculated.

The frequency response $H_K(f)$ of an ideal channel (eqn (7.7)) is unity, so that in this special case the useful receiver signal $\tilde{e}(t)$ is equal to the transmitted signal $s(t)$. It follows that with $F(f) = 1$ the sink signal-to-noise ratio is (see eqn (7.26))

$$\rho_{v,KE} \approx \frac{1}{4L_{th}} \min_k \min_{j \neq k} \int_{-\infty}^{+\infty} |S_k(f) - S_j(f)|^2 \, df \tag{7.28}$$

$S_k(f)$ and $S_j(f)$ are the spectra of the possible transmitter signals $s_k(t)$ and $s_j(t)$. Hence the channel efficiency (eqn (7.9)) is

$$\eta_K = \frac{\rho_{v,E}}{\rho_{v,KE}} \approx \frac{\min_k \min_{j \neq k} \int \{|H_K(f)|^2/F(f)\} |S_k(f) - S_j(f)|^2 \, df}{\min_k \min_{j \neq k} \int |S_k(f) - S_j(f)|^2 \, df} \tag{7.29}$$

In general the values of the pair (k,j) which define the minimum energy distance ΔE_{min} are different in the numerator and denominator, and so the calculation of the channel efficiency can be very lengthy.

It is clear from eqn (7.29) that the channel efficiency depends not only on the channel characteristics $H_K(f)$ and $F(f)$ but also on the transmitter parameters.

7.2.4 Transmitter efficiency

The maximum sink signal-to-noise ratio $\rho_{v,\mathrm{SKE}}$ which results from a digital system with an optimum transmitter (S), an ideal channel (K) and an optimum receiver (E) is needed to calculate the transmitter efficiency according to eqn (7.11). The problem is to define the normalized power spectrum $L_a(f)$ of the amplitude coefficients (as a transmission code characteristic) and the basic transmitter pulse $g_s(t)$ so that the sink signal-to-noise ratio $\rho_{v,\mathrm{KE}}$ in eqn (7.28) has the largest possible value. Then, using Parseval's theorem (eqn (6.25)), we have

$$\rho_{v,\mathrm{SKE}} = \max_{L_a(f),g_s(t)} \left[\frac{1}{4L_{\mathrm{th}}} \min_k \min_{j \neq k} \int_{-\infty}^{+\infty} \{s_k(t) - s_j(t)\}^2 \, \mathrm{d}t \right] \tag{7.30}$$

As the optimization operation cannot be achieved without further restrictions, a bipolar nonredundant M-level transmitter signal is assumed together with no intersymbol interference at the transmitter, i.e. the individual transmitter pulses do not mutually affect one another. Using eqn (2.11) it follows from eqn (7.30) that

$$\rho_{v,\mathrm{SKE}} = \max_{a_\nu^{(k)}, a_\nu^{(j)}, g_s(t)} \left[\frac{\int_{-\infty}^{+\infty} \{g_s(t)\}^2 \, \mathrm{d}t}{4L_{\mathrm{th}}} \min_k \min_{j \neq k} (a_\nu^{(k)} - a_\nu^{(j)})^2 \right] \tag{7.31}$$

It is plausible that the optimum transmitter signal has equidistant amplitude coefficients. Hence we obtain from eqns (2.10) and (2.15)

$$\rho_{v,\mathrm{SKE}} = \max_{M,H_\mathrm{S}(f)} \left\{ \frac{\hat{g}_s{}^2 T^2}{(M-1)^2 L_{\mathrm{th}}} \int_{-\infty}^{+\infty} |H_\mathrm{S}(f)|^2 \, \mathrm{d}f \right\} \tag{7.32}$$

In the calculation of the transmitter efficiency η_S it is necessary to distinguish between power-limited and peak-value-limited systems. In the case of power limitation the mean transmitter power (calculated from eqn (2.20))

$$S_s = \hat{g}_s{}^2 T \int_{-\infty}^{+\infty} L_a(f) |H_\mathrm{S}(f)|^2 \, \mathrm{d}f = \frac{\hat{g}_s{}^2 (M+1)}{3(M-1)} T \int_{-\infty}^{+\infty} |H_\mathrm{S}(f)|^2 \, \mathrm{d}f \tag{7.33}$$

is a given fixed value. Substituting this result in eqn (7.32) we have from the relationship $RT = \log_2 M$ and eqn (7.12)

$$\rho_\mathrm{L} = \rho_{v,\mathrm{SKE}} = \max_M \left\{ \frac{3 \log_2 M}{M^2 - 1} \frac{S_s}{L_{\mathrm{th}} R} \right\} = \frac{S_s}{L_{\mathrm{th}} R} \tag{7.34}$$

The maximum value of ρ_L occurs for level number $M = 2$. As the value $\rho_\mathrm{L} = S_s / L_{\mathrm{th}} R$ cannot be exceeded with redundant coding or unipolar amplitude coefficients, the reference value of the transmitter efficiency $\eta_{\mathrm{S,L}}$ under power limitation is determined.

Consequently, if we assume that both the amplitude coefficients and the decision values are equidistant, the transmitter efficiency given by eqn (7.13) is

$$\eta_{\mathrm{S,L}} = \frac{\rho_{v,\mathrm{KE}}}{\rho_{\mathrm{L}}} = \frac{RT}{(M-1)^2} \frac{\int |H_{\mathrm{S}}(f)|^2 \,\mathrm{d}f}{\int L_a(f)|H_{\mathrm{S}}(f)|^2 \,\mathrm{d}f} \tag{7.35}$$

The factor $1/(M-1)^2$ takes account of the fact that the distance between two neighboring amplitude values is smaller for multilevel transmission than for binary transmission. Hence the sink signal-to-noise ratio with an ideal channel and optimum receiver is smaller by the same factor. However, the factor RT describes the effect of the symbol rate which can be made smaller in multilevel transmission ($RT > 1$) than in nonredundant binary transmission ($RT = 1$) so that the channel bandwidth and hence the superimposed noise can be reduced. Finally, the last part of eqn (7.35) takes account of the fact that with constant transmitter power S_s the transmitter pulse amplitude $\hat{g}_s$ can also be increased if the nonredundant binary code is converted to a multilevel and/or redundant code. Although, as for RT, this term can be larger than unity, $\eta_{\mathrm{S,L}}$ cannot exceed unity.

In general $\eta_{\mathrm{S,L}}$ depends on the transmitter pulse shape, i.e. on $H_{\mathrm{S}}(f)$. However, in the special case of a nonredundant code ($L_{\mathrm{a}}(f) = \text{constant}$) $\eta_{\mathrm{S,L}}$ is independent of $H_{\mathrm{S}}(f)$ and we obtain (Fig. 7.4(a))

$$\eta_{\mathrm{S,L}} = \begin{cases} 3\dfrac{\log_2 M}{M^2-1} & \text{bipolar} \\ 3\dfrac{\log_2 M}{4M^2-6M+2} & \text{unipolar} \end{cases} \tag{7.36}$$

We obtain the following results for the redundant transmission codes described in Chapter 4 with a rectangular transmitter pulse shape (see eqn (7.35) and Table 4.2): for ternary partial-response codes

$$\eta_{\mathrm{S,L}} = \frac{1}{(3-1)^2}\frac{1}{0.5} = 0.5 \tag{7.37}$$

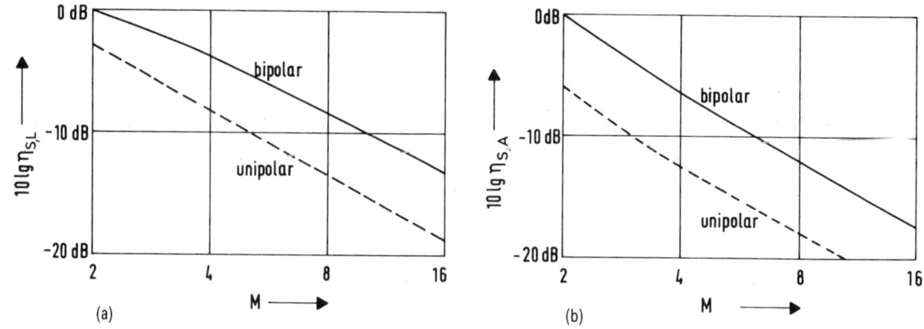

Fig. 7.4 Transmitter efficiency with nonredundant M-level NRZ rectangular signals: (a) with power limitation; (b) with peak-value limitation.

and for 4B3T codes

$$\eta_{S,L} \approx \frac{1.33}{(3-1)^2}\,\frac{1}{0.688} = 0.483 \tag{7.38}$$

Finally, we consider the transmitter efficiency with peak-value limitation. Here the following condition must be satisfied (see eqn (7.2)):

$$\max_{\langle a_v\rangle, t}\left\{\sum_{v=-\infty}^{+\infty} a_v g_s(t - vT)\right\} = s_{\max} \tag{7.39}$$

If we describe the intersymbol interference at the transmitter by the factor

$$\gamma_S = \frac{\hat{g}_s}{s_{\max}} = \gamma_S\{L_a(f), H_S(f)\} \tag{7.40}$$

we obtain from eqns (7.32) and (7.33)

$$\rho_A = \rho_{v,\mathrm{SKE}} = \max_{M, H_S(f)}\left\{\frac{\gamma_S{}^2 s_{\max}{}^2}{(M-1)^2 L_{\mathrm{th}}}\, T^2 \int_{-\infty}^{+\infty} |H_S(f)|^2\,\mathrm{d}f\right\} \tag{7.41}$$

This expression is a maximum for a nonredundant bipolar NRZ rectangular signal [8.33]. Then $\gamma_S = 1$ and the result of the integral is $1/T$, and so we have for the maximum sink signal-to-noise ratio with peak-value limitation

$$\rho_A = \max_{M}\left\{\frac{\log_2 M}{(M-1)^2}\,\frac{s_{\max}{}^2}{L_{\mathrm{th}} R}\right\} = \frac{s_{\max}{}^2}{L_{\mathrm{th}} R} \tag{7.42}$$

The maximum value here, as in the case of eqn (7.34), occurs for level number $M = 2$. Consequently for the transmitter efficiency with peak-value limitation (cf. eqn (7.14))

$$\eta_{S,A} = \frac{\rho_{v,\mathrm{KE}}}{\rho_A} = \gamma_S{}^2 \frac{RT}{(M-1)^2}\, T \int_{-\infty}^{+\infty} |H_S(f)|^2\,\mathrm{d}f \tag{7.43}$$

This is the equation for a nonredundant bipolar signal. For a unipolar signal $\eta_{S,A}$ is a factor of 4 smaller. γ_S is unity for a basic transmitter pulse limited to symbol duration T and is less than unity otherwise. The middle term takes account of the signal-to-noise ratio loss which is principally due to an M-level transmission with the co-condition of peak-value limitation. The final term defines the signal-to-noise ratio loss which is due to the fact that the rectangular NRZ transmitter pulse is not used and hence the potential transmitter power is not fully utilized.

Figure 7.4(b) shows the transmitter efficiency under peak-value limitation as a function of the number of levels of a nonredundant transmission system. Whereas the corresponding Fig. 7.4(a) for power limitation is valid for every type of transmitter pulse, Fig. 7.4(b) is only valid for the NRZ rectangular pulse.

7.2.5 Calculation and interpretation of the system efficiencies

The transmitter efficiencies $\eta_{S,L}$ and $\eta_{S,A}$ reflect the transmitter's performance capacity with an ideal channel and optimum receiver. They are components of the system efficiencies η_L and η_A, which describe the whole system and which are defined in such a way that they are suitable for the mutual optimization of transmitter and receiver. Therefore optimization of $\eta_{S,L}$ and $\eta_{S,A}$ for a given channel and a given (nonoptimum) receiver does not necessarily lead to the best transmitter. The system efficiencies η_L and η_A must always be considered in the solution of this optimization problem.

Substituting eqns (7.6), (7.9), (7.13) and (7.14) into eqns (7.16) and (7.17), we obtain for the system efficiency with power limitation

$$\eta_L = \eta_{S,L}\eta_K\eta_E = \frac{\rho_v}{\rho_L} \tag{7.44}$$

and for the system efficiency with peak-value limitation

$$\eta_A = \eta_{S,A}\eta_K\eta_E = \frac{\rho_v}{\rho_A} \tag{7.45}$$

ρ_L and ρ_A represent the mean and maximum power normalized by the reference noise power $N_{th} = L_{th}R$ which is available at the transmitter for transmission of the bit rate R (see eqns (7.34) and (7.42)). With a given regenerator-section length l, however, the sink signal-to-noise ratio ρ_v is greater, the higher is the mean and maximum power applied to the transmitter. If the mean bit error probability p_B is sufficiently small (e.g. $p_B < 10^{-6}$) and hence the sink signal-to-noise ratio is sufficiently large, then ρ_v is approximately proportional to the transmitter power. The ratio $\eta_L = \rho_v/\rho_L$ is therefore a measure of how well the digital system under consideration converts the mean transmitter power into the largest possible sink signal-to-noise ratio ρ_v. It is a "power transmission factor" which defines how much useful sink power a communication system obtains from the transmitter power with simultaneous suppression of the noise power.

If the actual noise affecting the system has a noise power density greater than L_{th}, this results in a smaller than achievable sink signal-to-noise ratio. In the present treatment this is counted as a deficiency of the system. Therefore the system comparison we aim for considers not only the value of the transmitted power but also the quality of the noise suppression. It is important for system comparison that the reference noise power $N_{th} = L_{th}R$ is the same for all systems, so that the comparison is based on the unavoidable thermal noise.

In principle the system efficiencies η_L and η_A are dependent on all the optimizable and nonoptimizable system parameters (Table 7.1). However, they are defined in such a way that the effect of the nonoptimizable system parameters is reduced as much as possible or completely removed. When the bit error probability

Table 7.1 Optimizable and nonoptimizable system parameters and their effect on system efficiencies

Row	*System parameter*	*Optimizable*	*Effect on* η_L	η_A	$\eta_{S,L}$	$\eta_{S,A}$	η_K	η_E
Transmitter								
1	Bit rate R [a]	No	No	No	No	No	No	No
2a	Transmitter peak value s_{max} [b]	No	Yes	No	Yes	No	No	No
2b	Mean transmitter power S_s [b]	No	No	Yes	No	Yes	No	No
2c	Transmitter pulse amplitude $\hat{g}_s$	No	Yes	Yes	Yes	Yes	No	No
3	Transmission code $L_a(f)$	Yes	Yes	Yes	Yes	Yes	Yes	Yes
4	Transmitter frequency response $H_S(f)$	Yes	Yes	Yes	Yes	Yes	Yes	Yes
Channel								
5	Channel frequency response $H_K(f)$	No	Yes	Yes	No	No	Yes	Yes
6a	Noise power spectrum $L_n(f)$	No	Yes	Yes	No	No	Yes	Yes
6b	Spectral noise factor $F(f)$	No	Yes	Yes	No	No	Yes	Yes
Receiver								
7a	Equalizer frequency response $H_E(f)$	Yes	Yes	Yes	No	No	No	Yes
7b	Pulse-shaper frequency response $H_I(f)$	Yes	Yes	Yes	No	No	No	Yes
8	Decision values $E_1 \ldots E_{M-1}$ [c]	Yes	Yes	Yes	No	No	No	Yes
9	Detection time T_D	Yes	Yes	Yes	No	No	No	Yes
10	DFE network $H_{DFE}(f)$	Yes	Yes	Yes	No	No	No	Yes

[a] In general the system efficiencies depend on the bit rate R. However, if the frequency responses are normalized by R, the efficiencies are approximately independent of R. A requirement of this approximation is a high sink signal-to-noise ratio and hence a small error probability.
[b] With peak-value limitation s_{max} is considered as an independent system parameter. The corresponding optimization criterion η_A is independent of this. Likewise, η_L is independent of S_s.
[c] The decision values are assumed throughout this chapter to be optimum. The effect of a threshold drift is treated in Section 8.7.

is low η_L is approximately independent of the mean transmitter power S_s. Therefore it is particularly suitable for use as the optimization and comparison criterion under power limitation. Similarly, η_A is independent of the transmitter peak power s_{max}^2 and hence is a suitable optimization criterion under the co-condition of peak-value limitation.

If the transmitter forms the transmitter signal $s(t)$ in such a way that it always emits the maximum possible transmitter power $S_s = s_{max}^2$ and if the transmission channel is also ideal and the receiver makes use of all the information available to it, both system efficiencies have the maximum value: $\eta_L = \eta_A = 1$. In contrast, $\eta_L = 0$ and

$\eta_A = 0$ corresponds to the sink signal-to-noise ratio $\rho_v = 0$ and hence the mean bit error probability $p_B = 0.5$. It is always the case that

$$0 \leqslant \eta_L \leqslant 1 \qquad \text{and} \qquad 0 \leqslant \eta_A \leqslant 1 \tag{7.46}$$

so that the term "system efficiency" is justifiable. It can be seen from the ratio of the two efficiencies

$$\frac{\eta_A}{\eta_L} = \frac{\rho_L}{\rho_A} = \frac{S_s}{s_{\max}^2} \leqslant 1 \tag{7.47}$$

that the system efficiency under peak-value limitation is always less than or equal to the system efficiency under power limitation. Hence it follows that

$$0 \leqslant \eta_A \leqslant \eta_L \leqslant 1 \tag{7.48}$$

This means that the demands of peak-value limitation (eqn (7.2)) are stronger than the demands of power limitation (eqn (7.1)).

In a digital system with threshold detection the sink signal-to-noise ratio is approximately equal to the worst-case signal-to-noise ratio ρ_U if the error probability is sufficiently small (see eqn (7.20) and Fig. 7.3). Hence, assuming a bipolar transmitter signal, we obtain the system efficiency

$$\eta_A = \frac{\rho_v}{\rho_A} = \frac{\{o(T_D)/2\}^2}{N_d} \frac{L_{th} R}{s_{\max}^2} \tag{7.49}$$

In the case of a low-pass system with direct signal transmission factors $|H_K(0)|$ and $|H_E(0)|$, where $|H_I(0)| = |H_K(0) H_E(0)|$, this can also be written

$$\eta_A = \left\{\frac{o(T_D)}{2 s_{\max} |H_I(0)|}\right\}^2 \frac{L_{th} R |H_E(0)|^2}{N_d} |H_K(0)|^2 \tag{7.50}$$

The first term corresponds to the square of the normalized eye opening given by eqn (3.25). The reciprocal of the second term is defined as the **normalized noise power**:

$$N_{\text{norm}} = \frac{N_d}{L_{th} R |H_E(0)|^2} = \frac{1}{R} \int_{-\infty}^{+\infty} F(f) \frac{|H_E(f)|^2}{|H_E(0)|^2} \, \mathrm{d}f \tag{7.51}$$

Hence we obtain for the system efficiency under peak-value limitation

$$\eta_A = \frac{o_{\text{norm}}^2(T_D)}{N_{\text{norm}}} |H_K(0)|^2 \tag{7.52}$$

This equation is for bipolar signals. The system efficiency for unipolar signals is smaller by a factor of 4. With the **transmitter peak-value factor (crest factor)**

$$\kappa_S = \frac{s_{\max}}{s_{\text{eff}}} = \frac{s_{\max}}{S_s^{1/2}} \tag{7.53}$$

this yields the system efficiency under power limitation:

$$\eta_L = \kappa_S{}^2 \frac{o_{norm}{}^2(T_D)}{N_{norm}} |H_K(0)|^2 \tag{7.54}$$

In the general case the transmitter peak-value factor κ_s has to be calculated from eqns (7.33) and (7.40). For the special case of nonredundant bipolar signals and transmitter pulses which do not overlap, the following is true:

$$\kappa_s = \left\{ \frac{M+1}{3(M-1)} T \int_{-\infty}^{+\infty} |H_S(f)|^2 \, df \right\}^{-1/2} \tag{7.55}$$

Figure 7.5 illustrates the relationship between the parameters ρ_v, ρ_A and η_A defined above and the mean bit error probability p_B. It shows the peak-value-limitation case: under power limitation ρ_A should be replaced by ρ_L and η_A by η_L. In this diagram the sink signal-to-noise ratio $10\lg\rho_v$ is plotted on the ordinate, the parameter $10\lg\rho_A$ is plotted as a measure of the available transmitter amplitude on the right-hand abscissa and the mean bit error probability p_B is plotted on the left-hand abscissa. Since

$$10\lg\rho_v = 10\lg\rho_A + 10\lg\eta_A \tag{7.56}$$

the lines of equal efficiency (η_A = constant) on the logarithmic scale of the right-hand side of the diagram are straight lines whose gradient is determined by the scale factors of ρ_v and ρ_A. As the system efficiency η_A can never be greater than unity the curves for all

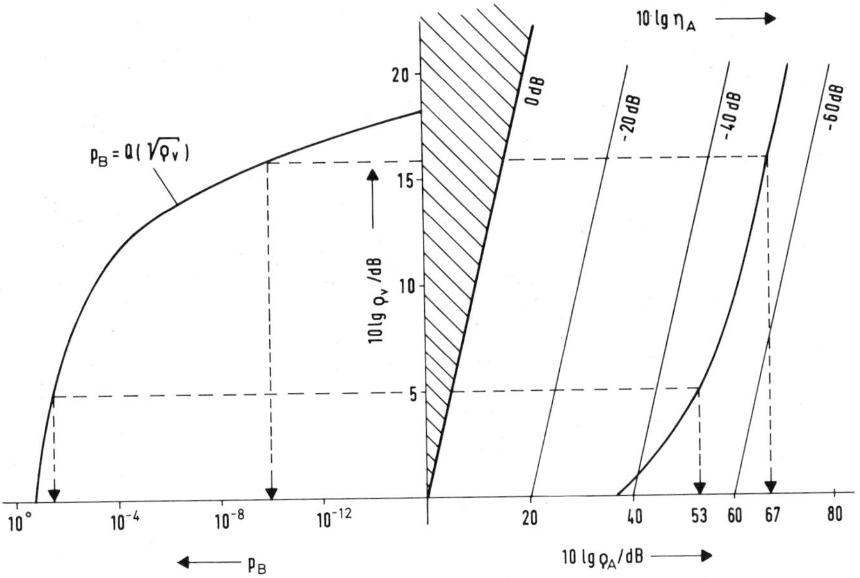

Fig. 7.5 Relationship between p_B, ρ_v, ρ_A and η_A.

systems lie below the hatched line through the origin. The bold curve is for the binary transmission system with a Gaussian pulse shaper which was discussed in Section 3.5.2. If the error probability is assumed to be 10^{-10}, this yields a power parameter $10\lg\rho_A \approx 67$ dB. The required transmitter amplitude can be calculated from this value by using eqn (7.42), and we obtain $s_{max}{}^2 = 10^{6.7} L_{th} R$ for the specified value. If the transmitter amplitude is larger than this, the sink signal-to-noise ratio ρ_v increases approximately in proportion to $s_{max}{}^2$. However, if s_{max}, and hence the power parameter ρ_A, are very much smaller than the value above, then the relationship between ρ_v and ρ_A is no longer linear. For example, when $10\lg\rho_A = 53$ dB, the sink signal-to-noise ratio is 5 dB, which corresponds to a mean bit error probability p_B of about 4.5%.

7.3 Maximum permissible bit rate and regenerator-section length

The system efficiencies η_L and η_A defined above correspond to a given channel frequency response and hence are valid for one specified regenerator-section length l. We shall now consider how large the regenerator-section length l can be for a given boundary condition without the error probability exceeding a predefined limit.

The following treatment is valid for transmission channels for which the attenuation constant

$$a_K(f) = 10\lg|H_K(f)| = \alpha l \tag{7.57}$$

increases in proportion to the regenerator-section length. This condition occurs in many transmission media, e.g. symmetric lines and coaxial cables. The example of a coaxial cable is considered below. According to Section 2.4.4 the amplitude response, to a good approximation, is given by

$$|H_K(f)| = \exp\left\{-a_*\left(\frac{2|f|}{R}\right)^{1/2}\right\} \tag{7.58}$$

where $a_* = \alpha_2(R/2)^{1/2} l$. The direct signal attenuation $\exp(-\alpha_0 l)$ is assumed to be negligible, which is permissible in the case of cables which are not extremely long ($l < 100$ km) [6.20]. The logarithmic system efficiency is plotted in Fig. 7.6 as a function of the characteristic cable attenuation value a_*. This figure is valid for white noise with a noise power density L_{th} (noise factor $F = 1$) and an optimum receiver, so that the receiver efficiency $\eta_E = 1$. In addition, a nonredundant code and a rectangular bipolar NRZ transmitter pulse are assumed, so that the system efficiencies under peak-value limitation and power limitation respectively are given by eqns (7.24), (7.25), (7.36) and (7.43):

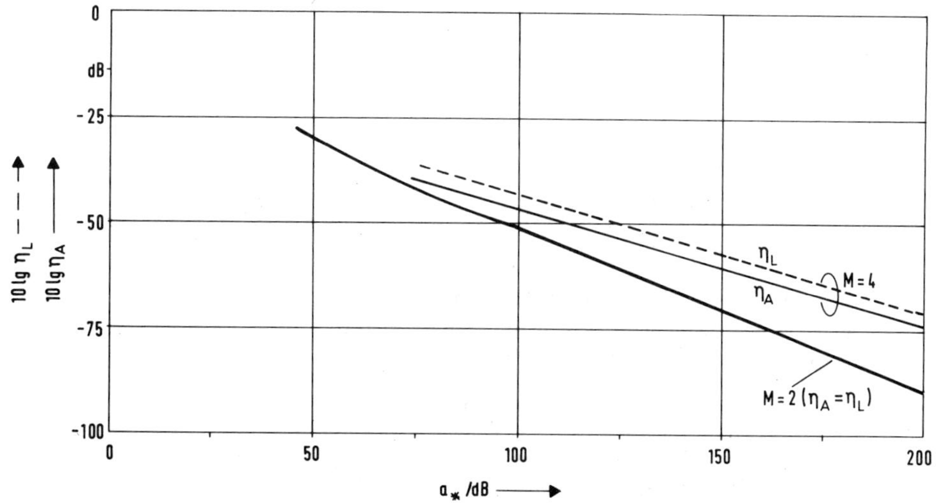

Fig. 7.6 System efficiency of a digital transmission system with optimum receiver as a function of the characteristic attenuation value a_* of a coaxial cable ($F(f) = 1$).

$$\eta_A(M, a_*) = \eta_{S,A}\eta_K = \frac{\log_2 M}{(M-1)^2} \frac{\Delta E_{\min}(M, a_*)}{4\hat{g}_s{}^2 T} \tag{7.59}$$

and

$$\eta_L(M, a_*) = \eta_{S,L}\eta_K = \frac{3\log_2 M}{M^2-1} \frac{\Delta E_{\min}(M, a_*)}{4\hat{g}_s{}^2 T} \tag{7.60}$$

where $\Delta E_{\min}$ is the minimum energy distance according to eqn (6.112). The nomenclature should make it clear that $\Delta E_{\min}$, and hence the system efficiencies η_A and η_L, depend on the level number M and the characteristic cable attenuation value a_*.

First, we consider a binary system ($M = 2$) in which the system efficiencies η_A and η_L are identical ($\eta_{S,A} = \eta_{S,L} = 1$). The minimum energy distance $\Delta E_{\min}$ for a binary coaxial cable system is shown in Fig. 6.14 as a function of the characteristic cable attenuation value. The system efficiencies $\eta_A = \eta_L$ can be determined from this.

Figure 7.6 shows that an approximately linear relationship exists between the logarithmic system efficiency and the cable attenuation value a_*, so that the curves for $a_* \geqslant 50$ dB can be approximated by straight lines:

$$10 \lg \eta_A \approx K_A - K_* a_* \tag{7.61}$$

$$10 \lg \eta_L \approx K_L - K_* a_* \tag{7.62}$$

This (approximately) linear relationship also exists in the case of multilevel systems and, as will be shown later, in systems with nonoptimum receivers. Calculating the system efficiency for two different values of the cable attenuation allows the constants K_A (K_L) and K_* to be determined. For example, for an optimum binary receiver with white noise of noise factor $F = 1$ we obtain

$$K_A = K_L = -13.6\,\text{dB} \qquad K_* = 0.38 \tag{7.63}$$

If we take into account additional noise (in the channel and the receiver) by means of a constant noise factor $F \geqslant 1$ we obtain

$$10\lg \eta_A = K_A - K_* a_* - 10\lg F \tag{7.64}$$

An analogous relation holds for power limitation.

To calculate the maximum bridgeable regenerator-section length under peak-value limitation we start with eqn (7.45). In logarithmic form this is

$$10\lg \eta_A = 10\lg \rho_v - 10\lg \rho_A \tag{7.65}$$

If the mean bit error probability p_B is not to exceed a defined limit p_{Gr}, then the system efficiency must obey the following:

$$10\lg \eta_A \geqslant 10\lg \rho_{Gr} - 10\lg \rho_A \tag{7.66}$$

The relationship between the **critical signal-to-noise ratio** $10\lg \rho_{Gr}$ and the corresponding **critical error probability** p_{Gr} is given by the complementary Gaussian error integral (see eqn (7.4)):

$$p_{Gr} = Q(\rho_{Gr}{}^{1/2}) \quad \text{or} \quad 10\lg \rho_{Gr} = 20\lg\{Q^{-1}(p_{Gr})\} \tag{7.67}$$

For example, if it is required that $p_{Gr} = 10^{-10}$ the necessary critical signal-to-noise ratio is $10\lg \rho_{Gr} = 16.1\,\text{dB}$. Using eqns (7.42) and (7.64) we obtain the maximum bridgeable cable attenuation value under peak-value limitation:

$$a_* \leqslant a_{*,\max} = \frac{1}{K_*}\left\{K_A + 10\lg\left(\frac{s_{\max}{}^2}{FL_{th}R}\right) - 10\lg \rho_{Gr}\right\} \tag{7.68}$$

From this we obtain the **maximum regenerator-section length** via eqn (7.58):

$$l \leqslant l_{\max} = \frac{1}{K_* \alpha_2 (R/2)^{1/2}}\left\{K_A + 10\lg\left(\frac{s_{\max}{}^2}{FL_{th}R}\right) - 10\lg \rho_{Gr}\right\} \tag{7.69}$$

By analogy we obtain for the co-condition of power limitation

$$l \leqslant l_{\max} = \frac{1}{K_* \alpha_2 (R/2)^{1/2}}\left\{K_L + 10\lg\left(\frac{S_s}{FL_{th}R}\right) - 10\lg \rho_{Gr}\right\} \tag{7.70}$$

Example With the constants as given by eqn (7.63) and the assumed values

$$s_{\max} = 3\,\text{V or } S_s = 9\,\text{V}^2 \qquad R = 1\,\text{Gbit s}^{-1} \qquad L_{th} = 1.5 \times 10^{-19}\,\text{V}^2\,\text{Hz}^{-1}$$
$$F = 6 \qquad p_{Gr} = 10^{-10} \rightarrow 10\lg \rho_{Gr} = 16.1\,\text{dB} \tag{7.71}$$

we obtain for the optimum binary system a maximum cable attenuation value a_* of

185 dB. At the given bit rate R this corresponds to a maximum regenerator-section length of 3.5 km of normal coaxial cable or 1.6 km of small coaxial cable (see Table 2.3).

The evaluation of eqns (7.69) and (7.70) shows that the maximum regenerator-section length l_{max} depends only slightly on the applied transmitter power. Changing the transmitter amplitude of s_{max} from 3 V to 1 V or 10 V results in maximum regenerator-section lengths of 3 km or 4 km respectively of normal coaxial cable. Similarly, the effect of the noise factor on the achievable regenerator-section length is relatively small.

The maximum regenerator-section length l_{max} is plotted against the bit rate R for normal coaxial cable (2.6 mm/9.5 mm) in Fig. 7.7. For small coaxial cable (1.2 mm/4.4 mm) the section length is shorter by a factor of about 0.45. It can be seen that in this log–log plot the curves can be approximated by straight lines. Using the abbreviations (units, dB)

$$A_0 = K_A + 10\lg\left(\frac{s_{max}^{\ 2}}{FL_{th}\,\text{MHz}}\right) - 10\lg\rho_{Gr} \tag{7.72}$$

and

$$A_1 = \frac{1}{2^{1/2}} K_* \alpha_2 \,\text{km MHz}^{1/2} \tag{7.73}$$

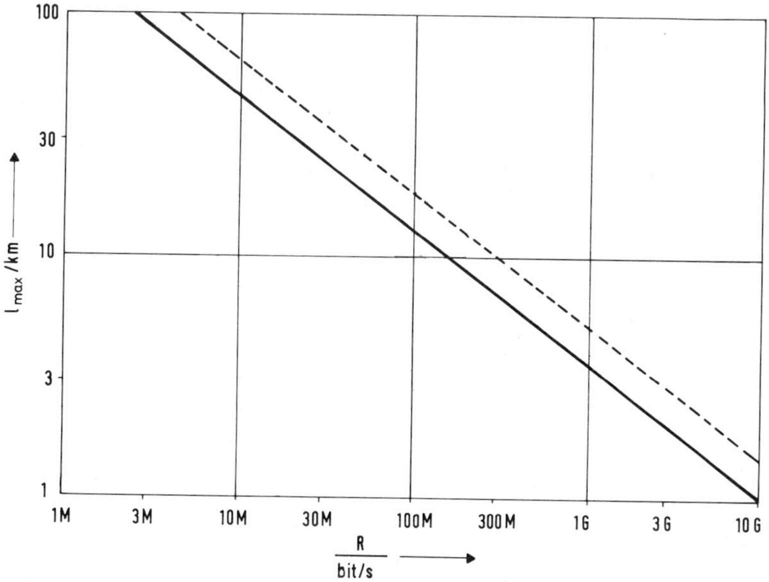

Fig. 7.7 Maximum regenerator-section length l_{max} of a normal coaxial cable (2.6 mm/9.5 mm) as a function of the bit rate R (for the values assumed in eqn (7.71)): ———, optimum binary receiver ($p_B = 10^{-10}$); —, channel capacity ($p_B = 0$).

we can rewrite eqn (7.70) as

$$\lg\left(\frac{l_{\max}}{\text{km}}\right)=\lg\left(\frac{A_0}{A_1}\right)-\frac{1}{2}\lg\left(\frac{R}{\text{MHz}}\right)+\lg\left\{1-\frac{10\lg(R/\text{MHz})}{A_0}\right\} \tag{7.74}$$

Under power-limitation conditions the constants K_A and $s_{\max}{}^2$ in eqn (7.72) replace K_L and S_s respectively. Using the approximation

$$\lg(1-x)\approx -x\lg e \tag{7.75}$$

which is valid at small values of x we finally obtain the maximum regenerator-section length

$$\lg\left(\frac{l_{\max}}{\text{km}}\right)\approx\lg\left(\frac{A_0}{A_1}\right)-\left(\frac{1}{2}+\frac{10\lg e}{A_0\,\text{dB}}\right)\lg\left(\frac{R}{\text{MHz}}\right) \tag{7.76}$$

It follows that the maximum transmittable bit rate $R_{\max}$ at a given regenerator-section length l is given by

$$\lg\left(\frac{R_{\max}}{\text{MHz}}\right)\approx\frac{\lg(A_0/A_1)-\lg(l/\text{km})}{0.5+(10\lg e)/(A_0/\text{dB})} \tag{7.77}$$

The channel capacity (under power limitation) is also drawn in Fig. 7.7 for comparison. This defines an information theory boundary for the error-free transmittable bit rate (see Section 7.4). It shows that, for a given regenerator-section length, the maximum transmittable bit rate $R_{\max}$ with an optimum binary receiver is about half the channel capacity. However, it should be remembered that the channel capacity is valid for error probabilities approaching zero, whereas the basis for the optimum binary receiver is an error probability of 10^{-10}.

The following approximations (upper boundaries) give the efficiencies η_A and η_L for multilevel nonredundant coding [6.20]:

$$\eta_A(M,a_*)\lesssim\frac{\log_2 M}{(M-1)^2}\eta_A\left\{M=2,\frac{a_*}{(\log_2 M)^{1/2}}\right\} \tag{7.78}$$

$$\eta_L(M,a_*)\lesssim\frac{3\log_2 M}{M^2-1}\eta_L\left\{M=2,\frac{a_*}{(\log_2 M)^{1/2}}\right\} \tag{7.79}$$

Both these efficiencies are plotted for the level number $M=4$ in Fig. 7.6. Over the whole range η_L is a factor $3(M-1)/(M+1)$ above the system efficiency η_A. Relative to the maximum regenerator-section length $l_{\max}$ and the maximum transmittable bit rate $R_{\max}$ the curves for the optimum multilevel receiver lie between those for the optimum binary receiver and the channel capacity. The corresponding values can be determined from eqns (7.72)–(7.77).

7.4 Information theory limits of digital transmission

The channel capacity introduced by Shannon [7.7, 7.8] is an upper boundary on the maximum transmittable information rate ϕ, derived from information theory, which, for a nonredundant source, is identical with the bit rate R.

7.4.1 Channel capacity of a discrete communication channel

To define the information theory quantities we first consider a **discrete communication channel** which can only have a finite number of input states (possible source symbols q_μ) and output states (possible sink symbols v_μ). The symbol set is assumed to be identical for source and sink and contains the M symbols $q_1, \ldots, q_M$. The mean information content of a source symbol is quantitatively determined by the source entropy:

$$H_q = \sum_{\mu=1}^{M} P(q_\nu = q_\mu)\log_2\left\{\frac{1}{P(q_\nu = q_\mu)}\right\} \tag{7.80}$$

In contrast with the general definition (2.3), statistically independent source symbols are assumed here. By analogy, the entropy of the sink is given by

$$H_v = \sum_{\mu=1}^{M} P(v_\nu = v_\mu)\log_2\left\{\frac{1}{P(v_\nu = v_\mu)}\right\} \tag{7.81}$$

With the M^2 different transition probabilities of the digital channel

$$P(v_m \mid q_\mu) = P(v_\nu = v_m \mid q_\nu = q_\mu) \qquad \begin{matrix} m = 1 \ldots M \\ \mu = 1 \ldots M \end{matrix} \tag{7.82}$$

the **irrelevance** (conditional entropy) $H_{v|q}$ can be defined as follows:

$$H_{v|q} = \sum_{\mu=1}^{M} P(q_\mu) \sum_{m=1}^{M} P(v_m \mid q_\mu)\log_2\left\{\frac{1}{P(v_m \mid q_\mu)}\right\} \tag{7.83}$$

The irrelevance is the mean conditional information content of the sink symbols v_ν if it is assumed that the source symbols are known. With the additive noise considered here the irrelevance is equal to the entropy (i.e. to the mean information content) of the noise signal $n(t)$. From this we can use eqn (7.81) to calculate the **synentropy**, i.e. the mean transinformation content of a symbol:

$$H_T = H_v - H_{v|q} \tag{7.84}$$

Hence H_T is the mean information content of a sink symbol reduced by the information content simulated by the noise. Dividing this value by the symbol duration T, we obtain the **mean transinformation rate** ϕ_T. The maximum value of this is the **channel capacity** introduced by Shannon which is to be maximized across all free parameters:

$$C = \max \phi_T = \max\left(\frac{H_T}{T}\right) \tag{7.85}$$

Hence the channel capacity C defines the maximum mean transinformation content which can be transmitted through the channel considered per unit time. Shannon has shown that an information rate ϕ emitted by a source can be transmitted over this channel without error, i.e. with a bit error probability p_B of zero, provided that $\phi \leqslant C$. Thus the channel capacity for a nonredundant source corresponds to the maximum bit rate $R_{max} = \phi_{max}$ for which error-free transmission is still possible. Optimum coding is assumed; the realization of this form of coding, which can be arbitrarily complicated, is not discussed here.

Example A binary symmetric channel ($M = 2$) is characterized by the following probabilities:

$$P(v_1 \mid q_2) = P(v_2 \mid q_1) = p$$

$$P(v_1 \mid q_1) = P(v_2 \mid q_2) = 1 - p$$

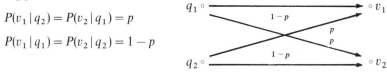

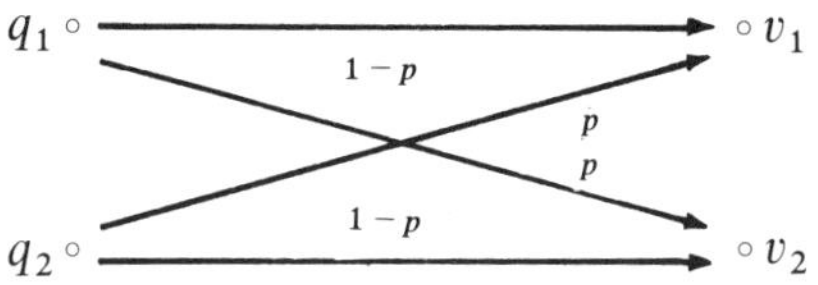

If the probabilities of occurrence of the source symbols are $p_1 = P(q_v = q_1)$ and $p_2 = P(q_v = q_2) = 1 - p_1$ the probabilities of occurrence at the receiver are

$$P(v_v = v_1) = P(q_v = q_1)P(v_1 \mid q_1) + P(q_v = q_2)\, P(v_1 \mid q_2) = p_1(1-p) + p_2 p \tag{7.86a}$$

$$P(v_v = v_2) = P(q_v = q_1)P(v_2 \mid q_1) + P(q_v = q_2)P(v_2 \mid q_2) = p_1 p + p_2(1-p) \tag{7.86b}$$

Using eqn (7.81) it follows from this that the mean information content of a sink symbol is given by

$$\begin{aligned} H_v = -\{p_1(1-p) + p_2 p\} \log_2\{p_1(1-p) + p_2 p\} - \\ \{p_1 p + p_2(1-p)\} \log_2\{p_1 p + p_2(1-p)\} \end{aligned} \tag{7.87}$$

In the case of a symmetric channel the irrelevance defined in (7.83) depends only on the noise characteristics and not on the source symbol probabilities p_1 and p_2:

$$\begin{aligned} H_{v|q} &= p_1[-P(v_1 \mid q_1)\log_2\{P(v_1 \mid q_1)\} - P(v_2 \mid q_1)\log_2\{P(v_2 \mid q_1)\}] \\ &\quad + p_2[-P(v_1 \mid q_2)\log_2\{P(v_1 \mid q_2)\} - P(v_2 \mid q_2)\log_2\{P(v_2 \mid q_2)\}] \\ &= p\log_2\left(\frac{1}{p}\right) + (1-p)\log_2\left(\frac{1}{1-p}\right) \end{aligned} \tag{7.88}$$

Therefore to maximize the mean transinformation content H_T it is sufficient to maximize the sink entropy H_v, as this is the only parameter that depends on the optimizable statistical characteristics of the source. By partial differential we obtain the optimum source symbol probabilities $\mathring{p}_1 = \mathring{p}_2 = 1/2$ and the maximum sink entropy $H_{v,\max} = 1$ bit. It then follows that the channel capacity of the binary symmetric channel is

$$C = \frac{1}{T}\left\{1 - p\log_2\left(\frac{1}{p}\right) - (1-p)\log_2\left(\frac{1}{1-p}\right)\right\} \tag{7.89}$$

7.4.2 Channel capacity of a continuous-value channel

The model shown in Fig. 7.8 is the basis of the calculation of the channel capacity of a continuous-value channel. The source is nonredundant, so that the information rate ϕ and the bit rate R are the same. The bit rate is increased to $R_c \geqslant R$ by means of a suitable coding, the realization of which is not discussed here and which can be arbitrarily complicated. However, the information rate ϕ to be transmitted remains unchanged. The attenuation-free channel considered is ideally band limited ("brick-wall channel"; see Table A3 in the Appendix):

$$H_K(f) = \begin{cases} 1 & |f| < B \\ 1/2 & |f| = B \\ 0 & |f| > B \end{cases} \tag{7.90}$$

where B is the one-sided bandwidth of the channel. According to the sampling theorem, a continuous information flow of mutually independent symbols is only possible over this channel if the symbol rate is no greater than twice the channel bandwidth. Hence the maximum symbol rate is

$$1/T = 2B \tag{7.91}$$

In the following only the discrete detection times $t_v = vT$ are considered, so that the continuous-time signals $s(t)$, $n(t)$ and $e(t)$ can be replaced by their sampled values

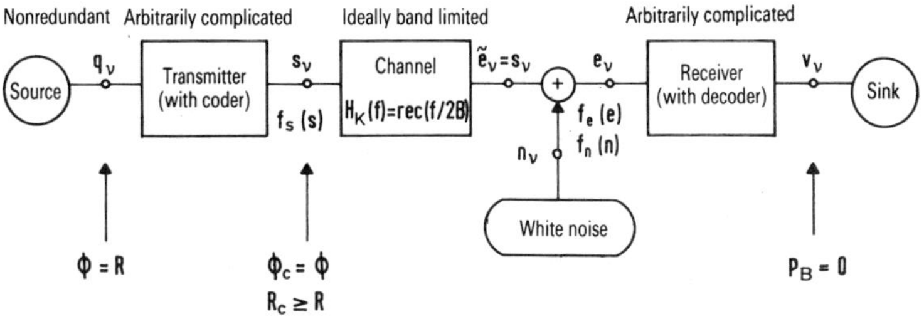

Fig. 7.8 Model for the calculation of the channel capacity of a frequency-independent ideally band-limited channel with white noise ($L_n(f) = L_0$).

s_ν, n_ν and e_ν. If eqn (7.91) is satisfied, the sampled values of the transmitted signal are not altered by transmission across the "brick-wall channel", and we obtain for the sample values of the received signal

$$e_\nu = s_\nu + n_\nu \tag{7.92}$$

By analogy with eqns (7.84) and (7.85) the capacity of the continuous-value discrete-time channel is given by

$$C = \max\left\{\frac{1}{T}(H_e - H_{e|s})\right\} \tag{7.93}$$

where H_e is the entropy of the continuous received signal $e(t)$ and of its sample values e_ν. The entropy can be calculated in the same way as for the discrete channel (see eqn (7.81)) by dividing the complete range of values of the received signal into small amplitude increments of width Δe:

$$H_e = \lim_{\Delta e \to 0}\left[\sum_{\mu=-\infty}^{+\infty} P\left(e_\nu = e_\mu \pm \frac{\Delta e}{2}\right)\log_2\left\{\frac{1}{P(e_\nu = e_\mu \pm \Delta e/2)}\right\}\right] \tag{7.94}$$

We now make Δe sufficiently small to enable the following approximation to be introduced:

$$P\left(e_\nu = e_\mu \pm \frac{\Delta e}{2}\right) = \int_{e_\mu - \Delta e/2}^{e_\mu + \Delta e/2} f_e(e)\,\mathrm{d}e \approx \Delta e f_e(e_\mu) \tag{7.95}$$

When this probability is inserted into eqn (7.94) and the limit $\Delta e \to 0$ is approached, the summation changes to an integration and Δe changes to $\mathrm{d}e$ and we obtain for the entropy of the received signal with the probability density function (PDF) $f_e(e)$

$$H_e = \int_{-\infty}^{+\infty} f_e(e)\log_2\left\{\frac{1}{f_e(e)}\right\}\mathrm{d}e + \lim_{\mathrm{d}e \to 0}\log_2\left(\frac{1}{\mathrm{d}e}\right) \tag{7.96}$$

The second term tends to infinity. However, the same expression occurs in the irrelevance $H_{e|s}$ so that it no longer needs to be considered when calculating the channel capacity by means of eqn (7.93).

When additive noise is present (eqn (7.92)) $H_{e|s}$ is equal to the entropy H_n of the noise signal, which by analogy with eqn (7.96) can be calculated as follows:

$$H_{e|s} = H_n = \int_{-\infty}^{+\infty} f_n(n)\log_2\left\{\frac{1}{f_n(n)}\right\}\mathrm{d}n + \lim_{\mathrm{d}n \to 0}\log_2\left(\frac{1}{\mathrm{d}n}\right) \tag{7.97}$$

If we allow $\mathrm{d}e$ and $\mathrm{d}n$ to approach zero (simultaneously), we obtain the mean transinformation content as defined in eqn (7.84):

$$H_\mathrm{T} = H_e - H_n = \int_{-\infty}^{+\infty} f_e(e)\log_2\left\{\frac{1}{f_e(e)}\right\}\mathrm{d}e - \int_{-\infty}^{+\infty} f_n(n)\log_2\left\{\frac{1}{f_n(n)}\right\}\mathrm{d}n \tag{7.98}$$

The only assumption made in this equation is that the sample values s_ν and n_ν of the transmitted signal and the noise signal are both mutually and reciprocally uncorrelated. Further restrictions are not necessary. With Gaussian-distributed noise the PDF $f_n(n)$ is given by eqn (2.46). The spectral components which do not fall into the transmitting range of the channel can be suppressed by a receiver filter and hence do not need to be considered. Therefore, assuming a constant power density L_0 (white noise), we obtain for the power of the noise signal limited to the band $\pm B$

$$N_B = 2BL_0 \tag{7.99}$$

Similarly, the sample values of the band-limited noise signal are mutually uncorrelated because the power spectrum has zeros at multiples of $\pm 2B$. Hence we obtain for the entropy of a noise signal of Gaussian distribution ideally band limited to $\pm B$ and with constant power density L_0 (see eqn (7.97))

$$\begin{aligned} H_n &= \frac{1}{(2\pi N_B)^{1/2}} \int_{-\infty}^{+\infty} \exp\left(-\frac{n^2}{2N_B}\right)\left[\log_2\{(2\pi N_B)^{1/2}\} + \frac{n^2}{2N_B \ln 2}\right] \mathrm{d}n \\ &= \tfrac{1}{2}\log_2(2\pi e N_B) \end{aligned} \tag{7.100}$$

Note Here and in the following, the entropy of a continuous signal $n(t)$ means only the first integral which has a finite value. The second term, $\lim \log_2(1/\mathrm{d}n)$, of eqn (7.97) is of no significance in the following and is therefore not considered further.

Channel capacity under power limitation As the sample values of the transmitted signal and the noise signal are uncorrelated both mutually and reciprocally, we have for the PDF of the received signal $e(t) = s(t) + n(t)$

$$f_e(e) = f_s(s) * f_n(n) = \int_{-\infty}^{+\infty} f_s(s) f_n(e-s)\,\mathrm{d}s \tag{7.101}$$

To calculate the channel capacity defined by eqn (7.85) the PDF $f_s(s)$ of the transmitted signal which gives the maximum transinformation rate must be calculated. With power limitation the following condition must be satisfied:

$$\int_{-\infty}^{+\infty} s^2 f_s(s)\,\mathrm{d}s \leqslant S_s \tag{7.102}$$

Shannon and Weaver [7.8] have shown that, in this special case with Gaussian-distributed noise, a transmitter signal which also has a Gaussian PDF (mean value zero, power S_s)

$$f_s(s) = \frac{1}{(2\pi S_s)^{1/2}} \exp\left(-\frac{s^2}{2S_s}\right) \tag{7.103}$$

leads to the maximum mean transinformation rate $\phi_{\mathrm{T,max}}$. Thus the received signal $e(t)$ also possesses a Gaussian PDF in which the power is $S_s + N_B$ (see eqn (7.99)). The entropy of the received signal can be calculated by analogy with eqn (7.100):

$$H_e = \tfrac{1}{2}\log_2\{2\pi e(S_s + N_B)\} \tag{7.104}$$

Hence with eqns (7.98) and (7.100) we obtain the channel capacity (maximum mean transinformation rate ϕ_T) of a continuous-valued discrete-time channel with additive white Gaussian noise (AWGN):

$$C_L = \max \phi_T = \frac{1}{2T} \log_2\left(1 + \frac{S_s}{N_B}\right) \tag{7.105}$$

The subscript L indicates that this result is only valid when power limitation is assumed. From eqns (7.91) and (7.99) the following is also true:

$$C_L = B \log_2\left(1 + \frac{S_s}{N_B}\right) = B \log_2\left(1 + \frac{S_s}{2BL_0}\right) \tag{7.106}$$

Therefore the channel capacity C_L depends only on the available transmitter power S_s, the noise power density L_0 (assuming white noise) and the bandwidth B. Figure 7.9 shows that, for $B \to \infty$, C_L tends towards a fixed limit:

$$C_\infty = \lim_{B \to \infty} C_L = \frac{1}{2 \ln 2} \frac{S_s}{L_0} \approx 0.721 \frac{S_s}{L_0} \tag{7.107}$$

It should be noted that this equation and the corresponding figure (Fig. 7.9) are only valid for a frequency-independent channel for which a very large optimum bandwidth results. This is not the case with real channels (see Section 7.4.3).

Shannon and Weaver [7.8] have further shown that, for a channel with channel capacity C_L, an infinitely long sequence of symbols with a Gaussian PDF $f_s(s)$ can be coded in such a way that the symbols can be transmitted error free with bit rate $R \leqslant C_L$. With realizable, and hence finite, coding the error rate can be signficantly reduced as long as $R \leqslant C_L$. However, an error-free transmission is not possible in principle.

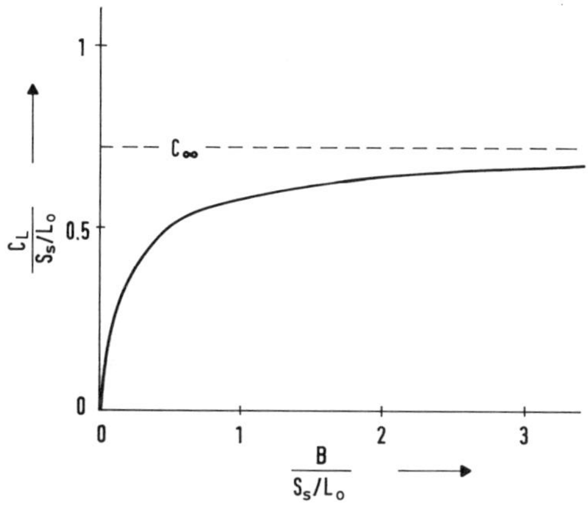

Fig. 7.9 Channel capacity (under power limitation) as a function of the one-sided channel bandwidth B.

If the transmitted signal $s(t)$ is a Gaussian-distributed analog signal and the channel is frequency independent, the bit rate R can be increased to the upper limit C_L without information loss. However, if the PDF $f_s(s)$ differs from the optimum Gaussian PDF, the maximum (error-free transmittable) bit rate is lower than the channel capacity C_L. In the following the maximum bit rate R_{max} will be determined for an M-level bipolar digital signal with rectangular pulses and equiprobable symbols. In this case the PDF $f_s(s)$ is a summation of the Dirac functions:

$$f_s(s) = \sum_{\mu=1}^{M} \frac{1}{M}\delta\left(s - \frac{2\mu - M - 1}{M - 1}\hat{g}_s\right) \tag{7.108}$$

Because of the power limitation the transmitter pulse amplitude $\hat{g}_s$ depends on the level number M (see eqn (7.33)):

$$\hat{g}_s = \left\{\frac{3(M-1)}{M+1}S_s\right\}^{1/2} \tag{7.109}$$

By substituting this PDF in eqns (7.101), (7.96) and (7.98) we obtain the mean transinformation content H_T of a symbol, from which the maximum transmittable bit rate $R_{max} = H_T/T$ of an M-level rectangular signal can be determined. R_{max} is plotted in Fig. 7.10(a) as a function of the signal-to-noise ratio $10\lg(S_s/N_B)$. Clearly at low signal-to-noise ratios the level number M has only a small effect on the maximum transmittable bit rate. At higher signal-to-noise ratios R_{max} approaches $(\log_2 M)/T$.

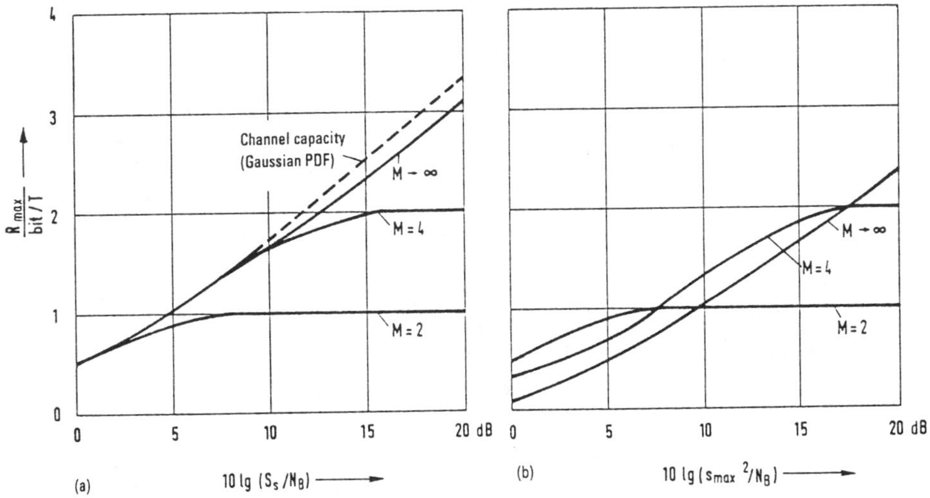

Fig. 7.10 Maximum transmittable bit rate in M-level digital signal transmission with NRZ rectangular pulses as a function of the signal-to-noise ratio at the channel output: (a) power limitation; (b) peak-value limitation.

The curve labeled $M \to \infty$ applies to an analog signal uniformly distributed between $+(3S_s)^{1/2}$ and $-(3S_s)^{1/2}$. In this special case the following approximation can be made for $S_s \gg 2BL_0$ [7.8]:

$$R_{\max} = \frac{1}{2T}\log_2\left(\frac{6}{\pi e}\frac{S_s}{N_B}\right) \tag{7.110}$$

Comparison of this result with that for a transmitted signal with Gaussian PDF (the broken curve in Fig. 7.10(a)), which leads to the maximum bit rate $R_{\max} = C_{\mathrm{L}}$, shows that a transmitter power which is a factor $\pi e/6$ (= 1.53 dB) higher has to be used here in order to transmit the same information (see eqn (7.106)).

Channel capacity under peak-value limitation In the case of power limitation the optimum transmitter signal is Gaussian and therefore can have arbitrarily high amplitude values (see eqn (7.103)). Under peak-value limitation this PDF is inappropriate.

The calculation of the channel capacity under the conditions of peak-value limitation and Gaussian noise is not possible analytically. Here the PDF

$$f_s(s) = \sum_{\mu=1}^{M} p_\mu \delta(s - s_\mu) \quad \text{with} \quad |s_\mu| \leqslant s_{\max} \tag{7.111}$$

has to be approximated by summing the Dirac functions, where the sum over all p_μ must produce the value unity. Hence, the PDF of a peak-value-limited transmitted signal can be approximated by letting M tend to the limit $M \to \infty$ and varying the probabilities p_μ. Therefore from eqn (7.101) we obtain the PDF of the received signal $e(t) = s(t) + n(t)$:

$$f_e(e) = \sum_{\mu=1}^{M} p_\mu f_n(e - s_\mu) \tag{7.112}$$

By inserting this PDF in eqn (7.98) and varying the probabilities p_μ the channel capacity C_{A} under peak-value limitation can be determined. Faerber [7.3] and Faerber and Appel [7.4] have shown that, in the case of peak-value limitation and for a frequency-independent channel (with uncorrelated noise), discrete probabilities lead to the maximum transinformation rate and hence to the channel capacity.

Figure 7.10(b) shows the maximum transmittable bit rate $R_{\max}$ for the condition that the modulation range of the transmitted signal is limited to $\pm s_{\max}$. The M amplitude levels are assumed to be equiprobable so that $p_\mu = 1/M$ in eqn (7.111). With a binary system $S_{\max} = s_{\max}{}^2$, so that for $M = 2$ the curves in Figs 7.10(a) and 7.10(b) are identical. However, with multilevel systems the maximum transmittable bit rate is lower if the transmitter peak value rather than the mean transmitter power is limited. In contrast with Fig. 7.10(a) (power limitation) the curves in Fig. 7.10(b) (peak-value limitation) are shifted to the right by an amount $10 \times \lg\{3(M-1)/(M+1)\}$ for a given M.

At large bandwidths (i.e. for a low signal-to-noise ratio at given noise power density) the binary signal ($M = 2$) is superior to the multilevel signals, whereas at small bandwidths (i.e. for a high signal-to-noise ratio) the uniformly distributed analog signal ($M \to \infty$) leads to the maximum transinformation rate. Hence for every bandwidth B there is an optimum level number M. Additional optimization of the symbol probabilities p_μ in accordance with ref. 7.3 only produces a slight increase in transinformation rate ϕ_T relative to the value with equiprobable symbols ($p_\mu = 1/M$) [7.1].

7.4.3 Channel capacity with frequency-dependent channel and correlated noise

If the channel frequency response $H_K(f)$ is frequency dependent or the noise sample values n_ν are mutually correlated ($L_n(f) \neq$ constant), treatment of the channel capacity proves to be more difficult than for a channel with constant frequency response and white noise. Whereas for a frequency-independent channel ideally band limited to $\pm B = 1/2T$ the received sample value $e_\nu = s_\nu + n_\nu$ and hence, with uncorrelated noise, the PDF $f_e(e) = f_s(s) * f_n(n)$ can be represented as a convolution integral, in the general case shown in Fig. 7.11(a) the following is true:

$$f_e(e) = f_{\tilde{e}}(\tilde{e}) * f_n(n) \tag{7.113}$$

The PDF $f_{\tilde{e}}(\tilde{e})$ of the useful received signal $\tilde{e}(t) = s(t) * h_K(t)$ can be determined from transmitter and channel characteristics in only a very few special cases. An example of such a case is a Gaussian transmitter signal which also results in a Gaussian PDF $f_{\tilde{e}}(\tilde{e})$ after linear filtering by $H_K(f)$.

Shannon has shown that a transmitter signal with Gaussian PDF as defined in eqn (7.103) is also the optimum with a frequency-dependent channel if the mean transmitter power is limited (power limitation). However, in contrast with Section

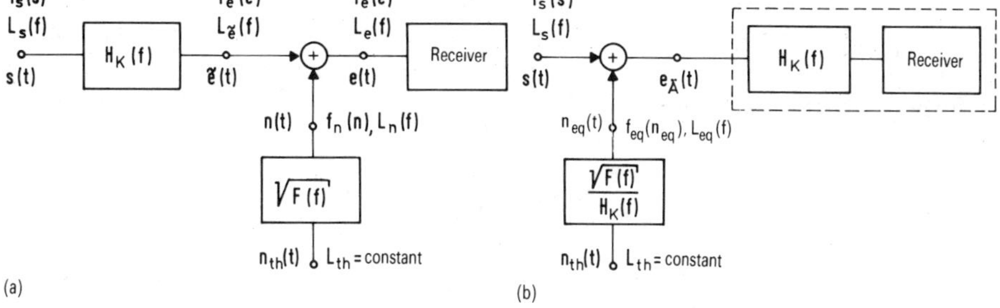

Fig. 7.11 Two models for calculating the capacity of a frequency-dependent channel with correlated noise.

7.4.2, for a frequency-dependent channel (or with correlated noise) the spectral distribution of the transmitter power, i.e. the spectral power density $L_s(f)$, also has to be optimized. For calculation of the channel capacity it is necessary to rearrange the block diagram (Fig. 7.11(a)). The noise-adding point is moved to the left-hand side of the channel frequency response $H_K(f)$, i.e. the noise is computed at the transmitter, and the block diagram shown in Fig. 7.11(b) is obtained. $n_{eq}(t)$ is referred to below as the **equivalent noise signal**. Its power spectrum is given by

$$L_{eq}(f) = \frac{L_n(f)}{|H_K(f)|^2} = \frac{F(f)}{|H_K(f)|^2} L_{th} \tag{7.114}$$

If the noise signal $n(t)$ has a Gaussian distribution, then so also does the noise signal $n_{eq}(t)$ filtered by $1/H_K(f)$. The noise power in the frequency range $-B$ to $+B$ is given by

$$N_B = \int_{-B}^{+B} L_{eq}(f)\,df = \int_{-B}^{+B} \frac{L_n(f)}{|H_K(f)|^2}\,df \tag{7.115}$$

This modification of the block diagram is only possible if the channel frequency $H_K(f)$ exhibits no zeros in the frequency range of interest. This is assumed to be so in the following. Now the effect of the channel on the useful signal does not have to be considered further in the calculation of the channel capacity. This effect can again be compensated for in the receiver without information loss (see Section 6.2.1). Therefore we can combine the channel $H_K(f)$ and the receiver in a new receiver for which the input signal is

$$e_{eq}(t) = s(t) + n_{eq}(t) \tag{7.116}$$

In the following treatment, therefore, it is sufficient to consider the channel capacity of a frequency-independent and attenuation-free channel in which the transmission is impaired by Gaussian coloured noise with power spectrum $L_{eq}(f)$. It was shown in Section 7.4.2 that, with power limitation and a frequency-independent channel, ideally band limited to $\pm 1/2T$, with uncorrelated noise, a transmitter signal with Gaussian PDF $f_s(s)$ leads to the maximum transinformation rate $\phi_{T,max} = H_{T,max}/T$. If we consider such a transmitter signal in a narrow frequency range (from f_κ to $f_\kappa + \Delta f$), the equivalent noise signal $n_{eq}(t)$ has an approximately frequency-independent power spectrum $L_{eq}(f_\kappa)$. Both the noise $n_{eq}(t)$ and the useful signal $s(t)$ have a Gaussian PDF so that in the narrow frequency range Δf the trans-information rate is a maximum and can be calculated by analogy with eqn (7.106) as follows:

$$\Delta\phi_\kappa = \frac{\Delta f}{2} \log_2\left\{1 + \frac{\Delta f\, L_s(f_\kappa)}{\Delta f\, L_{eq}(f_\kappa)}\right\} \tag{7.117}$$

Here $\Delta f\, L_s(f_\kappa)$ is the useful signal power in the frequency range $f_\kappa, \ldots, f_\kappa + \Delta f$ and $\Delta f\, L_{eq}(f_\kappa)$ is the corresponding noise power. Summing over all κ and approaching the limit $\Delta f \to 0$, we obtain the total transinformation rate

$$\phi = \frac{1}{2} \lim_{\Delta f \to 0} \sum_{\kappa=-\infty}^{\infty} \log_2\left\{1 + \frac{L_s(f_\kappa)}{L_{eq}(f_\kappa)}\right\}\Delta f = \frac{1}{2}\int_{-\infty}^{+\infty} \log_2\left\{1 + \frac{L_s(f)}{L_{eq}(f)}\right\} df \tag{7.118}$$

To determine the channel capacity $C_L = \phi_{T,max}$ we now have to calculate the transmitter power spectrum $\mathring{L}_s(f)$ for which the transinformation rate ϕ_T is a maximum. It is additionally required as a co-condition of the optimization that the mean transmitter power should not exceed the predefined value S_s (power limitation):

$$\int_{-\infty}^{+\infty} L_s(f)\, df \leqslant S_s \tag{7.119}$$

Shannon and Weaver [7.8] have demonstrated that the following is true for the optimum transmitter power spectrum:

$$\mathring{L}_s(f) = \begin{cases} L_{max} - L_{eq}(f) & \text{when } L_{eq}(f) \leqslant L_{max} \\ 0 & \text{when } L_{eq}(f) > L_{max} \end{cases} \tag{7.120}$$

This means that the sum of the power spectra of the transmitted signal and the equivalent noise signal must be constant:

$$\mathring{L}_s(f) + L_{eq}(f) = L_{max} \qquad \text{when } L_{eq}(f) \leqslant L_{max} \tag{7.121}$$

Substituting this result in eqn (7.118) we obtain the channel capacity of a frequency-independent channel $H_K(f)$ with coloured noise $L_n(f)$ under the co-condition of power limitation:

$$C_L = \phi_{max} = \int_0^{+\infty} \log_2\left\{1 + \frac{\mathring{L}_s(f)|H_K(f)|^2}{L_n(f)}\right\} df \tag{7.122}$$

The difficulty of numerical evaluation of this equation lies in the calculation of L_{max}. The following has to be true because of eqns (7.119) and (7.120):

$$\int_{-\infty}^{+\infty} \mathring{L}_s(f)\, df = \int_{L_{eq}(f) \leqslant L_{max}} \{L_{max} - L_{eq}(f)\}\, df = S_s \tag{7.123}$$

Figure 7.12(a) shows the optimum transmitter power spectrum $\mathring{L}_s(f)$ for the equivalent noise power spectrum $L_{eq}(f) = L_n(f)/|H_K(f)|^2$. The value of L_{max} was determined numerically so that the hatched area corresponds to the mean transmitter power S_s.

Example The channel capacity of a coaxial pair of length l is calculated. With thermal noise of noise factor F eqns (2.53) and (7.114) give

$$L_{eq}(f) = \frac{FL_{th}}{|H_K(f)|^2} = FL_{th} \exp(2\alpha_2 |f|^{1/2} l) \tag{7.124}$$

In this special case the equivalent noise power spectrum $L_{eq}(f)$ for positive frequencies is a monotonically increasing function (see Fig. 7.12(b)). According to

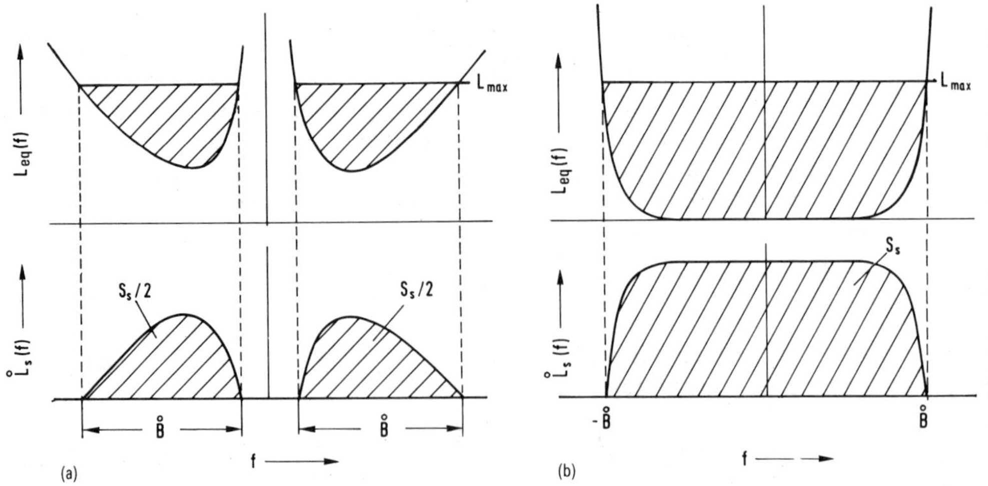

Fig. 7.12 Illustration of the calculation of the channel capacity of a frequency-dependent channel with correlated noise: (a) general case; (b) coaxial pair.

eqn (7.123) the frequency range in which the optimum transmitter power spectrum $\mathring{L}_s(f)$ lies is correspondingly spread from $-\mathring{B}$ to $+\mathring{B}$. Thus the optimum bandwidth $\mathring{B}$ and the value of $L_{\max} = L_{\text{eq}}(\mathring{B})$ can be analytically determined from the defined transmitter power S_s which corresponds to the hatched area in Fig. 7.12(b):

$$S_s = 2\mathring{B}L_{\text{eq}}(\mathring{B}) - \int_{-\mathring{B}}^{\mathring{B}} L_{\text{eq}}(f)\,\mathrm{d}f \tag{7.125}$$

Using eqn (7.124) and the approximation $\alpha_2 l B \gg 1$, we obtain from this [7.1]

$$\begin{aligned} S_s &= 2FL_{\text{th}}\mathring{B}\exp(2\alpha_2 l\mathring{B}^{1/2})\left(1 - \frac{1}{\alpha_2 l\mathring{B}^{1/2}}\right) \\ &\approx 2FL_{\text{th}}\mathring{B}\exp(2\alpha_2 l\mathring{B}^{1/2}) \end{aligned} \tag{7.126}$$

Hence the channel capacity emerges from eqns (7.120), (7.122) and (7.124):

$$\begin{aligned} C_{\mathrm{L}} &= \int_0^{\infty} \log_2\left\{1 + \frac{\mathring{L}_s(f)}{L_{\text{eq}}(f)}\right\}\mathrm{d}f \\ &= \int_0^{\mathring{B}} \log_2\left\{\frac{L_{\max}}{L_{\text{eq}}(f)}\right\}\mathrm{d}f = \frac{2\alpha_2 l}{3\ln 2}\mathring{B}^{3/2} \end{aligned} \tag{7.127}$$

Solving this equation for the optimum bandwidth and inserting $\mathring{B}$ from eqn (7.126) we obtain

$$S_s = 2FL_{\text{th}}\left(\frac{3\ln 2}{2\alpha_2 l}C_{\mathrm{L}}\right)^{2/3} \exp\{(2\alpha_2 l)^{2/3}(3\ln 2C_{\mathrm{L}})^{1/3}\} \tag{7.128}$$

After some nontrivial manipulation we obtain from this, using the approximation (7.75) [7.1],

$$\lg\left(\frac{C_L}{\text{Mbit s}^{-1}}\right) \approx \left(1-\frac{2}{K}\right)\lg K^3 - 2\left(1-\frac{3}{K}\right)\lg(2\alpha_2 l\,\text{MHz}^{1/2}) - \lg(3\ln 2) \quad (7.129)$$

where the constant K is a measure of the transmitter power used,

$$K = \ln\left(\frac{S_s}{2FL_{\text{th}}\,\text{MHz}}\right) \quad (7.130)$$

For cable lengths between 1 and 20 km the following is a further approximation producing a relative error of less than 10%:

$$\left(\frac{C_L}{\text{Mbit s}^{-1}}\right) \approx \left\{\frac{K^3}{(3\ln 2)(2\alpha_2 l)^2\,\text{MHz}}\right\}^{1-2/K} \quad (7.131)$$

Equations (7.129) and (7.131) are valid only under power limitation. For peak-value limitation and a frequency-dependent channel the calculation of channel capacity is more involved and leads to a somewhat smaller value [7.1]. It is shown in ref. 7.9 that an approximate value for the channel capacity C_A for a coaxial pair can be calculated similarly from eqns (7.129)–(7.131) if the value $2s_0^2/\pi e$ is substituted for S_s. Because attenuation increases with frequency we obtain a relatively small value for the optimum bandwidth $\mathring{B}$, and so the optimum transmitter signal with peak-value limitation is distributed equally between $-s_0$ and $+s_0$ (see the explanation of Fig. 7.10(b)). In a frequency-dependent channel this signal also results in the same mean transinformation rate as a Gaussian signal with power reduced by a factor of $\pi e/6$. Considering further that in the case of the uniformly distributed signal $S_s = s_0^2/3$, we obtain for the constant K with peak-value limitation

$$K = \ln\left(\frac{s_0^2}{\pi e F L_{\text{th}}\,\text{MHz}}\right) \quad (7.132)$$

Chapter 8

Optimization and Comparison of Digital Transmission Systems

Contents In this chapter the optimization of system parameters is explained by reference to examples in which the system efficiencies defined in Chapter 7 are used as criteria of quality. The optimization of transmitter and receiver parameters is described in Sections 8.1 and 8.2. Receiver optimization in systems without intersymbol interference (Nyquist systems) is discussed in Section 8.3. The mutual optimization of the transmitter and receiver, which can be performed analytically for systems without intersymbol interference, is treated in Sections 8.4 and 8.5. The variants of optimized systems are compared in Section 8.6, and finally some methods of examining the effects of tolerances are discussed in Section 8.7.

Assumptions The assumptions of Chapter 7 are maintained, i.e. the results defined below are valid only for additive Gaussian noise. The receiver uses symbol-wise detection and in addition can have decision feedback equalization (DFE). Except in Section 8.7 the system parameters (e.g. cut-off frequency, decision value and detection time) are assumed to be free from tolerances.

8.1 Optimization of the receiver parameters

One of the most important optimizable system parameters is the equalizer frequency response $H_E(f)$ or the pulse-shaper frequency response $H_I(f) = H_K(f)H_E(f)$. In the following some characteristics of the pulse shaper, e.g. the cut-off frequency and the roll-off factor, are optimized as functions of the other system parameters. The structure of the pulse-shaper frequency response remains unchanged. Hence the following assumptions are made: (a) the decision values are optimum ($\mathring{E}_1,\ldots,\mathring{E}_{M-1}$) in the centers of the eyes; (b) the detection time is optimum ($\mathring{T}_D$) at the maximum eye opening; (c) the receiver has no DFE or ideal DFE. In addition, the numerical results are based on a bipolar transmitter signal with NRZ rectangular pulses, a coaxial pair and white noise.

8.1.1 Nonredundant systems with a Gaussian pulse shaper

In many cases the pulse shaper can be approximated by a Gaussian low-pass filter, so that the pulse-shaper frequency response is given by (see the Appendix, Table A3)

$$H_I(f) = H_K(f)H_E(f) = \exp\left\{-\pi\left(\frac{f}{2f_I}\right)^2\right\} \tag{8.1}$$

The direct signal transmission factor $H_I(0)$ is assumed here to be unity. The only optimization parameter of this pulse shaper is its cut-off frequency f_I. The slope of the filter around the cut-off frequency is relatively moderate, and so this is referred to as "soft band limitation".

Eye opening for a system without decision feedback equalization If rectangular transmitter pulses are used, the basic detection pulse $g_d(t)$ is given by the response to a rectangular pulse defined in eqn (2.78) and Fig. 2.11(b). First, considering bipolar nonredundant M-level systems without DFE, we have for half the vertical eye opening (see eqns (3.11) and (3.51))

$$\frac{o(T_D)}{2} = \frac{|g_d(T_D)|}{M-1} - \sum_{\nu=1}^{\infty} |g_d(T_D - \nu T)| - \sum_{\nu=1}^{\infty} |g_d(T_D + \nu T)| \tag{8.2}$$

Absolute values need not be explicitly defined because the basic detection pulse considered here is nonnegative. After expansion of $\pm g_d(T_D)$ we obtain

$$\frac{o(T_D)}{2} = \frac{M}{M-1} g_d(T_D) - \sum_{\nu=-\infty}^{+\infty} g_d(T_D - \nu T) \tag{8.3}$$

If the basic transmitter pulse is an NRZ rectangular pulse ($T_s = T, \hat{g}_s = s_{max}$), which is assumed from now on, the infinite summation can be calculated as a direct-signal value and this yields the constant value $s_{max}H_I(0)$ independently of the detection time T_D. It follows from this that, for $H_I(0) = 1$,

$$\frac{o(T_D)}{2} = \frac{M}{M-1} g_d(T_D) - s_{max} \tag{8.4}$$

In a system without DFE the optimum detection time $\mathring{T}_D$ is zero. If we take

$$g_d(0) = \hat{g}_s[1 - 2Q\{(2\pi)^{1/2} f_I T\}] \tag{8.5}$$

according to eqn (2.79) and the relation $RT = \log_2 M$, which is valid for nonredundant systems, we obtain for the maximum half-eye opening

$$\frac{o(\mathring{T}_D = 0)}{2} = \frac{s_{max}}{M-1}\left[1 - 2MQ\left\{(2\pi)^{1/2} \log_2 M \frac{f_I}{R}\right\}\right] \tag{8.6}$$

$Q(x)$ is the complementary Gaussian error integral defined in the Appendix, Table

A2. From this the normalized eye opening according to eqn (3.25) is given by

$$o_{\mathrm{norm}}(\mathring{T}_{\mathrm{D}}=0)=\frac{o(\mathring{T}_{\mathrm{D}}=0)}{2s_{\max}H_{\mathrm{I}}(0)}=\frac{1}{M-1}\left[1-2MQ\left\{(2\pi)^{1/2}\log_2 M\frac{f_{\mathrm{I}}}{R}\right\}\right] \tag{8.7}$$

Equation (8.6) is valid for bipolar signals. The eye opening for unipolar signals is only half as large. However, because of the appropriate normalization, the right-hand side of eqn (8.7) is valid for both bipolar and unipolar signals.

Figure 8.1(a) shows the normalized eye opening plotted against the cut-off

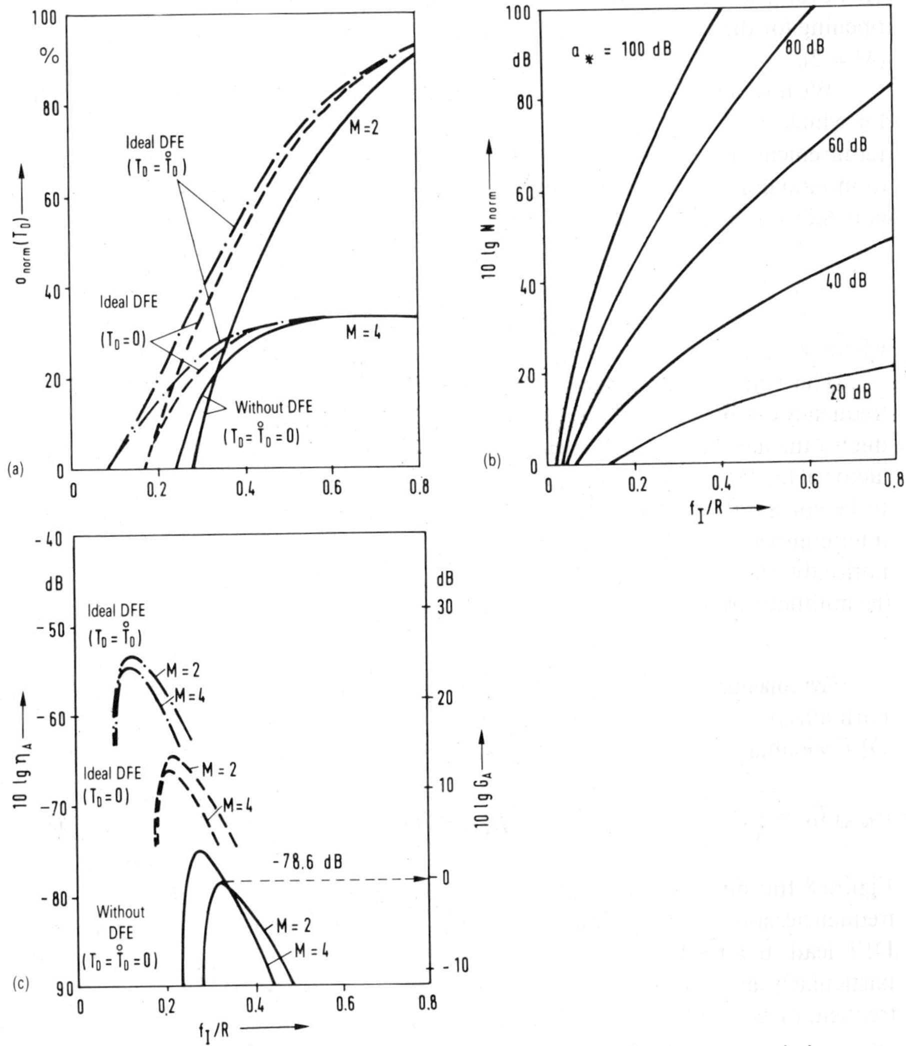

Fig. 8.1 Optimization of the cut-off frequency f_{I} of a nonredundant transmission system with a Gaussian pulse shaper: (a) normalized eye opening; (b) normalized detection noise power; (c) logarithmic system efficiency ($a_* = 80\,\mathrm{dB}$, $F(f) = 1$).

frequency f_{I} of the Gaussian low-pass filter for level numbers $M=2$ and $M=4$. The full curves are for systems without DFE. This figure shows that the binary system has a larger eye opening than the multilevel system as long as the cut-off frequency is sufficiently high. In the limiting case $f_{\mathrm{I}} \to \infty$ the intersymbol interference disappears so that the normalized eye opening for an M-level system reaches a maximum value

$$o_{\mathrm{norm}}(\mathring{T}_{\mathrm{D}}=0)|_{f_{\mathrm{I}}\to\infty} = \frac{1}{M-1} \tag{8.8}$$

However, if the cut-off frequency f_{I} of the pulse shaper is less than $0.35R$ the eye opening for the quaternary system ($M=4$) is larger than that for the binary system ($M=2$).

We now define the **minimum cut-off frequency** $\check{f}_{\mathrm{I}}$ of a Gaussian-type pulse shaper for which the eye is just opened as the characteristic parameter for the bandwidth requirement of individual codes. The minimum cut-off frequency for an M-level (nonredundant) transmission system without DFE can be calculated directly from eqn (8.7). It follows from the condition $o_{\mathrm{norm}}(\mathring{T}_{\mathrm{D}}=0)=0$ that

$$\check{f}_{\mathrm{I}} = \frac{1}{(2\pi)^{1/2}}\frac{R}{\log_2 M} Q^{-1}\left(\frac{1}{2M}\right) \tag{8.9}$$

where $Q^{-1}(x)$ is the inverse function of $Q(x)$ (see the Appendix, Table A2).

The values of $\check{f}_{\mathrm{I}}$ in Table 8.1, column 2, make it clear that a lower cut-off frequency can be chosen for the pulse shaper with increasing level number M. This is due to the fact that with a nonredundant M-level system the symbol rate $1/T$ is a factor of $\log_2 M$ smaller than that for the corresponding binary system (R is assumed to be constant), and hence the precursors and postcursors cause less intersymbol interference than in binary transmission. At low cut-off frequency f_{I} this effect is more noticeable than the loss due to the smaller eye opening (by a factor of $1/(M-1)$) of the multilevel systems.

Eye opening for a system with ideal decision feedback equalization For the normalized eye opening of a nonredundant M-level transmission system with ideal DFE we obtain from eqns (2.78), (3.25) and (5.12)

$$o_{\mathrm{norm}}(T_{\mathrm{D}}) = \frac{1}{M-1}\left[1 - MQ\left\{2(2\pi)^{1/2} f_{\mathrm{I}}\left(\frac{T}{2} - T_{\mathrm{D}}\right)\right\} - Q\left\{2(2\pi)^{1/2} f_{\mathrm{I}}\left(\frac{T}{2} + T_{\mathrm{D}}\right)\right\}\right] \tag{8.10}$$

Figure 8.1(a) shows that the vertical eye opening is increased by DFE at all cut-off frequencies and all level numbers. Even for the nonoptimum detection time $T_{\mathrm{D}}=0$ DFE leads to a marked increase in the eye opening. The gain due to (ideal) DFE is particularly large for a low pulse-shaper cut-off frequency. For the minimum cut-off frequency $\check{f}_{\mathrm{I}}$ we have for ideal DFE and $T_{\mathrm{D}}=0$

$$\check{f}_{\mathrm{I}} = \frac{1}{(2\pi)^{1/2}}\frac{R}{\log_2 M} Q^{-1}\left(\frac{1}{M+1}\right) \tag{8.11}$$

Table 8.1 Results of the optimization of systems with Gaussian pulse shapers

Code	$\check{f}_I/R$	$\mathring{f}_I/R$	$\mathring{T}_D/T$	S_s/s_{max}^2	o_{norm} (%)	N_{norm} (dB)	η_A (dB)	G_A (dB)	η_L (dB)	G_L (dB)
Without DFE ($T_D = \mathring{T}_D = 0$)										
$M = 2$[a]	0.27	0.32	0.00	1.00	15.5	62.4	−78.6	0.0	−78.6	0.0
$M = 3$[a]	0.24	0.30	0.00	0.67	12.7	58.0	−75.9	2.7	−74.1	4.5
$M = 4$[a]	0.23	0.28	0.00	0.55	11.9	56.4	−74.9	3.7	−72.3	6.3
$M = 8$[a]	0.20	0.24	0.00	0.43	6.2	50.0	−74.2	4.4	−70.5	8.1
$M = 16$[a]	0.18	0.22	0.00	0.38	3.7	46.7	−75.3	3.3	−71.1	7.5
AMI	0.36	0.40	0.00	0.50	11.3	73.6	−92.6	−14.0	−89.6	−11.0
Duo-binary	0.22	0.28	0.00	0.50	11.3	56.4	−75.3	3.3	−72.3	6.3
4B3T	0.29	0.35	0.00	0.69	13.8	66.7	−83.9	−5.3	−82.3	−3.7
With ideal DFE ($\mathring{T}_D = 0$)										
$M = 2$[a]	0.17	0.22	—	1.00	13.2	46.9	−64.5	14.1	−64.5	14.1
$M = 3$[a]	0.17	0.22	—	0.67	11.9	46.7	−65.2	13.4	−63.4	15.2
$M = 4$[a]	0.17	0.21	—	0.55	9.0	45.0	−65.9	12.7	−63.3	15.3
$M = 8$[a]	0.17	0.20	—	0.43	5.8	43.3	−68.0	10.6	−64.3	14.3
$M = 16$[a]	0.16	0.19	—	0.38	3.3	41.1	−70.7	7.9	−66.5	12.1
AMI	0.26	0.32	—	0.50	9.0	62.4	−83.3	−4.7	−80.3	−1.7
Duo-binary	0.20	0.24	—	0.50	7.7	50.0	−72.4	6.2	−69.4	9.2
4B3T	0.20	0.24	—	0.69	8.5	50.0	−71.4	7.2	−69.8	8.8
With ideal DFE ($T_D = \mathring{T}_D$)										
$M = 2$[a]	0.09	0.13	−1.64	1.00	6.6	29.7	−53.3	25.3	−53.3	25.3
$M = 3$[a]	0.09	0.12	−1.22	0.67	4.8	27.8	−53.8	24.8	−52.0	26.6
$M = 4$[a]	0.08	0.12	−0.96	0.55	4.4	27.5	−54.6	24.0	−52.0	26.6
$M = 8$[a]	0.08	0.12	−0.68	0.43	3.2	26.8	−56.7	21.9	−53.0	25.6
$M = 16$[a]	0.08	0.11	−0.52	0.38	2.2	26.4	−59.5	19.1	−55.3	23.3
AMI	0.08	0.13	−2.4	0.50	1.7	29.7	−64.8	13.8	−61.8	16.8
Duo-binary	0.08	0.13	−1.9	0.50	2.6	29.7	−61.3	17.3	−58.3	20.3
4B3T	0.08	0.13	−1.7	0.69	3.6	29.7	−58.5	20.1	−56.9	21.7

NRZ rectangular pulses, coaxial cable ($a_* = 80$ dB), white noise ($F = 1$) and optimum decision values are assumed.   [a] Nonredundant.

A comparison of the values given in Table 8.1 shows that for a nonredundant system with ideal DFE ($T_D = 0$) the minimum cut-off frequency $\check{f}_I \approx 0.17R$ is to a first approximation independent of the level number M.

However, the detection time $T_D = 0$ is not the optimum in a system with (ideal) DFE because the corrected basic detection pulse $g_k(t)$ is extremely asymmetrical (see Fig. 5.2). By differentiating eqn (8.10) with respect to T_D and setting to zero we obtain the following relationship for the optimum detection time $\mathring{T}_D$:

$$M \exp\left\{-4\pi f_I^2\left(\frac{T}{2} - \mathring{T}_D\right)^2\right\} \stackrel{!}{=} \exp\left\{-4\pi f_I^2\left(\frac{T}{2} + \mathring{T}_D\right)^2\right\} \tag{8.12}$$

The following assumptions are made in the differentiation of the Q function:

$$\frac{\mathrm{d}Q(x)}{\mathrm{d}x} = -\frac{1}{(2\pi)^{1/2}} \exp\left(-\frac{x^2}{2}\right) \tag{8.13}$$

It follows from this that the optimum detection time of a nonredundant M-level system with ideal DFE and a Gaussian pulse shaper is

$$\mathring{T}_D = \frac{-\ln 2}{8\pi \log_2 M\,(f_I/R)^2} T \approx \frac{-0.028}{\log_2 M\,(f_I/R)^2} T \tag{8.14}$$

As this equation shows, optimization of T_D produces a negative value. This means that in a system with (ideal) DFE it is advantageous to detect individual pulses before they reach their maximum. In this way the peak value $g_d(T_D)$ is reduced, but the perturbing effect of the precursors is reduced even more markedly. However, it must be remembered that the DFE must be realized more accurately for low cut-off frequency f_I and early pulse detection. Therefore shifts in the detection time are limited in practical systems for tolerance reasons (see Section 8.7).

Substitution of eqn (8.14) in eqn (8.10) gives the normalized eye opening at the optimum detection time $\mathring{T}_D$, with ideal DFE:

$$o_{\mathrm{norm}}(\mathring{T}_D) = \frac{1}{M-1}\left\{1 - MQ\left(\frac{f_I'}{2}\log_2 M + \frac{\ln 2}{f_I'}\right) - Q\left(\frac{f_I'}{2}\log_2 M - \frac{\ln 2}{f_I'}\right)\right\} \tag{8.15}$$

where $f_I' = 2(2\pi)^{1/2} f_I/R$.

The chain curves in Fig. 8.1(a) show that the cut-off frequency of the pulse shaper can be reduced to the limiting value $f_I \approx 0.09R$ by optimization of the detection time, without closing the eye. How this reduction of the cut-off frequency affects the signal-to-noise ratio, and hence the error probability, can be understood if the noise is taken into consideration.

Noise power and signal-to-noise ratio As an example we consider a coaxial pair with a characteristic attenuation value a_*. In the case of a Gaussian pulse shaper with cut-off frequency f_I we obtain for the normalized noise power at the decision device (see eqns (2.70) and (7.51))

$$N_{\mathrm{norm}} = \frac{1}{R}\int_{-\infty}^{+\infty} F(f) \exp\left\{2a_*\left(\frac{2|f|}{R}\right)^{1/2} - 2\pi\left(\frac{f}{2f_I}\right)^2\right\} \mathrm{d}f \tag{8.16}$$

In Fig. 8.1(b) the (logarithmic) noise power $10\lg N_{\text{norm}}$ is plotted as a function of the cut-off frequency of the pulse shaper and the characteristic attenuation value, where the noise factor $F(f)=1$. This figure is valid for all level numbers M and is also independent of whether or not the system has DFE. On a logarithmic scale it is clear that the noise power increases very markedly with increasing cable attenuation value a_* and increasing cut-off frequency f_I.

Because of the opposing effects of the cut-off frequency of the pulse shaper on the vertical eye opening and the noise power there is an optimum value for f_I which leads to a maximum value of system efficiency and hence a minimum error probability. The position and value of this maximum depends on the other system parameters, e.g. on the level number M and the characteristic attenuation value a_*.

In Fig. 8.1(c) the system efficiency under peak-value limitation (eqn (7.17))

$$10\lg \eta_A = 20\lg o_{\text{norm}}(T_D) - 10\lg N_{\text{norm}} \tag{8.17}$$

is plotted against the cut-off frequency f_I for the case where $a_* = 80$ dB and $F(f) = 1$. For a binary system without DFE the optimum cut-off frequency $\mathring{f}_I$ of the pulse shaper is about $0.32R$ under the assumptions made here. If f_I is less than $\mathring{f}_I$, the effect of intersymbol interference increases very strongly; the eye is closed when f_I is less than $\check{f}_I \approx 0.27R$. However, the noise is inadequately suppressed at higher pulse-shaper cut-off frequencies ($f_I > \mathring{f}_I$).

At the optimum cut-off frequency $\mathring{f}_I$ the system efficiency is about -78.6 dB. This means that the binary coaxial pair system ($a_* = 80$ dB) with an optimum Gaussian pulse shaper is 78.6 dB worse than a system with an ideal channel (eqn (7.7)) and an optimum receiver (Chapter 6).

If the (worst-case) error probability is not to exceed the limiting value 10^{-10}, the critical signal-to-noise ratio $10\lg \rho_{Gr}$ must be 16.1 dB or more, and correspondingly the following must apply (see eqns (7.42) and (7.66)):

$$\frac{s_{\max}^2}{L_{\text{th}} R} \geqslant 10^{9.47} \tag{8.18}$$

In this case the system efficiency η_L under power limitation is identical with η_A, as both transmitter efficiencies $\eta_{S,L}$ and $\eta_{S,A}$ are equal to unity when the transmitter pulses are binary NRZ rectangular (see Section 7.2.4).

This system (binary; nonredundant; NRZ rectangular transmitter pulse; Gaussian pulse shaper with optimized cut-off frequency; tolerance-free threshold detector without DFE) is used in the following as a **comparison system** and so the **gain in the signal-to-noise ratio under peak-value limitation** can be defined for a given system as follows:

$$G_A = \frac{\eta_A(\text{given system})}{\eta_A(\text{comparison system})} \tag{8.19}$$

In the same way we have for the gain in the signal-to-noise ratio under power limitation

$$G_L = \frac{\eta_L(\text{given system})}{\eta_L(\text{comparison system})} \tag{8.20}$$

It must be remembered that for these definitions the nonoptimizable system parameters have to be kept constant for the optimization and system comparison. This means, for example, that the gains G_A and G_L depend on the characteristic attenuation value a_*.

The gain G_A in the signal-to-noise ratio of a quaternary system without DFE is about 3.7 dB when the attenuation value a_* is 80 dB (see the right-hand scale in Fig. 8.1(c)), in which case the optimum cut-off frequency $\mathring{f}_I$ is rather lower than that of the binary system.

With the same pulse-shaper cut-off frequency ($f_I = 0.32R$) and the same detection time ($T_D = 0$) a signal-to-noise ratio improvement of approximately 7 dB can be achieved in a binary system ($M = 2$) by means of ideal DFE. When the cut-off frequency is optimized at a fixed detection time $T_D = 0$, an additional improvement of 7 dB is obtained for $f_I = 0.22R$. The mutual optimization of f_I and T_D finally gives a maximum improvement of 25.3 dB for $\mathring{f}_I \approx 0.13R$ and $\mathring{T}_D = -1.66T$.

The curves for the quaternary system ($M = 4$) with DFE show similar behavior. Here the maximum improvement relative to the (quaternary) system without DFE is about 20 dB.

The results of the optimization of a system with a Gaussian pulse shaper are summarized in Table 8.1, which is based on a characteristic attenuation value a_* of 80 dB. The gain G_L in the signal-to-noise ratio under power limitation is given in the final column. Under this co-condition the multilevel system is more satisfactory relative to the binary system than is the case under peak-value limitation because the amplitude of a multilevel transmitter is larger than that of a binary transmitter with the same power. The optimum receiver parameters $\mathring{f}_I$ and $\mathring{T}_D$, however, are independent of whether power or peak-value limitation of the transmitter is assumed.

Dependence of the optimum parameters on the characteristic attenuation value Figure 8.2(a) shows the optimum cut-off frequency $\mathring{f}_I$ of the Gaussian pulse shaper as a function of the characteristic attenuation value a_* of a coaxial pair. The larger a_* is, i.e. the larger the selected regenerator-section length l is, the smaller is the optimum value $\mathring{f}_I$ of the pulse-shaper cut-off frequency. This is accounted for by the fact that the detection noise power N_d increases strongly with increasing cable attenuation, which can be partially compensated by a lower cut-off frequency.

The system efficiency η_A under peak-value limitation is plotted against the characteristic attenuation value a_* in Fig. 8.2(b). The figure shows that, in the same way as for the optimum digital receiver, the logarithmic system efficiency $10 \lg \eta_A$ reduces approximately in proportion to the attenuation value a_* for typical attenuation values of 50 dB or more. Hence the maximum regenerator-section length can be determined using the equations introduced in Section 7.3. The slope of the curves is much smaller for the systems with DFE than for the systems without DFE, which leads to a greater regenerator-section length. This is due to the fact that in a system with DFE the bandwidth can be more markedly reduced than in a

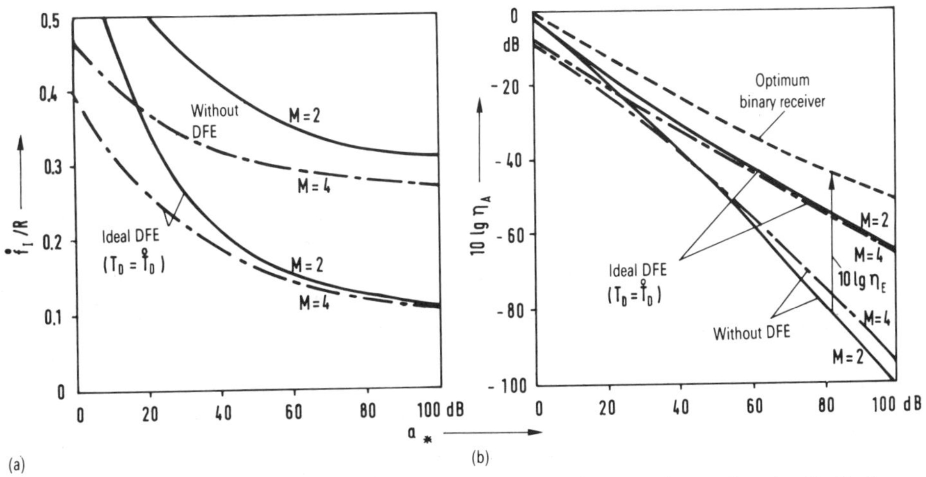

Fig. 8.2 (a) Optimum cut-off frequency and (b) system efficiency under peak-value limitation as a function of the characteristic attenuation value a_* of a coaxial pair and the level number M (NRZ rectangular transmitter pulse, white noise ($F = 1$) and Gaussian pulse shaper).

comparable system without DFE, and so the strong increase in the noise power with increasing attenuation value is partially compensated.

For comparison, the curve for the optimum binary receiver (the Viterbi receiver as defined in Chapter 6) is also plotted in Fig. 8.2(b). The curve for the (binary) threshold detector without DFE is about 35 dB below that for the optimum binary receiver for a characteristic attenuation value a_* of 80 dB. This means that the receiver efficiency is $10 \lg \eta_E = -35$ dB. However, the threshold detector with DFE is less than 10 dB away from the theoretical optimum.

8.1.2 Coded systems with Gaussian pulse shaper

The most important transmission codes were summarized in Chapter 4, and their coding algorithms and other parameters (e.g. redundancy, power spectra and eye opening) were defined. In the following section they are compared on the basis of the achievable signal-to-noise ratio. As in Section 8.1.1 the use of a coaxial pair and a Gaussian pulse shaper is assumed. As the noise power is not changed directly by coding, eqns (8.16) and Fig. 8.1(b) are still valid.

Figure 8.3(a) shows the normalized eye opening (at the optimum detection time $\mathring{T}_D = 0$) plotted against the cut-off frequency f_I for redundant coded systems without DFE. The curves for a nonredundant binary code and a nonredundant ternary code are also plotted for comparison. It can be seen that the widely used AMI code has a very small eye opening in this case. The curves, which can be calculated from eqns (3.25) and (4.15), are also good approximations for the HDB3 and B6ZS codes (see Section 4.2.4). The small eye opening of these ternary partial-response codes is due to the fact that despite the high redundancy of almost 37% the symbol sequences that are particularly adverse with regard to the eye opening are not completely excluded from the transmission. Because of the ternary detection (without simultaneous reduction of the symbol rate) the minimum cut-off frequency of these codes is very high ($\check{f}_I \approx 0.36R$). For a coaxial pair system with a characteristic attenuation value of 80 dB, optimization of the pulse-shaper cut-off frequency f_I results in the value $\mathring{f}_I \approx 0.4R$ (see Fig. 8.3(b)). The loss of signal-to-noise ratio with respect to the nonredundant binary system (comparison system) amounts to almost 14 dB if peak-value limitation is assumed. With power limitation there is only a loss of 11 dB because a ternary partial response transmitter signal has only half the transmitter power of the nonredundant binary signal (see Table 4.2 and Table 8.1).

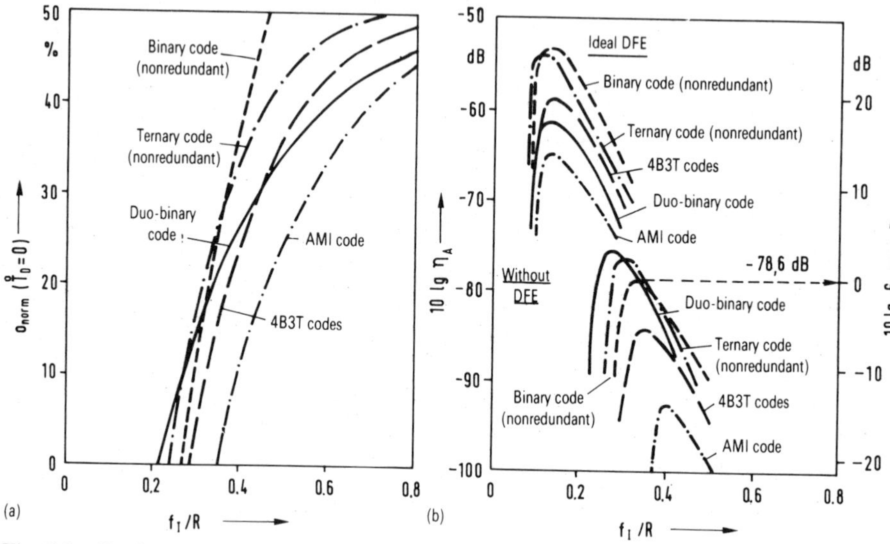

Fig. 8.3 Optimization of the cut-off frequency f_I of a redundant coded transmission system with a Gaussian pulse shaper: (a) normalized eye opening for a system without DFE; (b) logarithmic system efficiency ($a_* = 80$ dB; $F(f) = 1$).

The similarly "direct-signal-free" 4B3T codes are also 5–6 dB worse than the nonredundant binary code in this comparison under peak-value limitation. For $f_I < 0.29R$ the eye is closed with this coding. However, the duo-binary code is still slightly better than the nonredundant ternary code in a system without DFE. The reason for this is that symbol sequences that are particularly adverse with regard to intersymbol interference are excluded by the coding rules, and hence a very low bandwidth is required. The minimum cut-off frequency $\check{f}_I$ of the Gaussian pulse shaper can be determined from eqn (4.16). The smallest possible value ($\check{f}_I \approx 0.22R$) is obtained with the duo-binary code.

The corresponding values for the coded systems with DFE can be obtained from Table 8.1. All systems with redundant coding exhibit a smaller eye opening here than they do in nonredundant binary or ternary systems. Therefore, in a coaxial transmission channel ($a_* = 80$ dB), all nonredundant systems are 5–12 dB worse than the binary system (see Fig. 8.3(b)). The signal-to-noise ratio improvement obtained when ideal DFE is used is almost 28 dB for the AMI code and about 25 dB for the 4B3T code, but is only 14 dB for the duo-binary code. This is because in the case of the duo-binary code the particularly unfavorable symbol sequences are already excluded by the coding rules, so that the nonlinear DFE produces relatively little improvement.

8.1.3 Ternary partial response coding at the receiver

In a system without DFE the duo-binary code produces very encouraging results (see Fig. 8.3). However, the system efficiency can be improved even more if the coding is transferred from the transmitter side to the receiver side.

Like the AMI code the duo-binary code is a ternary partial response code. In Chapter 4, Section 4.2, it was shown that with this type of coding the coder, regardless of its realization, can be described by a nonlinear precoder (PC) as well as a linear coding network $H_C(f)$. Hence the block diagram shown in Fig. 8.4(a) is valid for a ternary partial response coded system. It follows that the detection signal is given by

$$d_S(t) = \sum_{\nu=-\infty}^{+\infty} a_\nu g_{d_S}(t - \nu T) + \overset{\times}{d}_S(t) \tag{8.21}$$

where the subscript S denotes the coding at the transmitter. The amplitude coefficients a_ν of the transmitter signal are $-1, 0$ or $+1$. The basic detection pulse is calculated from eqns (2.60) and (2.62):

$$g_{d_S}(t) = g_s(t) * h_K(t) * h_E(t) \tag{8.22}$$

Now if the linear coding network $H_C(f)$ is transferred from the transmitter to the receiver, we refer to ternary partial response coding at the receiver (see Fig. 8.4(b)). Nothing is changed with regard to the useful signal, i.e. it is given by

$$\tilde{d}_S(t) = \tilde{d}_E(t) = \sum_{\nu=-\infty}^{+\infty} b_\nu g_{d_E}(t - \nu T) \tag{8.23}$$

The amplitude coefficients $b_\nu \in \{-1; +1\}$ of the binary precoded signal are inserted in this equation, and the basic detection pulse is obtained from eqn (4.9):

$$g_{d_E}(t) = g_{d_S}(t) * h_C(t) = \tfrac{1}{2}\{g_{d_S}(t) - k_N g_{d_S}(t - NT)\} \tag{8.24}$$

This relationship is true for both the AMI code ($k_N = +1$) and the duo-binary code ($k_N = -1$). As $\tilde{d}_E(t) = \tilde{d}_S(t)$, equal (ternary) eye patterns are produced for the two different systems shown in Figs 8.4(a) and 8.4(b) (see Fig. 4.7). In particular, both systems have the same normalized eye opening, which is shown for a Gaussian pulse shaper in Fig. 8.3(a).

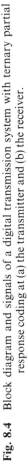

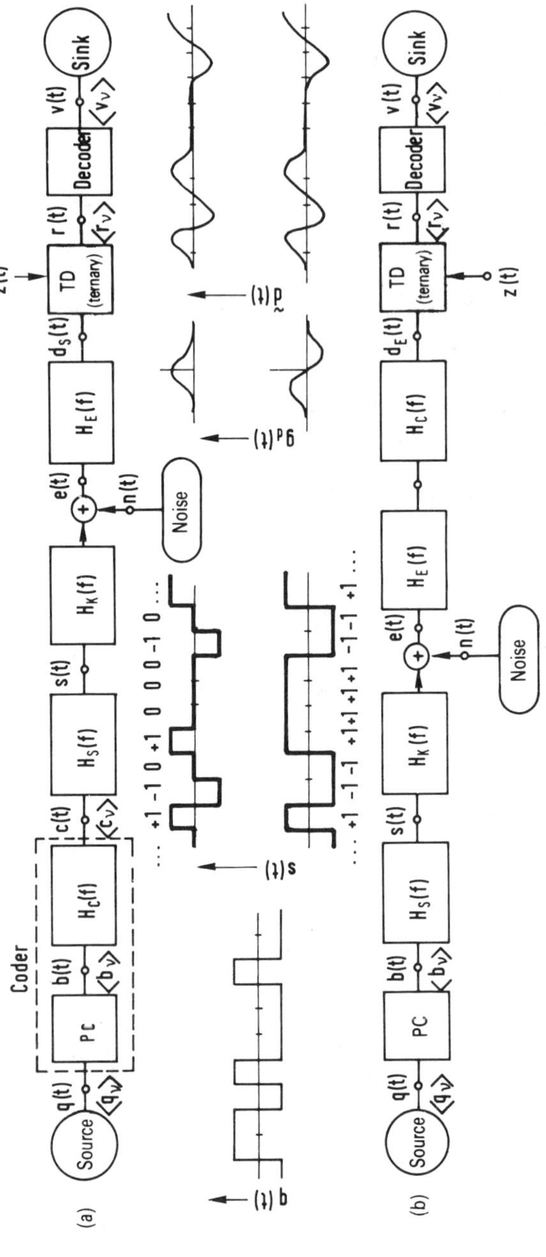

Fig. 8.4 Block diagram and signals of a digital transmission system with ternary partial response coding at (a) the transmitter and (b) the receiver.

However, the two systems shown in Fig. 8.4 differ with regard to the noise. The detection noise power with ternary partial response coding at the transmitter is obtained from eqn (2.69):

$$N_{d_S} = \int_{-\infty}^{+\infty} L_n(f)|H_E(f)|^2 \mathrm{d}f \tag{8.25}$$

In the case of coding at the receiver the following equation is obtained [4.13]:

$$N_{d_E} = \int_{-\infty}^{+\infty} L_n(f)|H_E(f)|^2|H_C(f)|^2 \mathrm{d}f \tag{8.26}$$

Since according to eqn (4.11) $|H_C(f)| \leqslant 1$ for all frequencies, N_{d_E} is always smaller than N_{d_S}. This means that the coding network $H_C(f)$, in addition to the equalizer $H_E(f)$, contributes to noise-power limitation if it is transferred from the transmitter to the receiver.

The normalized detection noise power $10 \lg N_{norm}$ is plotted against the cut-off frequency f_I of the Gaussian pulse shaper in Fig. 8.5(a). The curves are for various values of the characteristic attenuation value a_* of the coaxial pair. The continuous curves are valid for all nonredundant systems as well as for systems with ternary partial response coding at the transmitter (see Fig. 8.1(b)). However, application of AMI and duo-binary codes at the receiver results in a reduction of the noise power by up to 8 dB, i.e. by more than a factor of 6. The noise does not decrease

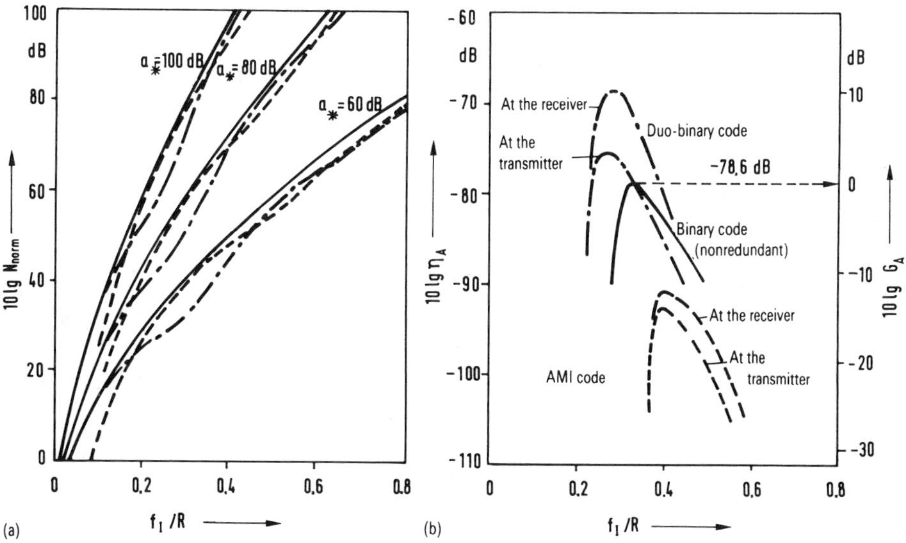

Fig. 8.5 Systems with a Gaussian pulse shaper and ternary partial response coding: (a) normalized detection noise power ($F(f) = 1$; —, nonredundant codes and ternary partial response codes at the transmitter; ---, AMI code at the receiver; —·—, duo-binary code at the receiver); (b) system efficiency with peak-value limitation ($a_* = 80$ dB, $F(f) = 1$).

monotonically with the pulse-shaper cut-off frequency but varies because of the cosine variation of $H_C(f)$ with cut-off frequency f_I. Whereas the AMI coding at the receiver causes a particularly marked noise reduction at very low cut-off frequencies and when $f_I \approx 0.5R$, duo-binary coding at the receiver is very worthwhile when $f_I = 0.2R–0.4R$. As the optimum cut-off frequency $\mathring{f}_I$ lies in this range duo-binary coding at the receiver leads to very favorable values for the system efficiency. Figure 8.5(b) clearly shows that an improvement in the signal-to-noise ratio of about 10 dB relative to the nonredundant binary system (comparison system) is possible for a Gaussian pulse shaper with "duo-binary coding at the receiver". Under the current assumptions this code is therefore more beneficial than any nonredundant code (see Fig. 8.1(c)). In addition it offers the advantage that error monitoring is possible using the signal redundancy.

Figure 8.5(b) is valid for peak-value limitation. With power limitation the improvement due to coding at the receiver is 3 dB lower than that for ternary partial response coding at the transmitter because the binary transmitter signal in Fig. 8.4(b) has twice the power of the redundant ternary transmitter signal in Fig. 8.4(a).

Note With ternary partial response coding at the receiver a binary signal is transmitted and a three-level signal is detected. In this case, as has just been shown, a signal-to-noise ratio improvement is possible. This gain in the signal-to-noise ratio can undoubtedly be increased still further if the possible amplitude levels of the transmitter and receiver are not predefined but are handled as optimizable system parameters (see Section 8.5).

8.1.4 Effect of the lower band limit

The use of redundant transmission codes is sensible if no direct signal can be transmitted through the channel. A nonredundant signal is generally only transmittable with acceptable error probability over such a band-pass channel if certain measures are taken at the receiver, e.g. direct-signal recovery as described in Section 5.3, to reconstruct the low-frequency spectral components.

A first-order high-pass filter (high-pass cut-off frequency f_{lo}) is now inserted into the transmission channel to examine the lower band limitation. We continue to use an equalizer with a Gaussian pulse shaper (low-pass cut-off frequency f_{up}) for noise-power limitation. Hence the pulse-shaper frequency response is given by (see Fig. 5.10(a))

$$H_I(f) = H_{HP}(f)H_{LP}(f) = \frac{-j\pi f/2f_{lo}}{1 + j\pi f/2f_{lo}} \exp\left\{-\pi\left(\frac{f}{2f_{up}}\right)^2\right\} \tag{8.27}$$

In this case the cut-off frequencies f_{lo} and f_{up} are not the 3 dB cut-off frequencies of the band-pass filter. Rather, f_{up} corresponds to the cut-off frequency of the Gaussian low-pass filter given by eqn (2.38), while f_{lo} defines the corresponding cut-off frequency of the low-pass filter $1\text{-}H_{HP}(f)$ that is equivalent to the high-pass filter $H_{HP}(f)$. Therefore digital signal transmission is also possible for $f_{lo} > f_{up}$.

The basic detection pulse $g_d(t) = g_s(t)*h_l(t)$ is illustrated in Fig. 8.6 where NRZ rectangular pulses are assumed at the transmitter. Curves are plotted here for various values of the high-pass cut-off frequency f_{lo}; the Gaussian low-pass cut-off frequency f_{up} always amounts to $0.5R$. If $f_{lo} \neq 0$, $g_d(t)$ becomes negative over a long period because of the missing direct component. The more extensive is the trailing edge the smaller is the high-pass cut-off frequency. However, a smaller main value $g_d(0)$ occurs for relatively large values of f_{lo}, while the amplitudes of the first postcursors are particularly large.

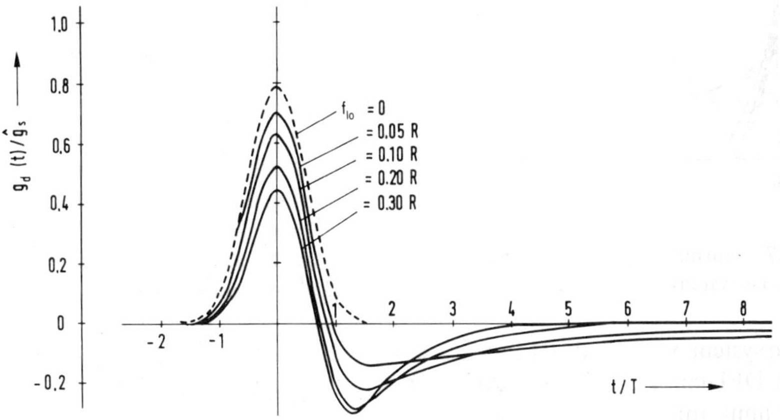

Fig. 8.6 Basic detection pulse for a binary band-pass system defined by eqn (8.27) for various lower cut-off frequencies f_{lo} ($f_{up} = 0.5R$).

Figure 8.7 shows the (normalized) eye opening which occurs with these band-pass-type pulse shapers. For nonredundant codes the normalized eye opening can be calculated from eqn (3.51) (without DFE) or eqn (5.12) (ideal DFE). The results for the coded system, however, were ascertained numerically.

If the lower cut-off frequency $f_{lo} \neq 0$, the eye for the nonredundant and duobinary codes is closed for a system without DFE (see Fig. 8.7(a)). However, the AMI code is relatively insensitive to a lower band limit because with this code the number of adjacent symbols of the same type (+ and −) is limited to 1. The 4B3T codes can also still be transmitted well over this band-pass channel, as long as the lower cut-off frequency is relatively small which can be assumed in general. The individual 4B3T codes, i.e. the MS43, FOMOT and Jessop–Waters 4B3T codes, differ only slightly.

As inspection of the power spectrum of Fig. 4.10 makes plain, the AMI code has very weak low-frequency spectral components. It is therefore understandable that this code is relatively insensitive to a lower band limit. The MS43, FOMOT and Jessop–Waters 4B3T codes exhibit stronger low-frequency spectral components. Therefore the lower band limitation produces increasing perturbation in the order in which these codes are listed above.

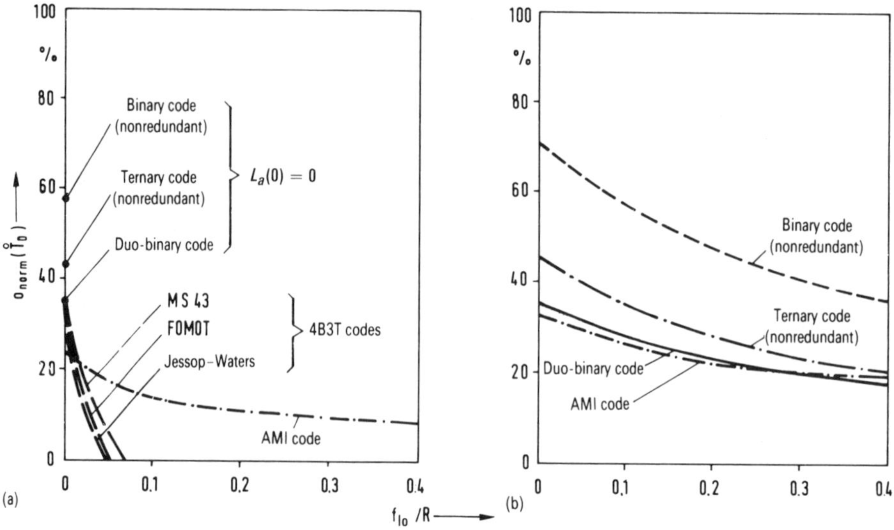

Fig. 8.7 Normalized eye opening as a function of the lower cut-off frequency f_{lo} of the band-pass system defined by eqn (8.27) ($f_{up} = 0.5R$) (a) without DFE and (b) with ideal DFE.

In systems with (ideal) DFE fundamentally different circumstances exist (see Fig. 8.7(b)). DFE compensates for the intersymbol interference caused by both the lower band limit (direct-signal recovery) and the upper band limit and therefore nonredundant signals also can be transmitted across the band-pass channel without a marked decline in the signal-to-noise ratio. For example the normalized eye opening of the nonredundant binary signal for channel parameters $f_{up} = 0.5R$ and $f_{lo} = 0.1R$ is about 57%, which is more than twice the size of the eye opening of the AMI-coded ternary signal. Hence the achievable signal-to-noise ratios of the two systems differ by more than 6 dB. With optimum dimensioning of the upper cut-off frequency f_{up} the loss in signal-to-noise ratio for the AMI code is even greater compared with the (nonredundant) binary code. This shows that the AMI code is also inferior to the nonredundant code in a band-pass system. However, it has the advantage that error monitoring is easier because of the redundancy.

A fixed upper cut-off frequency $f_{up} = 0.5R$ was assumed in Fig. 8.7. However, the dependence of the normalized eye opening and the system efficiency on the two variable system parameters f_{lo} and f_{up} is also of interest. Therefore we consider a nonredundant binary system with ideal DFE. Examination of the nonredundant multilevel system leads to qualitatively the same results.

The normalized eye opening $o_{norm}(\mathring{T}_D)$ is plotted in Fig. 8.8 against both the upper and the lower cut-off frequencies. It can be seen that the normalized eye opening approaches the limiting value of 100% for $f_{lo} \neq 0$ if the upper cut-off frequency is sufficiently high. In the limiting case $f_{up} \to \infty$ (high-pass system) the lower cut-off frequency can also be made very high (see Section 5.3). An upward shift in the lower cut-off frequency can therefore be compensated by a simultaneous

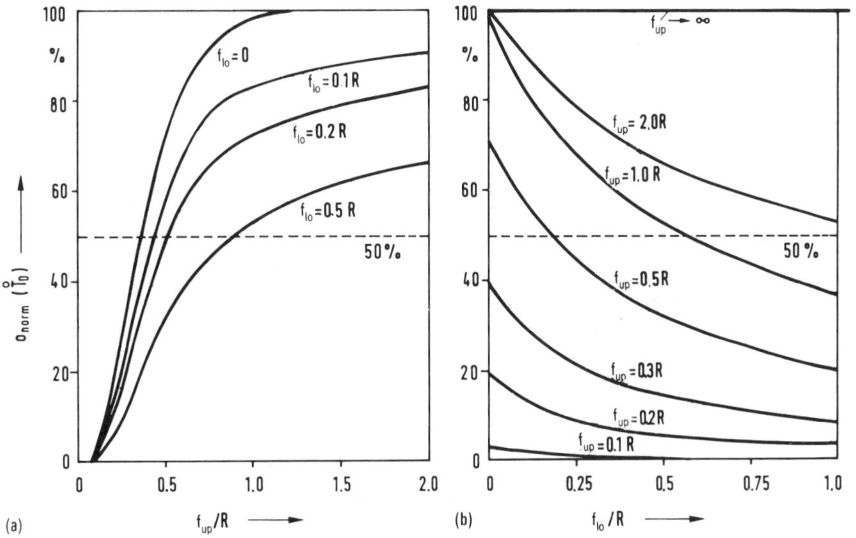

Fig. 8.8 Normalized eye opening for a nonredundant binary system with ideal DFE as a function of the lower and upper cut-off frequencies of the band-pass system defined by eqn (8.27).

upward shift in the upper cut-off frequency without changing the eye opening.

To clarify this we assume an eye opening $o_{norm}(\mathring{T}_D) = 50\%$, which is denoted by the horizontal broken line in Fig. 8.8. With a low-pass channel ($f_{lo} = 0$) the upper cut-off frequency f_{up} must be greater than $0.36R$ to satisfy the requirement $o_{norm}(\mathring{T}_D) = 50\%$. If $f_{lo} = 0.1R$ the upper cut-off frequency (with the same eye opening) has to be increased to $f_{up} = 0.43R$. For $f_{lo} = 0.5R$ the requirement is $f_{up} \geqslant 0.9R$.

Of course the increase in the upper cut-off frequency causes an increase in the noise power so that the system efficiency is reduced. In a frequency-independent transmission channel the noise power N_d increases proportionally to f_{up} and almost exponentially for a coaxial cable (see Fig. 8.1(b)). Therefore for every high-pass cut-off frequency f_{lo} there is an optimum pulse-shaper cut-off frequency $\mathring{f}_{up}$ which increases proportionally to f_{lo}. For clarification let us again consider a coaxial pair system.

The optimum upper cut-off frequency $\mathring{f}_{up}$ for several values of f_{lo} is plotted in Fig. 8.9(a) as a function of the characteristic attenuation value. If a_* is sufficiently large the optimum cut-off frequency f_{up} depends only slightly on the lower cut-off frequency. Figure 8.9(b) shows the system efficiency $10 \lg \eta_A$ as a function of a_*. The curves shown are with the lower cut-off frequency f_{lo} as a parameter; the upper cut-off frequency f_{up} has the optimum value $\mathring{f}_{up}$ given by Fig. 8.9(a). For $a_* = 80$ dB and $f_{lo} = 0.1R$ this results in a reduction in the signal-to-noise ratio of about 4 dB relative to the corresponding low-pass system ($f_{lo} = 0$). However, as in general f_{lo} can be given a lower value, the actual reduction in signal-to-noise ratio owing to the lower band limitation is less.

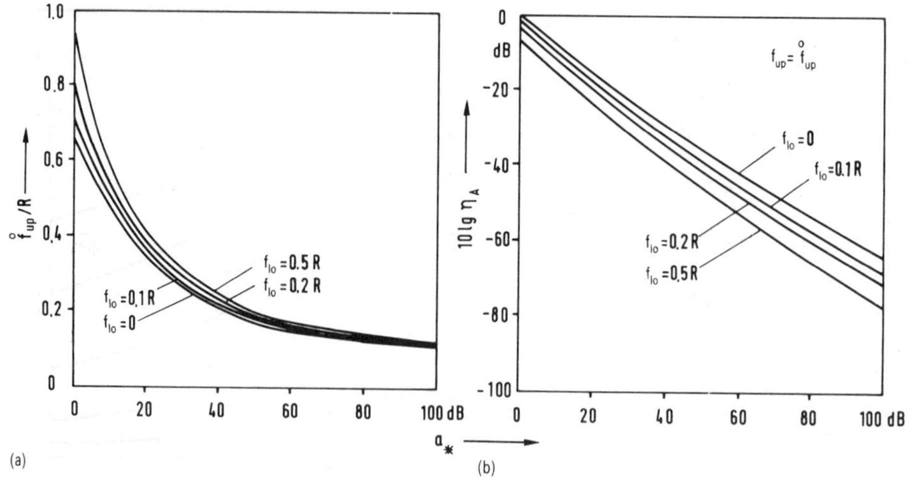

Fig. 8.9 (a) Optimum upper cut-off frequency and (b) system efficiency as a function of the lower cut-off frequency f_{lo} of the band-pass system and the characteristic attenuation value a_* of a coaxial pair (binary NRZ rectangular transmitter pulses, white noise ($F = 1$) and Gaussian pulse shaper).

Note The examples given above should explain the principal effect of the lower cut-off frequency. They assume an ideal direct-signal recovery. However, in a more extensive consideration the limited practical accuracy of DFE and direct-signal recovery must be taken into account.

8.1.5 Cosine roll-off pulse shaper

The Gaussian low-pass filter has only one variable parameter. To demonstrate the fundamental relationship between the cut-off frequency and the slope of the filter edges, we shall now consider the cosine roll-off low-pass filter (Fig. 8.10(a)). The pulse-shaper frequency response is given in the Appendix, Table A3 (row 10). At low frequencies $|f| \leqslant f_1$, $H_I(f) = 1$. Outside the frequency $\pm f_2$ the frequency response is zero. Between the frequencies f_1 and f_2 the variation is of cosine form. We define the **roll-off factor**

$$r_I = \frac{f_2 - f_1}{f_2 + f_1} \tag{8.28}$$

as a variable parameter in addition to the cut-off frequency $f_I = (f_1 + f_2)/2$. Special cases in the general group of cosine roll-off low-pass filters are the brick-wall low-pass filter ($r_I = 0$) and the raised cosine low-pass filter ($r_I = 1$).

Although the pulse response $h_I(t)$ of these low-pass filters has zero-crossings separated by equal distances $t = 1/2f_I$ (see Table A3), independent of the roll-off factor, the response to an NRZ rectangular pulse $g_d(t) = \hat{g}_s \operatorname{rec}(t/T) * h_I(t)$ does not have this characteristic. In contrast, the pulse precursors and postcursors cause intersymbol interference (Fig. 8.10(b)). This is larger the smaller the cut-off frequency f_I and the roll-off factor r_I are, i.e. the more rapidly the pulse-shaper frequency response decays. However, if the cut-off frequency is very high or the filter edges are not very steep, the noise is inadequately suppressed. Therefore there are optimum values of the parameters f_I and r_I.

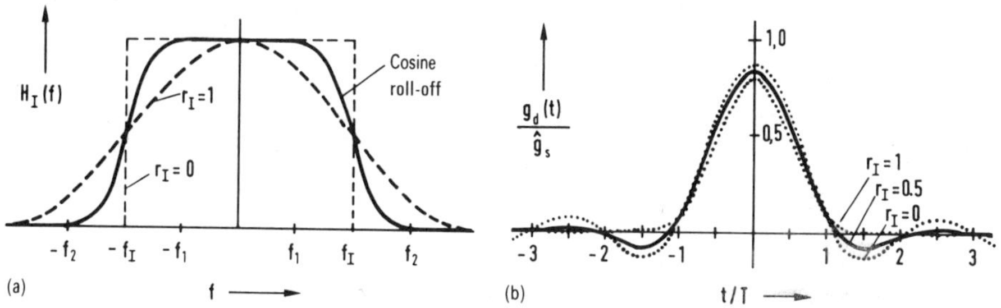

Fig. 8.10 (a) Pulse-shaper frequency response and (b) basic detection pulse in a system with an NRZ rectangular transmitter pulse and a cosine roll-off low-pass filter ($f_I = 0.5R$).

The observed gain in the signal-to-noise ratio for the example of a nonredundant binary coaxial cable system with a characteristic attenuation a_* of 80 dB is plotted against the cut-off frequency f_I and the roll-off factor r_I in Fig. 8.11. Figure 8.11(a) is for a system without DFE ($T_D = \mathring{T}_D = 0$). The optimum pulse-shaper parameters are $\mathring{f}_I \approx 0.5R$ and $\mathring{r}_I \approx 0.2$. In contrast with the comparison system (Gaussian low-pass filter) an improvement in the signal-to-noise ratio of about 5.8 dB can be gained by using these values.

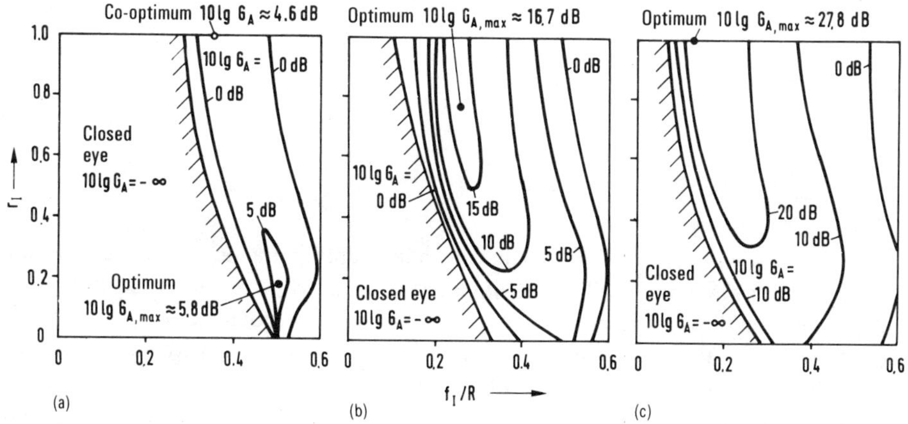

Fig. 8.11 Gain 10 lg G_A in the signal-to-noise ratio as a function of the cut-off frequency f_I and the roll-off factor r_I of a cosine roll-off low-pass filter ($M = 2$, $a_* = 80$ dB): (a) receiver without DFE; (b) receiver with ideal DFE ($T_D = 0$); (c) receiver with ideal DFE ($T_D = \mathring{T}_D$).

Varying the pulse-shaper cut-off frequency f_I at the optimum roll-off factor $\mathring{r}_I$ produces a considerable and very rapid loss in the signal-to-noise ratio relative to the optimum cut-off frequency $\mathring{f}_I$, as can be seen from the very densely packed contours in Fig. 8.11(a). The eye is closed for $f_I < 0.42R$, and the detection noise power increases substantially for $f_I > \mathring{f}_I$.

When the roll-off factor ($r_I > \mathring{r}_I$) is increased at optimum cut-off frequency $\mathring{f}_I$, a similar signal-to-noise ratio loss of up to 10 dB occurs relative to the optimum. This is due to the fact that the noise power increases very markedly because of the flat filter decay whereas the intersymbol interference decreases only slightly. When the roll-off factor is very small ($r_I < \mathring{r}_I$), the detection pulse decays very slowly so that the intersymbol interference increases negligibly. Moreover the temporal eye opening is very small, so that the perturbing effect of phase jitter also becomes stronger.

The fundamental relationship between the cut-off frequency f_I and the roll-off factor r_I is of interest. The less steep the filter edge is, i.e. the larger the value of r_I is, the smaller will be the value of the cut-off frequency. Figure 8.11(a) shows further that a local maximum exists for a raised cosine low-pass filter ($r_I = 1$) with cut-off frequency $f_I \approx 0.35R$. The intersymbol interference effects are significantly greater in this suboptimum system than in the optimum system. However, because of the lower cut-off frequency, the noise power is very much lower.

In a system with ideal DFE the suboptimum becomes the global optimum. At the nonoptimum detection time $T_D = 0$ the values of the optimum parameters are $\mathring{f}_I \approx 0.26R$ and $\mathring{r}_I \approx 0.7$. Here the maximum gain $G_{A,max}$ in the signal-to-noise ratio is about 16.7 dB (Fig. 8.11(b)). The pulse-shaper cut-off frequency can be further reduced by the additional optimization of the detection time T_D. In this case optimization leads to the following values: $\mathring{f}_I \approx 0.13R$; $\mathring{r}_I \approx 1$; $\mathring{T}_D = -1.8T$; $G_{A,max} \approx 27.8$ dB.

The results of the optimization of nonredundant multilevel systems using cosine roll-off low-pass filters can be deduced from Fig. 8.33 in Section 8.6. This figure shows that the signal-to-noise ratio of a system without DFE can be substantially improved by increasing the level number M. However, the binary system is best for a system with ideal DFE.

8.1.6 Effects of phase distortion

So far, for simplicity, we have generally considered frequency responses with no imaginary components and hence symmetrical pulses. The results of Sections 8.1.1–8.1.5 are therefore valid only when **ideal phase equalization** is assumed. i.e. when the pulse-shaper phase response $b_I(f)$ increases linearly with frequency and hence the group delay $\tau_I(f)$ is constant:

$$b_I(f) = kf \tag{8.29a}$$

$$\tau_I = \frac{1}{2\pi}\frac{db(f)}{df} = \frac{k}{2\pi} \tag{8.29b}$$

In the realization of a digital transmission system, ideal phase equalization is only possible at infinite expense. The disturbing effect of phase distortion is discussed in refs 5.17 and 8.9, for example. The aim of the work reported in these publications was to design the equalizer of a nonredundant 140 Mbit s^{-1} binary system ($a_* = 96$ dB) so that the resulting pulse-shaper frequency response $H_I(f)$ quantitatively approximated a cosine roll-off low-pass filter ($f_I = 0.5R$, $r_I = 0.25$) as closely as possible.

Figure 8.12(a) shows that the absolute value of the pulse-shaper frequency response approximates the desired curve very well. However, the group delay time is not ideal ($\tau_I(f) \neq$ constant) (Fig. 8.12(b)). This has the effect that the basic detection pulse is not symmetrical and the postcursors are very much larger than the precursors (Fig. 8.12(c)). The normalized eye opening, which can be calculated from eqns (3.24) and (3.25), thus amounts to only about 13% (Fig. 8.12(d)).

By means of ideal phase equalization, e.g. by using all-pass filters, the symmetrical basic detection pulse shown in Fig. 8.12(e) is obtained. The normalized eye opening (Fig. 8.12(f)) increases to about 73% and thus is almost as large as in the case of the ideal cosine roll-off low-pass filter. This means that the remaining deviation relative to $|H_I(f)|$ is not noticeable in the eye opening.

A second possible method of increasing the eye opening is offered by DFE which can be effected very easily with this basic detection pulse. Figure 8.12(h) shows

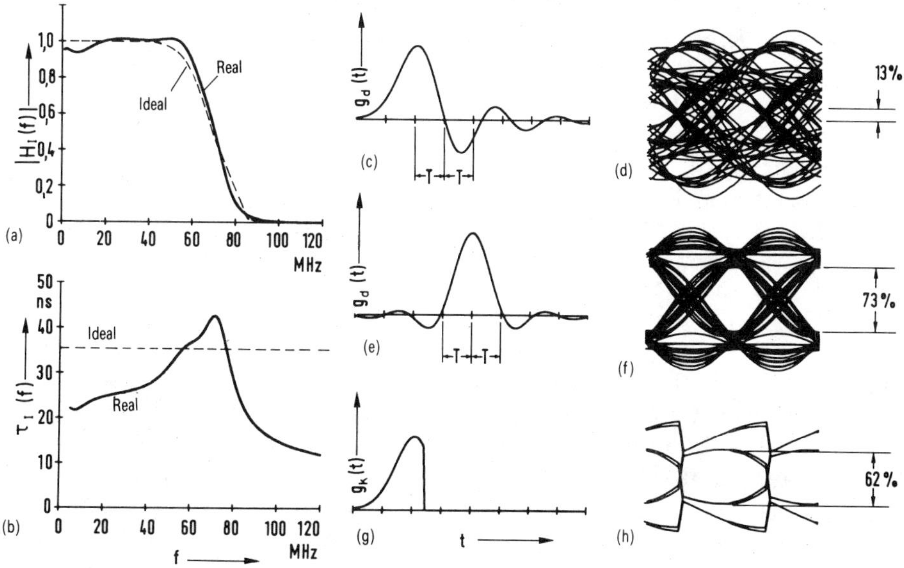

Fig. 8.12 Effect of phase distortion on the digital signal transmission [8.9]: (a) amplitude response and (b) group delay of the pulse shaper under consideration; (c) basic detection pulse and (d) eye pattern without phase equalization or DFE; (e) basic detection pulse and (f) eye pattern with ideal phase equalization and without DFE; (g) corrected basic detection pulse and (h) eye pattern without phase equalization and with ideal DFE.

that with ideal DFE the normalized eye opening is about 63%. As the detection noise power N_d is changed by neither phase equalization nor DFE, the normalized eye opening is an equivalent measure of the system efficiency and hence the error probability. In the present case the signal-to-noise ratio is improved by 15 dB by ideal phase equalization and by about 13.7 dB by ideal DFE.

Phase equalization is always necessary in a system without DFE. However, complete phase equalization is not sensible if a receiver has DFE. A further improvement can be achieved by using a special phase equalization matched to the DFE which differs from the linear phase response. The system with DFE is then clearly superior to the system without because fewer precursors and more postcursors are produced by the special phase distortion and the latter are compensated for by DFE.

8.1.7 Optimization of the pole and zero distribution

Practical equalizer circuits can be fully described by defining the zeros $p_N^{(j)}$ and the poles $p_P^{(k)}$ [1.17]. With the complex frequency $p = \sigma + j2\pi f$ we have for the **complex equalizer frequency response**

$$H_E(p) = K_E \frac{\prod_{j=1}^{J}(p - p_N^{(j)})}{\prod_{k=1}^{K}(p - p_P^{(k)})} = K_E \frac{(p - p_N^{(1)})(p - p_N^{(2)})\dots(p - p_N^{(J)})}{(p - p_P^{(1)})(p - p_P^{(2)})\dots(p - p_P^{(K)})} \tag{8.30}$$

The constant K_E is required for normalizing. However, it has no effect on the system efficiency and hence the error probability. If the numbers J and K of zeros and poles are given, the optimum values of these can be determined numerically. As the zeros $p_N^{(j)}$ and poles $p_P^{(k)}$ are either real or complex conjugates, there are exactly $J + K$ independent (real) parameters. The following principles can be advanced for optimizing these poles and zeros with regard to minimum error probability [8.7, 8.9 and 8.32].

(a) Fix the number of zeros and poles. To begin with it is advisable to keep J and K (and hence the number of independent parameters) as small as possible.

(b) Fix the starting values for the poles and zeros. The zeros $p_N^{(1)} \dots p_N^{(J)}$ and the poles $p_P^{(1)} \dots p_P^{(K)}$ should be chosen at the start of the optimization so that the channel frequency response is equalized as well as possible at low frequencies ($H_E(f) \approx 1/H_K(f)$). However, at higher frequencies the equalizer frequency response must decay more sharply than the channel frequency response, so that the resulting pulse shaper $H_I(f) = H_K(f)H_E(f)$ has low-pass characteristics.

(c) Calculate the normalized noise power according to eqn (7.51). With white noise the normalized noise power N_{norm} is completely specified by the equalizer frequency response $H_E(f)$. This is obtained from the complex frequency response $H_E(p)$ by substituting $p = j2\pi f$.

(d) Calculate the normalized eye opening (eqns (3.25) and (3.51)). The normalized eye opening $o_{\text{norm}}(T_D)$ depends only on the transmission code and the basic detection pulse (eqn (2.62)) which can be calculated from the frequency responses $H_S(f)$, $H_K(f)$ and $H_E(f)$ using the inverse (fast) Fourier transform

$$g_d(t) = \hat{g}_s T \int_{-\infty}^{+\infty} |H_S(f) H_K(f) H_E(f)| \exp(j2\pi f t)\,df \tag{8.31}$$

The Laplace transform offers another elegant method of determining $g_d(t)$. The equalizer pulse response $h_E(t)$ can then be derived directly from a decomposition of the complex frequency response $H_E(p)$ to partial fractions [1.17, 1.29]. The basic detection pulse $g_d(t)$ can also be determined from this by a convolution operation.

(e) Calculate the system efficiency from eqn (7.52). The system efficiency $\eta_A = o_{norm}{}^2(T_D)/N_{norm}$ is a measure of the error probability. The optimum detection time $\mathring{T}_D$ is obtained by maximizing η_A.

(f) Vary the poles and zeros (continuation of (c)). The poles and zeros are varied so that the system efficiency becomes a maximum. Various optimization algorithms (e.g. gradient methods or stochastic searches) can be used for this purpose. In addition, various co-conditions can be considered, e.g. the condition that all poles should be negative and real.

(g) Increase the number of poles and zeros (continuation of (b)). Increasing the values of J and K produces a general improvement. The values determined in (f) can be used as new starting values for the poles and zeros. When a further increase in J and K provides no significant improvement, the iterative optimization process can be terminated.

Even with the fast computers available today, optimization can require very long computation times if J and K are very large or if inappropriate starting values have been chosen. Therefore the choice of suitable starting values is of great importance.

Example We want to design an equalizer for a coaxial cable system. The channel frequency response $H_K(f)$ is given by eqn (2.53). In ref. 8.20 it is shown that $H_K(f)$ can be approximated by a series expansion:

$$H_K(f) = \exp\{-\alpha_2 l (2jf)^{1/2}\} \approx \frac{1 + jf/f_0}{1 + 2jf/f_0} \frac{1}{\Pi_{\kappa=1}^{Z}(1 + jf/f_\kappa)} \tag{8.32a}$$

with

$$f_0 = \frac{0.231}{\alpha_2{}^2 l^2} \quad \text{and} \quad f_\kappa = \frac{\pi^2}{8\alpha_2{}^2 l^2}(2\kappa - 1)^2 \tag{8.32b}$$

The cable-specific coefficient α_2 is in nepers (see Table 2.3) and l is the regenerator-section length.

The resulting pole–zero configuration consists of Z real poles at the points $p_\kappa = -2\pi f_\kappa$, $\kappa = 1, \ldots, Z$ and one extra pole–zero pair to improve the approximation at low frequencies. It is shown in ref. 8.32 that even when $Z = 8$ the frequency response of a normal coaxial cable of length $l = 5$ km is reproduced well up to about 100 MHz.

According to eqn (8.32) the starting value of the complex equalizer frequency response can be given as follows:

$$H_E(p) = K_E \frac{(p - 0.5p_0)\Pi_{\kappa=1}^{Z}(p - p_\kappa)}{p - p_0} \frac{\Pi_j(p - p_N^{(j)})}{\Pi_k(p - p_P^{(k)})} \tag{8.33}$$

Hence the complex pulse-shaper frequency response $H_I(p)$ determining the shape of the basic detection pulse is defined solely by the last term. It is usually sufficient for the starting function of the optimization that the pulse-shaper frequency response only has poles. If all poles are chosen to be equal and real ($p_P^{(k)} = -\sigma_P$) the pulse shaper is determined from the two independent parameters K and $\sigma_P > 0$:

$$H_I(p) = \frac{1}{\Pi_{k=1}^{K}(p - p_P^{(k)})} = \frac{1}{(p + \sigma_P)^K} \tag{8.34}$$

Rather better results are obtained with multiple complex conjugate pole pairs ($p_P^{(k)} = -\sigma \pm j\omega_P$) [8.17]. In this case the starting function of the pulse shaper has three independent parameters:

$$H_I(p) = \frac{1}{p^2 + \sigma_P p + (\sigma_P{}^2 + \omega_P{}^2)\}^{K/2}} \tag{8.35}$$

A comparison of a number of papers which deal with pole–zero configuration optimization shows that an improvement in the signal-to-noise ratio relative to that of an equalizer with a Gaussian pulse shaper can be achieved with four zeros and seven poles.

8.2 Optimization of the transmitter parameters

The basic transmitter pulse $g_s(t)$ represents another optimizable system parameter in addition to the equalizer frequency response. In this section a selection of the most common transmitter pulses (rectangular, Gaussian, raised cosine and trapezoidal) is examined and their characteristics are optimized.

The optimum basic transmitter pulse $\mathring{g}_s(t)$ depends on the transmission code and the characteristics of the channel and the receiver among other things. For reasons of illustration we assume here, as in Section 8.1.1, a nonredundant code (M levels) and an equalizer with a Gaussian pulse shaper (cut-off frequency f_I). Rather different results are obtained for other types of equalizer (see Section 8.4).

8.2.1 Effect of the duration of the transmitter pulse

First we consider the rectangular basic transmitter pulse

$$g_s(t) = \hat{g}_s \operatorname{rec}\left(\frac{t}{\Delta t_s}\right) \tag{8.36}$$

Δt_s is the (equivalent) transmitter pulse duration given by eqn (2.13) which affects both the mean transmitter power (eqn (2.20)) and the normalized eye opening (eqn (3.25)). The normalized eye opening can be calculated from eqns (2.78), (3.51) and (5.12). It is plotted in Fig. 8.13 as a function of the transmitter pulse duration Δt_s. Figure 8.13(a) is for systems without DFE so that the optimum detection time $\mathring{T}_D$ is zero.

The full curves are for the binary system ($M = 2$). If the pulse-shaper cut-off frequency is sufficiently high (e.g. $f_I = 0.5R$) the duration of the NRZ transmitter pulse is optimum, i.e. $\Delta\mathring{t}_s = T$. If a rectangular RZ transmitter pulse ($\Delta t_s < T$) is used, the main value $g_d(0)$ of the basic detection pulse is smaller, which also leads to a reduction in the eye opening (see eqn (2.79)). However, if $\Delta t_s > T$ adjacent transmitter pulses overlap and so the modulation range is twice as large for a given transmitter pulse amplitude $\hat{g}_s$.

If, in contrast, the pulse-shaper cut-off frequency is reduced (e.g. $f_I = 0.3R$) the optimum transmitter pulse duration $\Delta\mathring{t}_s$ has lower values. The reason for this is that, because of the narrow basic transmitter pulse width, the basic detection pulse is also less wide than for NRZ transmitter pulses. The associated reduction in the intersymbol interference becomes more noticeable than the reduction in the main value $g_d(0)$, which as a first approximation is proportional to Δt_s.

It can also be seen from Fig. 8.13(a) that, for a binary system, the increase in eye opening owing to reduction of the transmitter pulse duration is relatively small. However, in a quaternary system ($M = 4$) the eye opening can be substantially increased by using RZ transmitter pulses in a system without DFE.

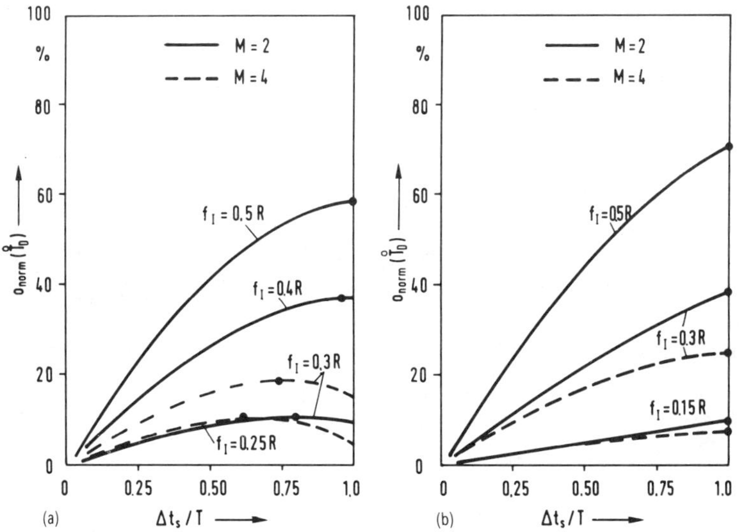

Fig. 8.13 Normalized eye opening as a function of the pulse duration Δt_s of a rectangular transmitter pulse for a system (a) without and (b) with ideal DFE.

Figure 8.13(b) is for a system with ideal DFE. In this case the optimum transmitter pulse duration $\Delta\mathring{t}_s$ is equal to T and is independent of the level number M and the cut-off frequency f_I. This means that the NRZ rectangular pulse is always the optimum transmitter pulse when DFE is used. Therefore we consider only systems without DFE in the remainder of this section.

The efficiency $10\lg\eta_\mathrm{A}$, which is valid for peak-value limitation, is plotted against the transmitter pulse duration Δt_s in Fig. 8.14(a). It is calculated according to eqn (8.17) from the normalized eye opening (Fig. 8.13(a)) and the normalized noise power (Fig. 8.1(b)). It is assumed here that the cut-off frequency of the Gaussian pulse shaper is optimally matched to the transmitter pulse duration. Apart from this the same assumptions as in the comparable Fig. 8.1(c) apply (nonredundant code with $M = 2$ or $M = 4$; coaxial pair system with $a_* = 80$ dB; white noise with $F(f) = 1$; systems without DFE).

In a binary system a gain in signal-to-noise ratio of only 0.2 dB relative to the NRZ rectangular pulse is obtained from an RZ pulse of duration $\Delta\mathring{t}_s = 0.9T$. Compared with this, a gain in signal-to-noise ratio of about 3.8 dB can be achieved in a quaternary system ($M = 4$) owing to a reduction of the transmitter pulse duration to $\Delta\mathring{t}_s = 0.6T$.

A much larger signal-to-noise ratio gain G_L is achieved under the co-condition of power limitation (Fig. 8.14(b)). The figure shows further that the optimum values of the transmitter pulse duration are smaller under the co-condition of power limitation than under peak-value limitation. This is because the mean transmitter power S_s with rectangular transmitter pulses increases in proportion to Δt_s. Since, according to eqn (7.54), the system efficiency η_L depends on S_s, a smaller value of $\Delta\mathring{t}_s$ is obtained than is the case under peak-value limitation.

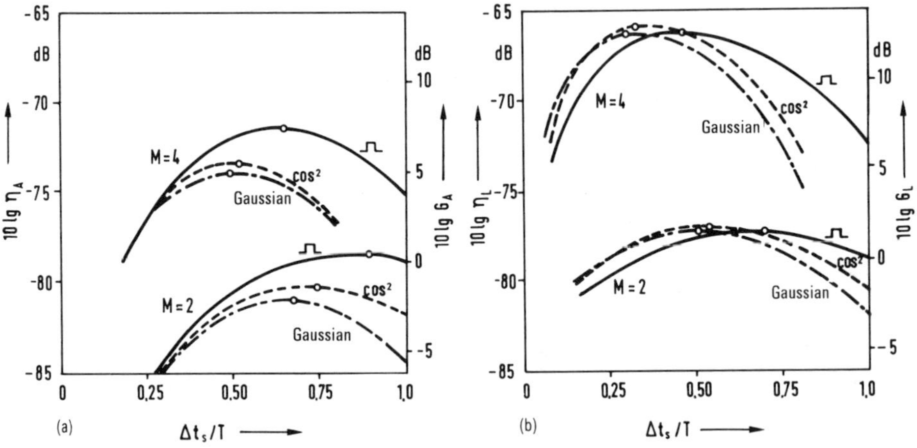

Fig. 8.14 System efficiency and gain in the signal-to-noise ratio as a function of the equivalent pulse duration of rectangular, Gaussian and raised cosine transmitter pulses under (a) peak-value limitation and (b) power limitation.

Note For completeness Fig. 8.14 also contains the curves for the raised cosine transmitter pulse (broken curves) and the Gaussian transmitter pulse (chain curves). In contrast with peak-value limitation, under power limitation a further, albeit small, improvement in signal-to-noise ratio can be achieved from a raised cosine or Gaussian transmitter pulse.

8.2.2 Effect of the steepness of the leading and trailing edges

In high-speed pulse transmission precisely rectangular transmitter pulses are often not achieved. To assess the effect of the steepness of the leading and trailing edges of real transmitter pulses, the additional case of a basic transmitter pulse in the form of the trapezoidal pulse shown in Fig. 8.15(a) is considered. The **roll-off factor** r_s is used to measure the steepness of the leading and trailing edges of the transmitter pulse, where

$$r_s = \frac{T_s - \Delta t_s}{\Delta t_s} \qquad 0 \leqslant r_s \leqslant 1 \tag{8.37}$$

This can be computed from the absolute and equivalent transmitter pulse durations T_s and Δt_s (eqn (2.13)). The special cases of a rectangular pulse ($r_s = 0$) and a triangular pulse ($r_s = 1$) are also included in this figure.

The gain G_A in the signal-to-noise ratio under peak-value limitation (see eqn (8.19)) is plotted against the parameters Δt_s and r_s in Fig. 8.15(b). This figure is for the level number $M = 2$ only; apart from this the same assumptions as for Fig. 8.14 are valid. It can be seen that under peak-value limitation the optimum roll-off factor $\mathring{r}_s$ of the transmitter pulse is zero. This means that, for this co-condition, the rectangular pulse is more beneficial than a pulse with edges of finite steepness. A triangular pulse

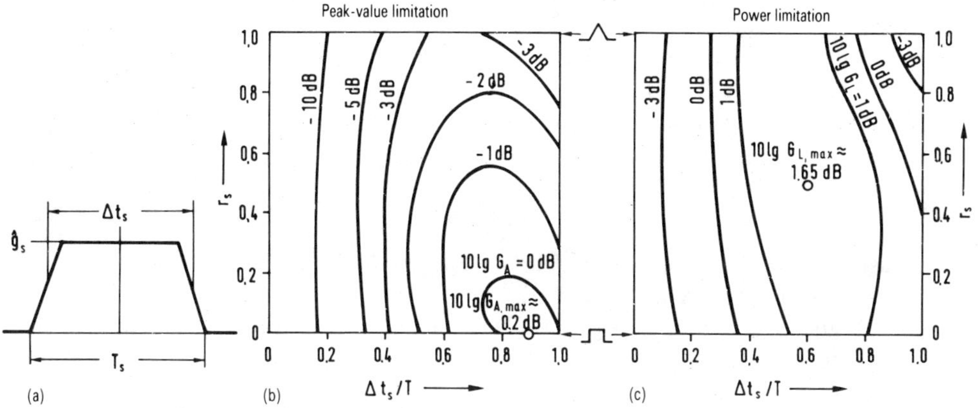

Fig. 8.15 (a) Basic trapezoidal transmitter pulse; (b), (c) gain in the signal-to-noise ratio under peak-value limitation ($10 \lg G_A$) and power limitation ($10 \lg G_L$) as a function of the equivalent transmitter pulse duration Δt_s and the roll-off factor r_s for the pulse shown in (a).

($r_s = 1$) can be expected to give a loss of signal-to-noise ratio of more than 3 dB compared with the rectangular pulse ($r_s = 0$).

Under power limitation, with optimum pulse duration ($\Delta \mathring{t}_s < T$) and optimum roll-off factor ($\mathring{r}_s > 0$) for the transmitter pulse, a slight improvement relative to the NRZ rectangular pulse is obtained (Fig. 8.15(c)). The optimum parameter values of the basic transmitter pulse are $\Delta \mathring{t}_s \approx 0.6T$ and $\mathring{r}_s \approx 0.5$. By varying the roll-off factor between the extreme values of zero (rectangular) and unity (triangular), with fixed transmitter pulse duration ($\Delta t_s = 0.6T =$ constant), we obtain a loss in signal-to-noise ratio of less than 0.5 dB relative to the optimum roll-off factor ($\mathring{r}_s = 0.5$).

Examination of trapezoidal pulses for multilevel systems and systems with DFE leads to quantitatively similar results.

8.2.3 Asymmetric basic transmitter pulse with a negative component

It is shown in ref. 8.18 that a substantial improvement in the signal-to-noise ratio can be achieved using a basic transmitter pulse with a negative component. Additionally we have to assume here a receiver without DFE and a Gaussian pulse shaper.

For a given transmitter pulse amplitude $\hat{g}_s$ this asymmetrical pulse is defined by the three independent parameters T_+, T_- and α. The rectangular pulse is included as a special case, $\alpha = 0$, in this general illustration.

The basic detection pulse can be determined from the step response of the Gaussian low-pass filter (see the Appendix, Table A3) by using the shift theorem. We obtain (Fig. 8.16(b))

$$g_d(t) = \hat{g}_s[\phi\{f_I'(t + T_+)\} - (1 + \alpha)\phi(f_I't) + \alpha\phi\{f_I'(t - T_-)\}] \tag{8.38}$$

where $f_I' = 2(2\pi)^{1/2} f_I$. $\phi(x)$ denotes the Gaussian error integral. The eye opening can be computed from this using eqn (3.51). Further, as the noise power is given by eqn (8.16) and Fig. 8.1(b), the system efficiency η_A can be determined from eqn (8.17).

The optimization of transmitter parameters under peak-value limitation (i.e. the maximizing of η_A) leads in all cases to the values

$$T_+ = T - T_- \tag{8.39a}$$

$$\alpha = 1 \tag{8.39b}$$

This means that, with peak-value limitation, the optimum basic transmitter pulse always has the maximum energy $\hat{g}_s^{\,2} T$.

Figure 8.16(c) shows the gain G_A in the signal-to-noise ratio for a binary system as a function of the parameter T_-. Otherwise the assumptions are the same as for Figs 8.14 and 8.15. If $f_I = 0.32R$ (optimum pulse-shaper cut-off frequency with NRZ rectangular transmitter pulse) and $T_- = 0$, the gain G_A in the signal-to-noise ratio is by definition 0 dB. With equal cut-off frequency the optimum value of T_- is about $0.3T$. However, by using the asymmetric basic transmitter pulse with a negative

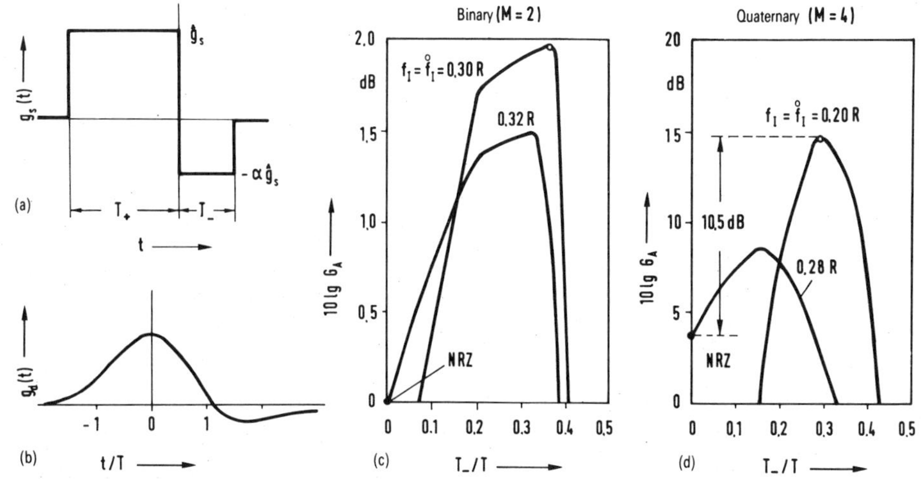

Fig. 8.16 (a) Basic transmitter pulse with a negative component [8.19]; (b) the corresponding basic detection pulse; (c), (d) gain 10 lg G_A in the signal-to-noise ratio under peak-value limitation as a function of the duration T_- of the negative component for $M = 2$ and $M = 4$ ($a_* = 80$ dB).

component it is also possible to reduce the pulse-shaper cut-off frequency f_I which favorably affects the limitation of the detection noise power. An improvement of about 2 dB in the signal-to-noise ratio relative to the optimum system with NRZ rectangular transmitter pulses can be achieved for the optimum parameter values $\mathring{f}_I = 0.3R$ and $\mathring{T}_- = 0.35T$. Figure 8.16(d) (which is on a different scale) makes it clear that for a multilevel system a very much larger improvement is possible using the negative component. For example, for $M = 4$ the improvement is more than 10 dB [3.12].

The negative component of the basic transmitter pulse provides some compensation for the postcursors of the basic detection pulse. Therefore such a basic transmitter pulse is only sensible in a system without DFE. In a system with DFE the NRZ rectangular transmitter pulse is the optimum. It can also be shown that with an optimum equalizer the NRZ rectangular pulse is also optimum without DFE if peak-value limitation is assumed (see Section 8.4).

8.3 Optimization of the equalizer in Nyquist systems

It is possible to eliminate intersymbol interference completely by suitable shaping of the equalizer frequency response. Such a system is often referred to as a "Nyquist system". In this section the Nyquist criteria are defined, and the optimum Nyquist equalizer is computed for systems with and without DFE.

8.3.1 Nyquist criteria in the time and frequency domains

If the basic detection pulse $g_d(t)$ has zero-crossings separated by equal intervals νT, where T is the symbol duration, it is referred to as a **Nyquist pulse**:

$$g_d(t) = g_N(t) \quad \text{when} \quad g_d(\nu T) = 0 \quad \text{for} \quad \text{all } \nu \neq 0 \tag{8.40}$$

where the subscript N indicates a Nyquist pulse. A detection time T_D of zero is assumed in this definition. However, the following expressions can be adapted for the case $T_D \neq 0$ by using the shift theorem.

The first Nyquist criterion For many applications it is necessary to define the Nyquist criteria in the frequency domain. If the basic detection pulse

$$g_d(t) = \hat{g}_s T\{h_S(t) * h_K(t) * h_E(t)\}$$

is a Nyquist pulse as defined above, the overall frequency response of the transmitter, the channel and the equalizer is called the **Nyquist frequency response** $H_N(f) \bullet\!\!-\!\!\circ\, g_N(t)/\hat{g}_s T$:

$$H_S(f)H_K(f)H_E(f) = H_S(f)H_I(f) \stackrel{!}{=} H_N(f) \tag{8.41}$$

The condition which a Nyquist frequency response must satisfy was defined by Nyquist in 1928 [8.28] and is known as the first **Nyquist criterion**:

$$\sum_{\kappa=-\infty}^{+\infty} H_N\left(f - \frac{\kappa}{T}\right) = K_N = \text{constant} \tag{8.42}$$

T is the time interval between the equidistant zero-crossings of the Nyquist pulse $g_N(t)$. The constant K_N is calculated using eqn (8.48).

Proof Using eqn (8.41) it emerges from the Nyquist criterion in the time domain that

(a) $$g_N(\nu T) = \hat{g}_s T \int_{-\infty}^{+\infty} H_N(f) \exp(j2\pi f \nu T)\, df \stackrel{!}{=} 0 \qquad \nu \neq 0$$

If the Fourier integral is decomposed into partial integrals of width $1/T$, the equations become

(b) $$\sum_{\kappa=-\infty}^{+\infty} \int_{(\kappa-1/2)/T}^{(\kappa+1/2)/T} H_N(f) \exp(j2\pi f \nu T)\, df \stackrel{!}{=} 0 \qquad \nu \neq 0$$

Substituting $f' = f + \kappa/T$ it follows that

(c) $$\sum_{\kappa=-\infty}^{+\infty} \int_{-1/2T}^{1/2T} H_N\left(f' - \frac{\kappa}{T}\right) \exp\left\{j2\pi\left(f' - \frac{\kappa}{T}\right)\nu T\right\} df' \stackrel{!}{=} 0 \qquad \nu \neq 0$$

For integer values of κ and ν, $\exp(-j2\pi\kappa\nu) = 1$. Exchanging summation and integration we obtain for $f' \to f$

(d) $$\int_{-1/2T}^{1/2T} \sum_{\kappa=-\infty}^{+\infty} H_N\left(f - \frac{\kappa}{T}\right) \exp(j2\pi f \nu T)\, df \overset{!}{=} 0 \qquad \nu \neq 0$$

This requirement can only be fulfilled for all $\nu \neq 0$ if the sum to infinity (eqn (8.42)) is independent of f and thus has a constant value. □

The frequency response with the smallest bandwidth which satisfies the first Nyquist criterion is that of the rectangular (brick-wall) low-pass filter with cut-off frequency $f_N = 1/2T$ (see the Appendix, Table A3):

$$H_N(f) = K_N \operatorname{rec}(fT) = K_N \operatorname{rec}\left(\frac{f}{2f_N}\right) \tag{8.43}$$

For the corresponding Nyquist pulse the following is true:

$$g_N(t) = K_N \hat{g}_s \operatorname{si}\left(\pi \frac{t}{T}\right) \tag{8.44}$$

The oscillations of this pulse decay very slowly between zero-crossings (asymptotically with $1/t$), and therefore the eye opening narrows to an infinitely small slit, i.e. $o(T_D) = 0$ for $T_D \neq 0$. Hence a transmission with a sufficiently small error probability is only possible with an ideal (jitter-free) clock signal.

If the absolute (one-sided) bandwidth of a channel is less than half the symbol rate (**Nyquist frequency**)

$$f_N = \frac{1}{2T} \tag{8.45}$$

then the Nyquist criterion in eqn (8.42) cannot be satisfied and a transmission free from intersymbol interference is not possible. Further, it can also be shown that, even in the absence of noise, not all symbol sequences (e.g. the long LOLOLO sequence) can be transmitted error-free over such a channel. This means that for a channel bandwidth which is smaller than the Nyquist frequency f_N the eye is closed if special measures, e.g. DFE or Viterbi detection, are not taken. Other Nyquist frequency responses which utilize a larger frequency band can be derived from the rectangular Nyquist frequency response.

Example The real rectangular Nyquist frequency response in Fig. 8.17(a) is divided into six partial-frequency responses. If these partial-frequency responses are displaced along the frequency axis in integer multiples of the symbol rate $1/T = 2f_N$, a Nyquist frequency response again occurs (Fig. 8.17(b)). As only real-time functions are considered here, the rectangular frequency response has to be decomposed and the parts have to be displaced (according to eqn (8.42)) such that the resulting real part is even.

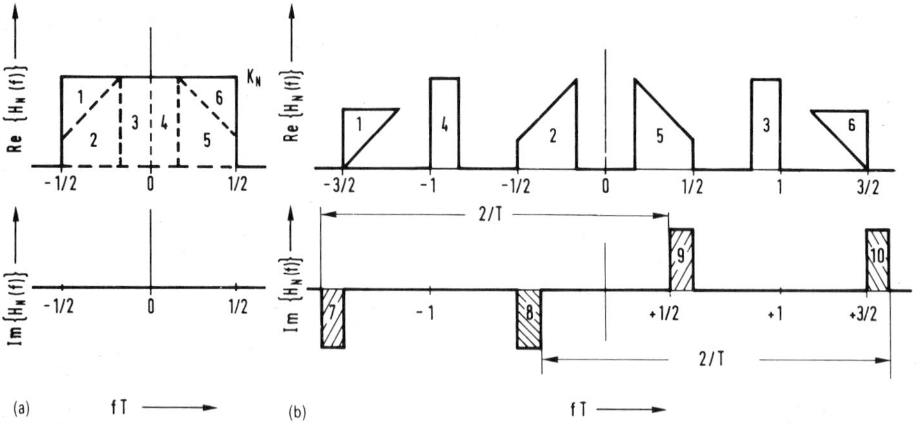

Fig. 8.17 Examples of Nyquist frequency responses (real and imaginary components): (a) rectangular real; (b) arbitrary complex.

In general the Nyquist frequency response can also have an imaginary component as long as the following requirement is not violated:

$$\sum_{\kappa=-\infty}^{+\infty} \mathrm{Im}\left\{H_N\left(f-\frac{\kappa}{T}\right)\right\}=0 \tag{8.46}$$

The Nyquist pulse $g_N(t)$ corresponding to the complex Nyquist frequency response in Fig. 8.17(b) is asymmetric. From eqn (8.42) we obtain for time $t=0$

$$g_N(0)=\hat{g}_s T\int_{-\infty}^{+\infty} H_N(f)\,\mathrm{d}f=\hat{g}_s T\int_{-1/2T}^{1/2T}\sum_{\kappa=-\infty}^{+\infty} H_N\left(f-\frac{\kappa}{T}\right)\mathrm{d}f=K_N\hat{g}_s \tag{8.47}$$

This means that the constant

$$K_N=\sum_{\kappa=-\infty}^{+\infty} H_N\left(\frac{\kappa}{T}\right)=\frac{g_N(0)}{\hat{g}_s} \tag{8.48}$$

which has not yet been determined, specifies the ratio of the amplitudes of the basic detection pulse and the transmitter pulse.

$1/T$ Nyquist frequency responses Of particular significance to digital transmission systems are the Nyquist frequency responses which are limited to the frequency range $-1/T \leqslant f \leqslant +1/T$ and are continuous. These are often referred to in the literature as **$1/T$ Nyquist frequency responses**.

As can be seen from eqn (8.42) the following must be true for the real and imaginary components of a $1/T$ Nyquist frequency response in the case of real pulses (first Nyquist criterion):

$$\mathrm{Re}\{H_N(f_N+f)\}+\mathrm{Re}\{H_N(f_N-f)\}=H_N(0)=K_N \tag{8.49a}$$

$$\mathrm{Im}\{H_N(f_N+f)\}+\mathrm{Im}\{H_N(f_N-f)\}=0 \tag{8.49b}$$

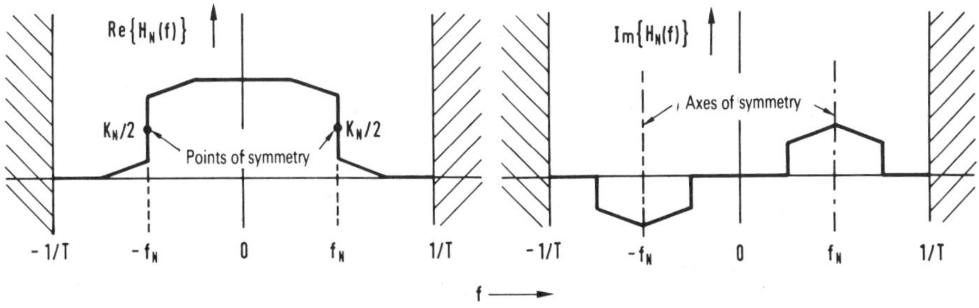

Fig. 8.18 Symmetry characteristics of the real and imaginary components of a $1/T$ Nyquist frequency response.

These two equations specify that the real component of a $1/T$ Nyquist frequency response is centrosymmetric about the two points $\{f = \pm f_N; H_N(\pm f_N) = K_N/2\}$, whereas the imaginary component is axisymmetric about the frequencies $f = \pm f_N$ (Fig. 8.18).

Example The trapezoidal low-pass filter and the cosine roll-off low-pass filter in Table 8.2 satisfy the condition above as long as their cut-off frequency is equal to

Table 8.2 Nyquist frequency responses and the corresponding eye patterns

Trapezoidal low-pass	*Cosine roll-off low-pass*	*Cosine low-pass*	*Raised cosine low-pass*
Frequency response 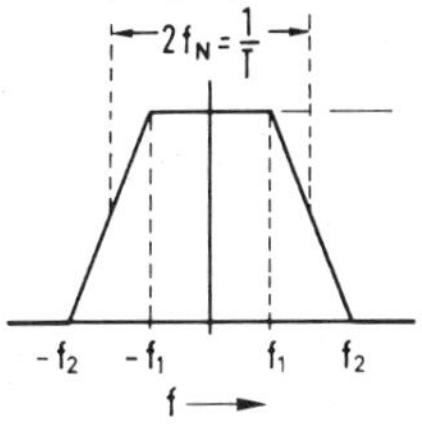	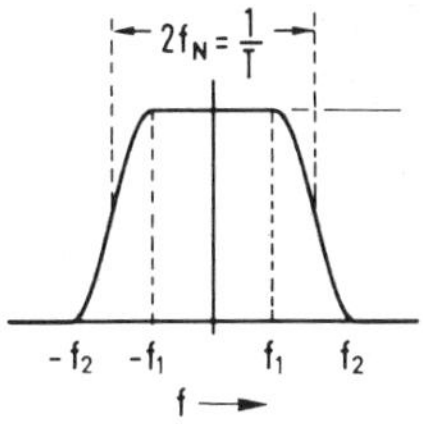	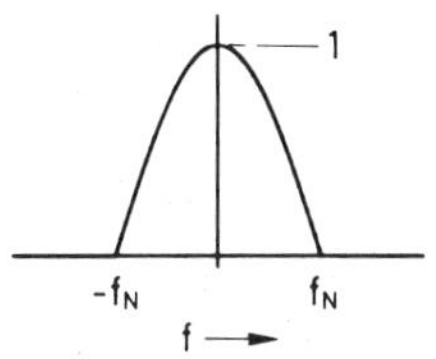	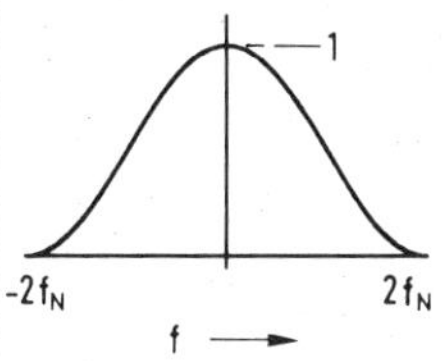
Nyquist pulse $\hat{g}_s\, \mathrm{si}(\pi r_N \frac{t}{T})\, \mathrm{si}(\pi \frac{t}{T})$	$\hat{g}_s \frac{\cos(\pi r_N t/T)}{1-(2r_N t/T)^2} \mathrm{si}(\pi \frac{t}{T})$	$\hat{g}_s \frac{2}{\pi} \frac{\cos(\pi t/T)}{1-(2t/T)^2}$	$\hat{g}_s \frac{\cos(\pi t/T)}{1-(2t/T)^2} \mathrm{si}(\pi \frac{t}{T})$
Binary eye diagram 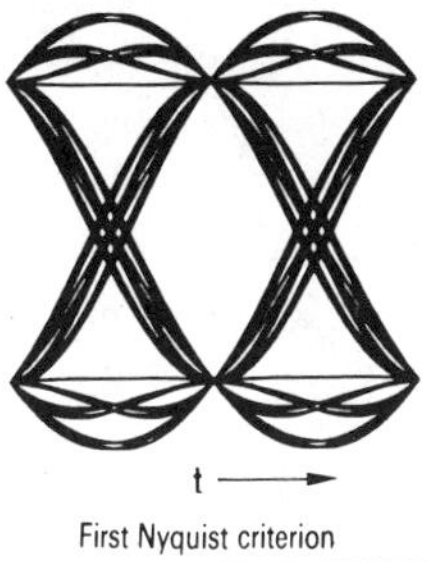 First Nyquist criterion	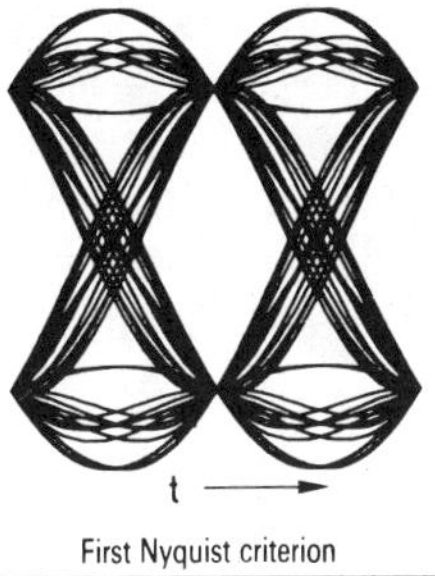 First Nyquist criterion	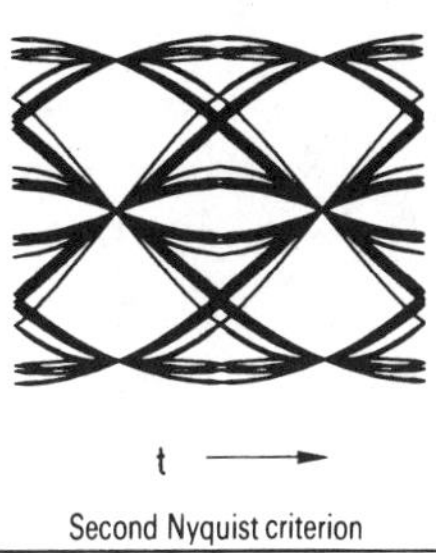 Second Nyquist criterion	 First and second Nyquist criteria

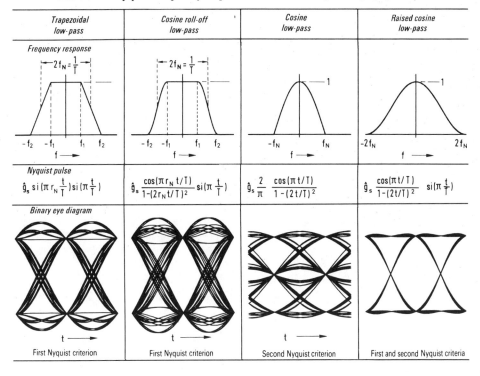

the Nyquist frequency $f_N = 1/2T$. The **roll-off factor**

$$r_N = \frac{f_2 - f_1}{f_2 + f_1} \tag{8.50}$$

can then have any arbitrary value between zero and unity. The pulse responses of both low-pass filters have equidistant zero-crossings at intervals T. The oscillations between the times νT are smaller the larger the roll-off factor r_N becomes. The pulse response of the trapezoidal low-pass filter decays asymptotically with $1/t^2$ which, in contrast with the si pulse ($r_N = 0$), leads to a temporal eye opening. This improved response is due to the more moderate slope of the trapezoidal low-pass filter compared with the rectangular Nyquist frequency response. The pulse response of the cosine roll-off low-pass filter decays asymptotically with $1/t^3$ as, in contrast with the trapezoidal low-pass filter, its frequency response no longer exhibits kinks.

The second Nyquist criterion The first Nyquist criterion is satisfied if the basic detection pulse $g_d(t)$ has zero-crossings at times $\pm T$, $\pm 2T$etc. In this case the eye is completely open vertically (see Table 8.2, columns 1 and 2).

However, the basic detection pulse is defined as a **Nyquist-2 pulse** $g_{N2}(t)$ if it has zero-crossings at times $\pm 1.5T$, $\pm 2.5T$, $\pm 3.5T$etc. In the case of a Nyquist-2 pulse the zero-crossings of the detection signal are not displaced from their true position $(\nu + 1/2)T$, and therefore the horizontal eye opening is maximum and equal to the symbol duration T of the transmitter (see Table 8.2, columns 3 and 4). However, with this type of equalization the pulse detection at times νT can be affected by the precursors and postcursors of the neighboring pulses.

A Nyquist-2 pulse can always be represented by the sum of two (possibly different) Nyquist-1 pulses displaced by $\pm T/2$. If we restrict our attention to symmetric pulses we obtain the following relationship in the time domain:

$$g_{N2}(t) = g_N\left(t + \frac{T}{2}\right) + g_N\left(t - \frac{T}{2}\right) \tag{8.51}$$

Figure 8.19(a) illustrates this result for the case where the Nyquist-1 pulse $g_N(t)$ is a si

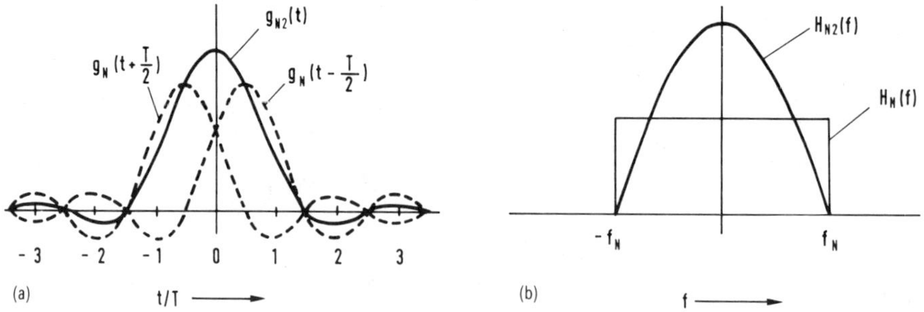

Fig. 8.19 Illustration of the second Nyquist criterion in (a) the time domain and (b) the frequency domain.

pulse. The resulting Nyquist-2 pulse $g_{N2}(t)$ exhibits zero-crossings, as required, at times $\pm 1.5T, \pm 2.5T, \ldots$. At time $t = \pm T/2$, $g_d(t) = g_N(0)$ is not zero. The maximum value $2g_N(T/2)$ occurs when $t = 0$.

To derive the second Nyquist criterion in the frequency domain we again assume symmetric pulses so that eqn (8.51) can be used. For the corresponding Nyquist-2 frequency response $H_{N2}(f) \bullet\!\!-\!\!\circ\, g_{N2}(t)/\hat{g}_s T$ we obtain from the shift theorem

$$H_{N2}(f) = H_N(f)\exp(\mathrm{j}\pi f T) + H_N(f)\exp(-\mathrm{j}\pi f T) = 2H_N(f)\cos(\pi f T) \quad (8.52)$$

However, as $H_N(f) \bullet\!\!-\!\!\circ\, g_N(t)/\hat{g}_s T$ must satisfy the first Nyquist criterion defined in eqn (8.42), we obtain the second Nyquist criterion for symmetric pulses as follows:

$$\sum_{\kappa=-\infty}^{+\infty} \frac{H_{N2}(f-\kappa/T)}{\cos(\pi f T - \kappa\pi)} = \frac{\sum_{\kappa=-\infty}^{+\infty}(-1)^{\kappa} H_{N2}(f-\kappa/T)}{\cos(\pi f T)} = \text{constant} \quad (8.53)$$

A si pulse was assumed for the Nyquist-1 pulse in Fig. 8.19, so that the corresponding Nyquist-1 frequency response $H_N(f)$ is a rectangular low-pass filter with cut-off frequency $f_N = 1/2T$. It follows from this that the Nyquist-2 frequency response is (see eqn (8.52))

$$H_{N2}(f) = 2K_N \operatorname{rec}(fT)\cos(\pi f T) \quad (8.54)$$

The cosine-squared low-pass filter has particular significance for digital transmission systems (see Table 8.2, column 4). In the general representation of the cosine roll-off low-pass filter (Table 8.2, column 2) this comprises a special case with $r_N = 1$ and it satisfies both the first and second Nyquist criteria. Hence in the case of this pulse shaper neither the symbol detection (error probability) nor the timing recovery (phase jitter) is adversely affected by the precursors and postcursors of the neighboring pulses.

8.3.2 Optimum Nyquist equalizer for a system without decision feedback equalization

In this section the optimum equalizer frequency response $\mathring{H}_E(f)$ is determined for systems without DFE. The system efficiency defined in Section 7.2 is used as the optimization criterion. Optimization is performed under the co-condition that the overall frequency response $H_S(f)H_K(f)H_E(f)$ is a Nyquist frequency response. It can indeed be shown that in the transmission channels commonly used the optimum Nyquist system is better than all the systems which have intersymbol interference if symbol-by-symbol detection is assumed (see Section 8.3.3).

The transmitter characteristics (frequency response $H_S(f)$, level number M) are assumed to be given. Therefore the optimization obtained here is valid under both power limitation and peak-value limitation. This means that both system efficiencies η_L (eqn (7.16)) and η_A (eqn (7.17)) can be used in the same way as the optimization criterion. Both criteria produce the same optimum equalizer frequency response $\mathring{H}_E(f)$.

In the following the system efficiency η_A under peak-value limitation is assumed. As only bipolar Nyquist systems are considered, $o(T_D) = 2g_d(T_D)/(M-1)$, and we obtain from eqn (7.49)

$$\eta_A = \frac{\{g_d(0)\}^2}{(M-1)^2 s_{\max}{}^2} \frac{L_{th}R}{N_d} = \frac{\gamma_S{}^2 K_N{}^2 L_{th} R}{(M-1)^2 N_d} \tag{8.55}$$

Here the detection time $\mathring{T}_D = 0$ which is optimum for a symmetric pulse is assumed. The factor $\gamma_S = \hat{g}_s/s_{\max}$ given by eqn (7.40) takes account of intersymbol interference at the transmitter and, like the term $1/(M-1)^2$, is a constant for receiver optimization. $K_N = g_d(0)/\hat{g}_s$ is the ratio of pulse amplitudes according to eqn (8.48). With normalized noise power (eqn (7.51)) we obtain

$$\eta_A = \frac{\gamma_S{}^2}{(M-1)^2} \frac{K_N{}^2}{|H_E(0)|^2} \frac{1}{N_{\text{norm}}} \tag{8.56}$$

for the system efficiency. For receiver optimization with a given transmitter and channel the ratio

$$\alpha = \frac{K_N}{|H_E(0)|} \tag{8.57}$$

can be considered to be a constant. Hence the maximization of η_A is equivalent to minimization of the normalized detector input noise power according to eqn (7.51):

$$N_{\text{norm}} = \frac{1}{|H_E(0)|^2 R} \int_{-\infty}^{+\infty} F(f)|H_E(f)|^2 \, df \stackrel{!}{=} \text{minimum} \tag{8.58}$$

It should be noted here that the overall frequency response

$$H_S(f) H_K(f) H_E(f) = H_N(f)$$

must satisfy the first Nyquist criterion (eqn (8.42)). Therefore using the constant α from eqn (8.57) the above optimization condition can also be written as

$$N_{\text{norm}} = \frac{\alpha^2}{K_N{}^2 R} \int_{-\infty}^{+\infty} \frac{F(f)|H_N(f)|^2}{|H_S(f)|^2 |H_K(f)|^2} \, df \stackrel{!}{=} \text{minimum} \tag{8.59}$$

The optimum Nyquist frequency response $\mathring{H}_N(f)$ which satisfies this condition can be calculated by variational calculus and is [1.15, 8.22]

$$\mathring{H}_N(f) = K_N \frac{|H_S(f)|^2 |H_K{}'(f)|^2}{\sum_{\kappa=-\infty}^{+\infty} |H_S(f-\kappa/T)|^2 |H_K{}'(f-\kappa/T)|^2} \tag{8.60}$$

For simplicity the channel frequency response

$$H_K{}'(f) = \frac{H_K(f)}{\{F(f)\}^{1/2}} \tag{8.61}$$

based on the noise factor $F(f)$ is used here, so that eqn (8.60) is valid for both white and colored noise.

Proof Every Nyquist pulse (see eqn (8.40)) can be written in the form

(a)
$$g_N(t) = \hat{g}_s \, \mathrm{si}\left(\pi \frac{t}{T}\right) b(t)$$

The equidistant zero-crossings at times $\nu T \neq 0$ are governed by the si function. At other times $g_N(t)$ is defined by the arbitrary dimensionless function $b(t)$. For $t = 0$ it must be true that $b(0) \neq 0$. Using the convolution theorem we obtain for every arbitrary Nyquist frequency response $H_N(f)$ •——○ $g_N(t)/\hat{g}_s T$

(b)
$$H_N(f) = \mathrm{rec}(fT) * B(f) = \int_{f-1/2T}^{f+1/2T} B(f')\,\mathrm{d}f'$$

where $B(f)$ is the Fourier transform of $b(t)$. Inserting this into eqn (8.59) produces the optimization condition

(c)
$$N_{\mathrm{norm}} = \frac{\alpha^2}{K_N{}^2 R} \int_{-\infty}^{+\infty} \frac{\{\mathrm{rec}(fT) * B(f)\}^2}{|H_S(f)|^2 |H_K'(f)|^2} \mathrm{d}f \overset{!}{=} \text{minimum}$$

Now, according to the method of variational calculus, we insert the function

(d)
$$B(f) = \mathring{B}(f) + \varepsilon B_\varepsilon(f)$$

in this equation. $\mathring{B}(f)$ is the desired optimum function which fulfils the condition (c), while $\varepsilon B_\varepsilon(f)$ represents an arbitrary deviation from this optimum function. If the variable ε is non-zero, the resultant frequency function $B(f)$ deviates from the optimum $\mathring{B}(f)$ so that the noise power given by (c) cannot be a minimum. It thus follows that for each arbitrary function $B_\varepsilon(f)$ the following must be true:

(e)
$$\lim_{\varepsilon \to 0} \frac{\mathrm{d}N_{\mathrm{norm}}}{\mathrm{d}\varepsilon} \overset{!}{=} 0$$

When this operation is completed, the following condition emerges from (c):

(f)
$$\int_{-\infty}^{+\infty} \frac{\{\mathrm{rec}(fT) * \mathring{B}(f)\}\{\mathrm{rec}(fT) * B_\varepsilon(f)\}}{|H_S(f)|^2 |H_K'(f)|^2} \mathrm{d}f \overset{!}{=} 0$$

In addition, it should be noted that $\mathrm{rec}(fT) * \mathring{B}(f) = \mathring{H}_N(f)$. Furthermore, it follows from (b) that

(g)
$$\int_{-\infty}^{+\infty} \frac{\mathring{H}_N(f)}{|H_S(f)|^2 |H_K'(f)|^2} \int_{f-1/2T}^{f+1/2T} B_\varepsilon(f')\,\mathrm{d}f'\mathrm{d}f \overset{!}{=} 0$$

For abbreviation we now introduce the frequency response

(h)
$$H_{\mathrm{TF}}(f) = \frac{\mathring{H}_N(f)}{K_N |H_S(f)|^2 |H_K'(f)|^2}$$

and the indefinite integral $B_I(f) = \int B_\varepsilon(f)\,df$. Hence condition (g) can be written

$$\text{(i)} \qquad \int_{-\infty}^{+\infty} H_{TF}(f) B_I\left(f + \frac{1}{2T}\right) df \stackrel{!}{=} \int_{-\infty}^{+\infty} H_{TF}(f) B_I\left(f - \frac{1}{2T}\right) df$$

Substitution of $f - 1/2T$ for the integration variable f on the left-hand side of the equation and $f + 1/2T$ for f on the right-hand side yields

$$\text{(j)} \qquad \int_{-\infty}^{+\infty} H_{TF}\left(f - \frac{1}{2T}\right) B_I(f)\,df \stackrel{!}{=} \int_{-\infty}^{+\infty} H_{TF}\left(f + \frac{1}{2T}\right) B_I(f)\,df$$

For every arbitrary frequency function $B_\varepsilon(f)$—and hence also for every arbitrary integral function $B_I(f)$—conditions (j) can only be satisfied if $H_{TF}(f - 1/2T) = H_{TF}(f + 1/2T)$. However, this means that $H_{TF}(f)$ is a periodic function of the frequency with the symbol rate $1/T$. Considering also the first Nyquist criterion (eqn (8.42)), we obtain from (h) the condition

$$\text{(k)} \quad \sum_{\kappa=-\infty}^{+\infty} \mathring{H}_N\left(f - \frac{\kappa}{T}\right) = K_N \sum_{\kappa=-\infty}^{+\infty} H_{TF}\left(f - \frac{\kappa}{T}\right) \left|H_S\left(f - \frac{\kappa}{T}\right)\right|^2 \left|H_K'\left(f - \frac{\kappa}{T}\right)\right|^2 \stackrel{!}{=} K_N$$

As $H_{TF}(f)$ is periodic with $1/T$ this function can be extracted from the summation and we obtain

$$\text{(l)} \qquad H_{TF}(f) = \left\{ \sum_{\kappa=-\infty}^{+\infty} \left|H_S\left(f - \frac{\kappa}{T}\right)\right|^2 \left|H_K'\left(f - \frac{\kappa}{T}\right)\right|^2 \right\}^{-1}$$

The result shown in eqn (8.60) is obtained directly from (h) and (l). □

Thus the optimum Nyquist frequency response $\mathring{H}_N(f)$ is determined. For the frequency response $\mathring{H}_E(f) = \mathring{H}_N(f)/H_S(f)H_K(f)$ of the optimum Nyquist equalizer we obtain from eqn (8.61)

$$\mathring{H}_E(f) = \underbrace{K_N \frac{H_S^*(f) H_K^*(f)}{F(f)}}_{\text{matched filter } H_{MF}(f)} \underbrace{\left\{ \sum_{\kappa=-\infty}^{+\infty} \frac{|H_S(f-\kappa/T)|^2 |H_K(f-\kappa/T)|^2}{F(f-\kappa/T)} \right\}^{-1}}_{\text{transverse filter } H_{TF}(f)} \qquad (8.62)$$

Interpretation of the optimum Nyquist equalizer The first component of $\mathring{H}_E(f)$ is the same as the matched filter frequency response $H_{MF}(f)$ for colored noise (see eqn (6.51)). The second component is a function which is periodic with the symbol rate $1/T$ and hence can be written as a Fourier series:

$$H_{TF}(f) = \sum_{\lambda=-\infty}^{+\infty} k_\lambda \exp(-j2\pi f \lambda T) \qquad (8.63)$$

where the k_λ are the coefficients of the Fourier series. Comparison with Fig. 5.6 shows that $H_{TF}(f)$ corresponds to the frequency response of a (noncausal) transverse filter with filter coefficients k_λ.

The optimum Nyquist equalizer for a system without DFE is thus a combination of a matched filter and an infinite transverse filter (Fig. 8.20(a)):

$$\mathring{H}_E(f) = H_{MF}(f)\mathrm{H}_{TF}(f) \tag{8.64}$$

The characteristics of a matched filter were discussed in Section 6.3. If an input pulse $g_e(t) = g_s(t)*h_K(t)$ is applied to the input of the matched filter, then its output pulse is given by eqn (6.54):

$$g_m(t) = g_e(t)*h_{MF}(t) = K_N\hat{g}_s T\int_{-\infty}^{+\infty} \frac{|H_S(f)|^2|H_K(f)|^2}{F(f)} \exp(-j2\pi f t)\,df \tag{8.65}$$

If the noise signal has the power spectrum $L_n(f) = F(f)L_{th}$, the signal-to-noise ratio

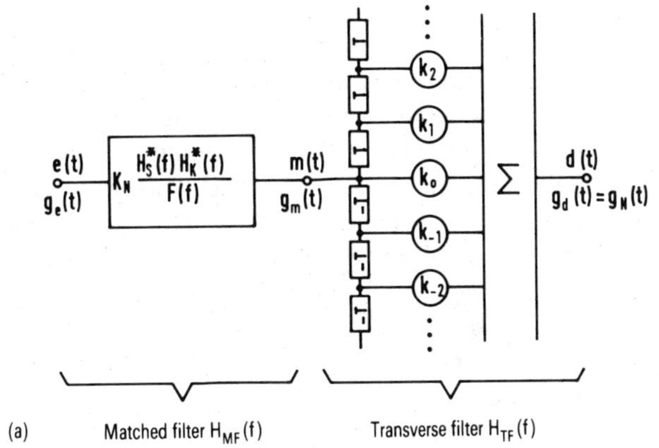

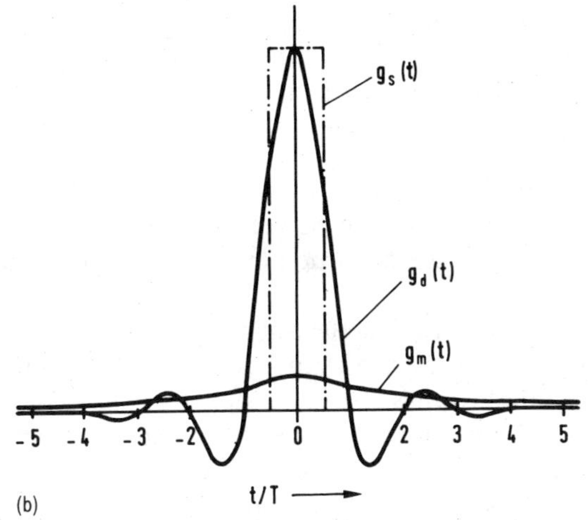

Fig. 8.20 (a) Block diagram of the optimum Nyquist equalizer; (b) basic pulses $g_m(t)$ and $g_d(t)$ of a coaxial cable system ($M = 2$, $a_* = 20$ dB and $F(f) = 1$).

at the detection time $\mathring{T}_D = 0$ is a maximum:

$$\rho_m = \frac{g_m^2(0)}{N_m} = \frac{\hat{g}_s^2 T}{L_{th}} T \int_{-\infty}^{+\infty} \frac{|H_S(f)|^2 |H_K(f)|^2}{F(f)} df \tag{8.66}$$

No other (linear) equalizer achieves a larger signal-to-noise ratio at the time point considered. However, the output pulse $g_m(t)$ of the matched filter is often very wide, so that large intersymbol interferences occur in digital transmission and the eye of the signal $m(t)$ is closed.

The purpose of the transverse filter is to eliminate completely the intersymbol interference that exists in the signal $m(t)$. However, this linear compensation for the intersymbol interference, in comparison with nonlinear compensation using DFE or a Viterbi detector, often involves a considerable increase in the noise power, i.e. the detection noise power N_d is generally very much larger than the noise power N_m in the signal $m(t)$.

Example To demonstrate the above results, consider a binary coaxial cable system with an optimum Nyquist equalizer. Rectangular NRZ pulses are transmitted and the noise factor $F(f)$ is assumed to be unity. Figure 8.20(b) shows the pulse $g_m(t)$ at the matched filter output; it is very wide and has a very small amplitude even with the characteristic attenuation value considered here ($a_* = 20$ dB). However, the basic detection pulse at the output of the transverse filter is a Nyquist pulse with a maximum value $g_d(0) = \hat{g}_s$ and equidistant zero-crossings.

The amplitude responses of the matched filter and the transverse filter are shown for various characteristic attenuation values in Fig. 8.21(a). It is apparent that the bandwidth of the matched filter becomes increasingly narrow with increasing characteristic attenuation value a_* (i.e. with increasing cable length). This has the effect that the output pulse $g_m(t)$ of the matched filter becomes increasingly amplitude attenuated and time dilated as a_* increases. For example, for a characteristic attenuation value of 80 dB, the pulse amplitude $g_m(0)$ is only about 0.6% of the transmitter pulse amplitude $\hat{g}_s$ and a single pulse extends over hundreds of symbols.

It can be seen that the output pulse $g_m(t)$ of the matched filter is symmetrical, in contrast with the input pulse $g_e(t)$. This is because the matched filter completely compensates the channel phase distortion because of the complex conjugate frequency responsc $II_K{}^*(f)$.

The frequency response $H_{TF}(f)$ of the transverse filter is real and periodic, and its maximum lies exactly at the Nyquist frequency. The greater is the characteristic attenuation value a_*, the wider is the input signal $g_m(t)$ of the transverse filter and, correspondingly, the greater the power the filter must employ to achieve complete Nyquist equalization. For $a_* = 80$ dB, $20 \lg H_{TF}(f_N) \approx 160$ dB.

Figure 8.21(b) makes it clear that the maximum value of the optimum equalizer frequency response $\mathring{H}_E(f) = H_{MF}(f) H_{TF}(f)$ lies closer to the Nyquist frequency f_N as the characteristic attenuation value a_* increases. The optimum pulse-shaper

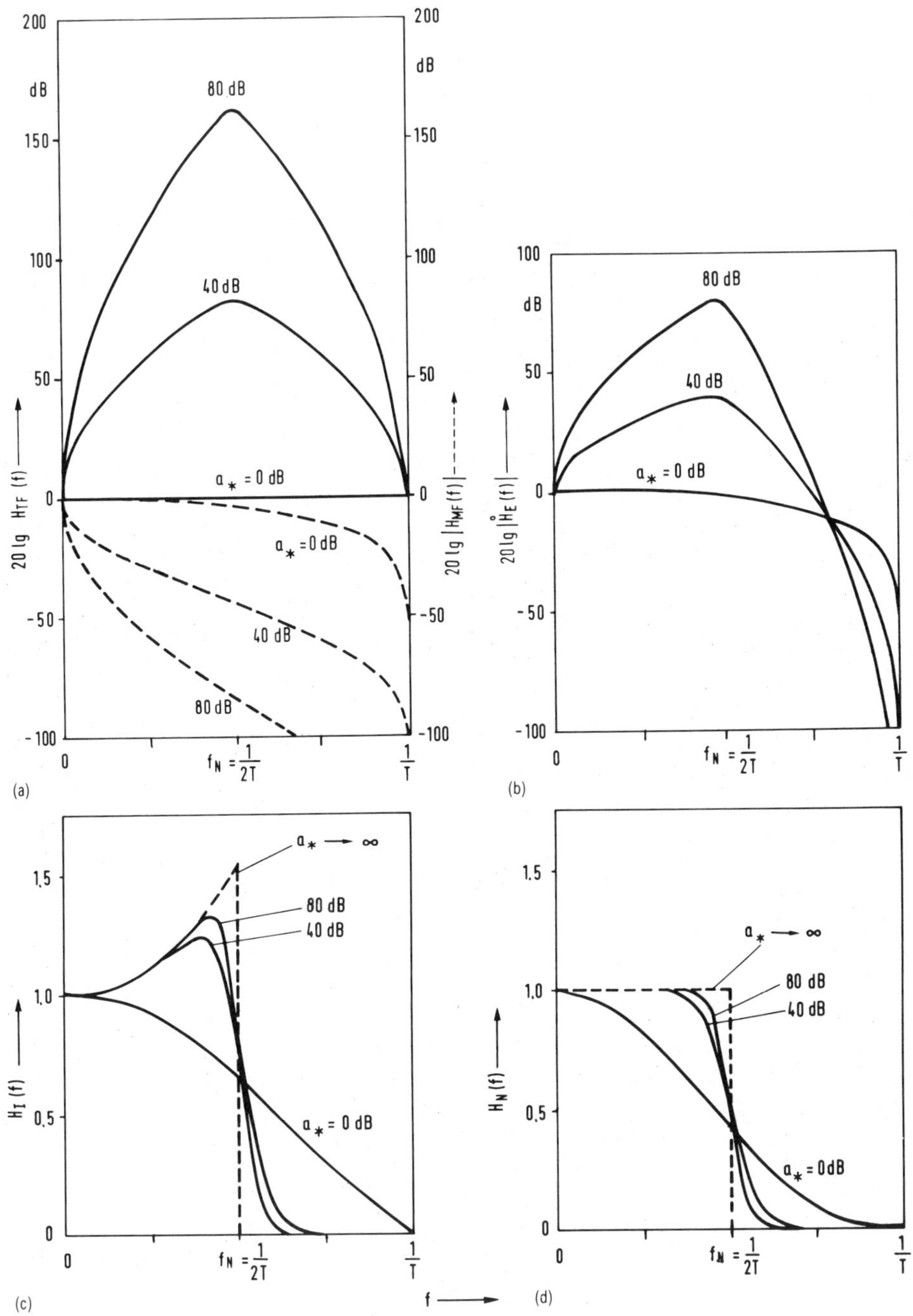

Fig. 8.21 (a) Frequency response of a matched filter and a transverse filter (logarithmic); (b) equalizer frequency response (logarithmic); (c) pulse-shaper frequency response (linear); (d) Nyquist frequency response (linear).

frequency response $\mathring{H}_I(f) = H_K(f)\mathring{H}_E(f)$ increases slightly in the region of the Nyquist frequency before decaying at higher frequencies (see Fig. 8.21(c)). However, the optimum Nyquist frequency response $\mathring{H}_N(f)$ is constant over a wide frequency range (see Fig. 8.21(d)). The slope increases with increasing characteristic attenuation value a_*, which means that noise is better suppressed.

Approximation to the optimum Nyquist equalizer The expression for the frequency response of the optimum Nyquist equalizer (eqn (8.62)) is complicated. However, it is apparent from comparison of Fig. 8.21(d) and Table 8.2 that the optimum Nyquist frequency response $\mathring{H}_N(f)$ for transmission channels in which attenuation increases with frequency can be very closely approximated by a cosine roll-off low-pass filter with cut-off frequency $f_N = 1/2T$, by which the optimum equalizer frequency response $\mathring{H}_E(f) = \mathring{H}_N(f)/H_S(f)H_K(f)$ is also defined. In making this approximation it is assumed that the condition

$$\left|H_S\left(\frac{\kappa}{T}\right)\right|^2 \left|H_K\left(\frac{\kappa}{T}\right)\right|^2 \ll |H_S(0)|^2 |H_K(0)|^2 \qquad (8.67)$$

is satisfied for all $\kappa \neq 0$, which is the case for most transmission channels.

The optimum roll-off factor $\mathring{r}_N$ is still to be determined. This depends on the system parameters M, $H_S(f)$ and $H_K(f)$. It is plotted for a coaxial pair in Fig. 8.22, which shows that $\mathring{r}_N$ decreases approximately hyperbolically with increasing characteristic attenuation value a_*. In the limiting case ($a_* \to \infty$) a rectangular (brick-wall) low-pass filter provides the optimum Nyquist frequency response (see Fig. 8.21(d)). Figure 8.22 also makes it clear that the filter edges must be steeper for a binary system ($M = 2$) than for a multilevel system, and steeper for NRZ transmitter pulses than for RZ pulses.

Finally, consider a frequency-independent channel, which is a good approximation for a radio link. Figure 8.21 shows this special case of $H_K(f) = 1$ for $a_* = 0$ dB.

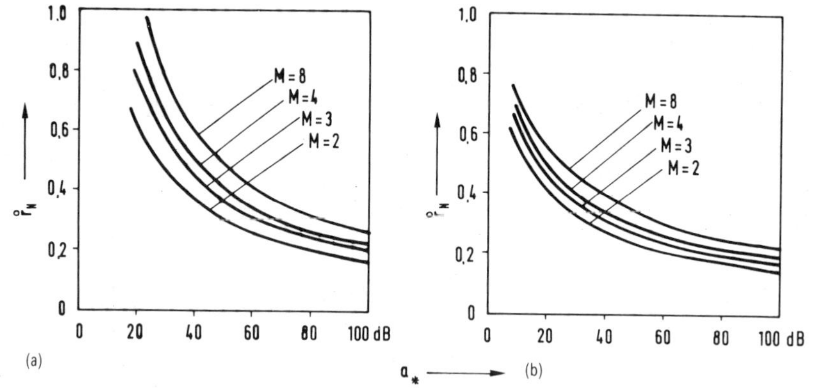

Fig. 8.22 Optimum roll-off factor of the approximate Nyquist frequency response as a function of the level number M and the characteristic attenuation value a_* of a coaxial pair: (a) RZ rectangular transmitter pulse ($\Delta t_s = 0.5T$); (b) NRZ rectangular transmitter pulse.

With white noise ($F(f) = 1$) we obtain from eqn (8.62) the following expression for the frequency response of the matched filter:

$$H_{MF}(f) = K_N H_S^*(f)$$

Therefore the output pulse $g_m(t)$ of the matched filter has the same shape as the energy autocorrelation function $l_{gs}^{\bullet}(t)$ of the basic transmitter pulse defined in eqn (2.23). If $g_s(t)$ is limited to the time interval $|t| \leqslant T/2$, then $g_m(t) = 0$ for $|t| \geqslant T$ and there is no intersymbol interference. Hence the transverse filter can be omitted ($H_{TF}(f) = 1$). For example, a triangular basic detection pulse is obtained from rectangular transmitter pulses.

Calculation of the system efficiencies The structure of the optimum Nyquist equalizer is given by eqn (8.64) and Fig. 8.20(a). The coefficients k_λ of the transverse filter $H_{TF}(f)$ still have to be suitably determined. A comparison of step (l) in the proof of eqn (8.60) and eqn (8.63) leads to the following condition:

$$\sum_{\mu=-\infty}^{+\infty} k_\mu \exp(-j2\pi f \mu T) \stackrel{!}{=} \frac{1}{\sum_{\kappa=-\infty}^{+\infty} |H_S(f-\kappa/T)|^2 |H_K'(f-\kappa/T)|^2} \tag{8.68}$$

In contrast with eqn (8.63) the variable λ has been substituted by μ. Multiplying both sides of the equation by $\exp(2\pi f \lambda T)$, integrating over a period of $1/T$ and interchanging the summation and the integration, we obtain

$$\sum_{\mu=-\infty}^{+\infty} k_\mu T \int_{-1/2T}^{+1/2T} \exp\{j2\pi f(\lambda-\mu)T\}\mathrm{d}f$$

$$\stackrel{!}{=} T \int_{-1/2T}^{+1/2T} \frac{\exp(j2\pi f \lambda T)}{\sum_{\kappa=-\infty}^{+\infty} |H_S(f-\kappa/T)|^2 |H_K'(f-\kappa/T)|^2} \mathrm{d}f \tag{8.69}$$

The integral on the left-hand side has a value of unity for $\mu = \lambda$; all other integrals ($\mu \neq \lambda$) are zero. Since the denominator of the integrand on the right-hand side is an even function of frequency, we obtain the expression

$$k_\lambda = T \int_{-1/2T}^{1/2T} \frac{\cos(2\pi f \lambda T)}{\sum_{\kappa=-\infty}^{+\infty} |H_S(f-\kappa/T)|^2 |H_K'(f-\kappa/T)|^2} \mathrm{d}f \tag{8.70}$$

for the optimum filter coefficients. Thus the optimum Nyquist equalizer is fully defined. Inserting $\mathring{H}_E(f)$ from eqn (8.62) into eqn (8.58) we obtain

$$N_{\text{norm}} = \frac{\alpha^2}{K_N^2 R} \int_{-\infty}^{+\infty} F(f)|H_{MF}(f)|^2 |H_{TF}(f)|^2 \mathrm{d}f \tag{8.71}$$

for the (minimum) normalized noise power. It follows from step (h) of the proof of eqn (8.60) and eqn (8.61) that

$$N_{\text{norm}} = \frac{\alpha^2}{K_N R} \int_{-\infty}^{+\infty} \mathring{H}_N(f) H_{TF}(f) \mathrm{d}f \tag{8.72}$$

On decomposing this integral into partial integrals of width $1/T$, as in the proof of eqn (8.60), interchanging the summation with the integration and taking account of the periodicity of $H_{\mathrm{TF}}(f)$, we obtain

$$N_{\mathrm{norm}} = \frac{\alpha^2}{K_N R} \int_{-1/2T}^{1/2T} H_{\mathrm{TF}}(f) \sum_{\kappa=-\infty}^{+\infty} \mathring{H}_N\left(f - \frac{\kappa}{T}\right) \mathrm{d}f = \frac{\alpha^2}{R} \int_{-1/2T}^{1/2T} H_{\mathrm{TF}}(f) \mathrm{d}f \quad (8.73)$$

In this case the first Nyquist criterion (eqn (8.42)) was used, which states that the infinite summation in the integrand is equal to K_N. Considering further that the filter coefficient k_0 can also be represented as a Fourier coefficient (eqn (8.63))

$$k_0 = T \int_{-1/2T}^{1/2T} H_{\mathrm{TF}}(f) \mathrm{d}f \quad (8.74)$$

we obtain from eqn (8.73) the normalized noise power for the case when an optimum Nyquist equalizer is used:

$$N_{\mathrm{norm}} = \frac{\alpha^2}{RT} k_0 \quad (8.75)$$

It follows from this and eqns (8.56), (8.57) and (7.53) that the two system efficiencies, under peak-value or power limitation, are given by

$$\eta_A = \frac{RT}{(M-1)^2} \gamma_S^2 \frac{1}{k_0} = \frac{\log_2 M}{(M-1)^2} \gamma_S^2 \frac{1}{k_0} \quad (8.76)$$

$$\eta_L = \frac{RT}{(M-1)^2} \kappa_S^2 \frac{1}{k_0} = \frac{3 \log_2 M}{(M^2 - 1)} \frac{1}{T \int |H_S(f)|^2 \mathrm{d}f} \frac{1}{k_0} \quad (8.77)$$

Both the equations on the left-hand side are generally valid, but those on the right-hand side only hold for nonredundant transmitter signals ($RT = \log_2 M$). γ_S represents the intersymbol interference at the transmitter (see eqn (7.40)); for rectangular pulses $\gamma_S = 1$. κ_S is the transmitter peak-value factor given by eqn (7.53). The centre filter coefficient k_0 of the transverse filter can be calculated using eqn (8.70).

Example The optimum filter coefficients k_λ for rectangular NRZ transmitter pulses and a coaxial channel are given in Table 8.3. As the characteristic attenuation value a_* is suitably normalized, the table is valid for binary and multilevel systems as well as for nonredundant and redundant systems.

Figure 8.23 shows the corresponding system efficiencies η_A and η_L. Even with optimum Nyquist equalization a first approximation to the decrease with increasing a_* is linear, and so the maximum regenerator-section length can again be calculated from Section 7.3. This is true for both power and peak-value limitation.

The curves for the ternary partial response codes (AMI code and duo-binary code) are parallel to those for $M = 2$. The reason for this is that the symbol rate $1/T$ is not changed relative to the nonredundant binary system, so that the same noise power results. Under peak-value limitation the ternary partial response codes are about 6 dB worse than the nonredundant binary code because the eye opening is

Table 8.3 Coefficients k_λ of the transverse filter for a coaxial pair with characteristic attenuation value a_* and white noise ($F = 1$)

$a^*/(RT)^{1/2}$ (dB)	$10 \lg k_0$ (dB)	k_1/k_0	k_2/k_0	k_3/k_0	k_4/k_0	k_5/k_0	k_6/k_0	k_7/k_0	k_8/k_0	k_9/k_0
0	0	0	0	0	0	0	0	0	0	0
20	16.61	−0.590	0.168	−0.060	0.013	−0.007	0	−0.002	−0.001	−0.001
40	34.60	−0.793	0.438	−0.212	0.100	−0.047	0.022	−0.010	0.005	−0.001
60	53.26	−0.878	0.617	−0.381	0.224	−0.129	0.074	−0.042	0.024	−0.014
80	72.25	−0.919	0.727	−0.518	0.348	−0.228	0.148	−0.095	0.061	−0.039
100	91.44	−0.943	0.797	−0.621	0.458	−0.328	0.231	−0.172	0.112	−0.078

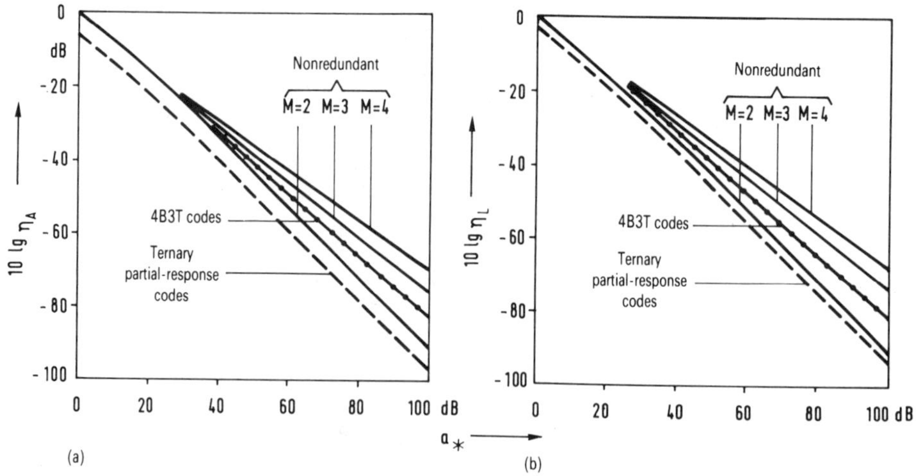

Fig. 8.23 (a) System efficiency $10\lg\eta_A$ and (b) system efficiency $10\lg\eta_L$ with optimum Nyquist equalization as a function of the characteristic attenuation value a_* of a coaxial pair ($F(f) = 1$).

only half as large. With power limitation the distance between the two curves is only 3 dB because the transmitter amplitude with ternary partial response coding can be a factor of $2^{1/2}$ greater than that with nonredundant binary coding (S_s is assumed to be constant).

However, with the nonredundant multilevel system the symbol rate is lower than in binary transmission, and so the curves become flatter. If the characteristic attenuation value is sufficiently large, the multilevel system is superior to the binary system even with optimum Nyquist equalization. Under the co-condition of power limitation the gain in signal-to-noise ratio is greater than that with peak-value limitation. This statement is also qualitatively valid for systems with 4B3T coding.

8.3.3 Optimum Nyquist equalizer for a system with decision feedback equalization

It was shown in Section 8.1 that a considerable improvement in signal-to-noise ratio can generally be achieved by using DFE. In a Nyquist system the linear equalizer already forms a pulse free from intersymbol interference, and so the vertical eye opening and hence the system efficiency cannot be increased further by using DFE which merely increases the temporal eye opening in this case [8.25]. An improvement can only be achieved by DFE if an equalizer is used whose bandwidth is narrower than that of the Nyquist equalizer $\mathring{H}_E(f)$ which is optimum for the system without DFE. This produces a smaller noise power, but the basic detection pulse $g_d(t)$ no longer satisfies the Nyquist condition (eqn (8.40)) and so intersymbol interference occurs. However, this can be reduced to zero by suitable design of the feedback network.

Figure 8.24(a) shows the structure of an optimum equalizer for a system with DFE. An additional transverse filter with n positive and v negative time-delay elements is inserted after the Nyquist equalizer $\mathring{H}_E(f) = H_{MF}(f)H_{TF}(f)$ which is

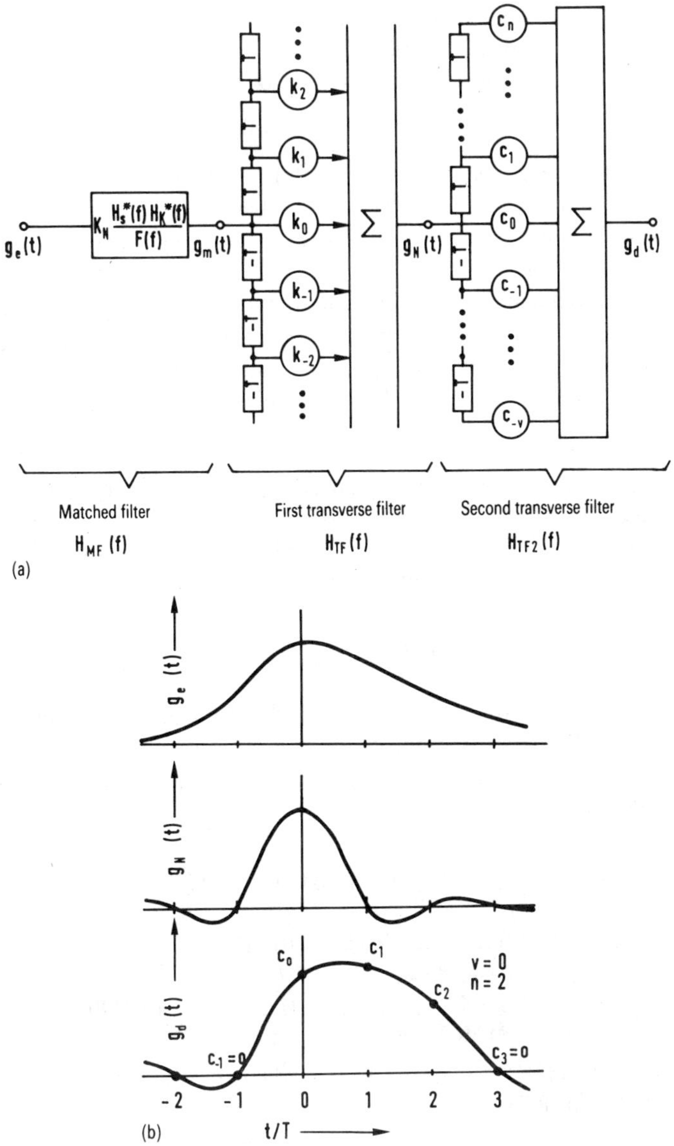

Fig. 8.24 Optimum equalizer for a digital system with DFE: (a) block diagram; (b) basic pulses $g_e(t)$, $g_N(t)$ and $g_d(t)$.

optimum for the system without DFE. The coefficients of this filter are $c_{-v} \ldots c_v \ldots c_n$. Hence the frequency response of the second transverse filter is (see eqn (8.63))

$$H_{\mathrm{TF}_2}(f) = \sum_{v=-v}^{n} c_v \exp(-\mathrm{j}2\pi f v T) \tag{8.78}$$

Firstly let $v = 0$. The basic pulses for this special case are shown for various points in the block diagram in Fig. 8.24(b). $g_{\mathrm{N}}(t)$ is the Nyquist pulse, which would be the optimum in a system without DFE. The "asymmetrical" transverse filter $H_{\mathrm{TF}_2}(f)$ forms from this the basic detection pulse $g_d(t)$ for which the sample values are

$$g_d(vT) = c_v g_{\mathrm{N}}(0) \tag{8.79}$$

If we assume that the first n postcursors $g_d(T) \ldots g_d(nT)$ of the basic detection pulse are fully compensated by DFE, the corrected basic detection pulse $g_k(t)$ is again a Nyquist pulse and accordingly the eye is completely open at the optimum detection time $\mathring{T}_{\mathrm{D}} = 0$ (see Section 5.1).

If the filter coefficients c_v are suitably dimensioned, the basic detection pulse $g_d(t)$ has fewer high-frequency components than the Nyquist pulse $g_{\mathrm{N}}(t)$(see Fig. 8.24(b)). The spectral components in the region of the Nyquist frequency f_{N} are strongly attenuated by the filter $H_{\mathrm{TF}_2}(f)$ and so the noise power is markedly reduced.

Optimization of the coefficients of the asymmetric transverse filter Now $v = 0$, and for the purpose of this discussion it is also assumed that the coefficient $c_0 = 1$. Hence according to eqn (8.79) the pulse amplitude $g_d(0) = g_{\mathrm{N}}(0)$, and therefore the system efficiency which is to be maximized is given by (see eqn (8.56))

$$\eta_{\mathrm{A}} = \frac{1}{(M-1)^2} \frac{1}{N_{\mathrm{norm}}} \tag{8.80}$$

The constants γ_{S} and α are put equal to unity here. For a given level number M maximization of the system efficiency η_{A} is the same as minimization of the normalized noise power N_{norm}, which is given by the following expression rather than by eqn (8.71):

$$N_{\mathrm{norm}} = \frac{1}{K_{\mathrm{N}}^2 R} \int_{-\infty}^{+\infty} F(f) |H_{\mathrm{MF}}(f)|^2 |H_{\mathrm{TF}}(f)|^2 |H_{\mathrm{TF}_2}(f)|^2 \, \mathrm{d}f \tag{8.81}$$

By analogy with eqn (8.72) this can also be written

$$N_{\mathrm{norm}} = \frac{1}{K_{\mathrm{N}} R} \int_{-\infty}^{+\infty} \mathring{H}_{\mathrm{N}}(f) H_{\mathrm{TF}}(f) |H_{\mathrm{TF}_2}(f)|^2 \, \mathrm{d}f \tag{8.82}$$

As the function $H_{\mathrm{TF}}(f)|H_{\mathrm{TF}_2}(f)|^2$ is also periodic with the symbol rate $1/T$, we have by analogy with eqn (8.73)

$$N_{\mathrm{norm}} = \frac{1}{RT} T \int_{-1/2T}^{1/2T} H_{\mathrm{TF}}(f) |H_{\mathrm{TF}_2}(f)|^2 \, \mathrm{d}f \tag{8.83}$$

This means that, up to the factor $1/RT$, the normalized noise power N_{norm} is identical with the zeroth Fourier coefficient of the periodic function

$$H_{\text{TF}}(f)|H_{\text{TF}_2}(f)|^2 = \underbrace{\sum_{\lambda=-\infty}^{+\infty} k_\lambda \exp(-j2\pi f \lambda T)}_{H_{\text{TF}}(f)} \underbrace{\sum_{\nu=0}^{n} c_\nu \exp(-j2\pi f \nu T)}_{H_{\text{TF}_2}(f)} \underbrace{\sum_{\mu=0}^{n} \exp(j2\pi f \mu T)}_{H_{\text{TF}_2}{}^*(f)} \tag{8.84}$$

where the k_λ are the filter coefficients of the transverse filter $H_{\text{TF}}(f)$ which can be determined from eqn (8.70).

The zeroth Fourier coefficient is calculated from the above equation at the condition $\mu - \lambda - \nu = 0$ [8.25]. Hence we obtain

$$N_{\text{norm}} = \frac{1}{RT} \sum_{\nu=0}^{n} \sum_{\mu=0}^{n} k_{\mu-\nu} c_\nu c_\mu \qquad c_0 = 1 \tag{8.85}$$

for the normalized noise power at the equalizer output in Fig. 8.24(a). By setting the n differential ratios $dN_{\text{norm}}/dc_\lambda$ equal to zero we obtain n linearly independent equations for the optimum filter coefficients $\mathring{c}_1, \ldots, \mathring{c}_n$:

$$\sum_{\nu=1}^{n} k_{\lambda-\nu} \mathring{c}_\nu \overset{!}{=} -k_\lambda \qquad \text{for} \quad \lambda = 1 \ldots n \qquad \mathring{c}_0 = 1 \tag{8.86}$$

This means that the filter coefficients $c_1, \ldots, c_n$ can be optimized by solving a linear set of equations of degree n.

Example For $n = 1$ this set of linear equations reduces to a single equation from which the filter coefficient $\mathring{c}_1 = -k_1/k_0$ can be determined. Inserting this in eqn (8.85) we obtain

$$N_{\text{norm}} = \frac{k_0}{RT}\left(1 - \frac{k_1^2}{k_0^2}\right) \tag{8.87}$$

This means that the (normalized) noise power is already reduced by a factor of $1 - k_1^2/k_0^2$ with an additional transverse filter of only first order. For a binary coaxial cable system with characteristic attenuation value $a_* = 80$ dB the ratio $k_1/k_0 = -0.919$ (see Table 8.3). It thus follows that for this system the noise power is reduced by about a factor of 6.4 (about 8.1 dB) by the second transverse filter ($n = 1$), and hence the system efficiency is increased in the same way.

The larger is the chosen value of n, the greater is the gain which can be achieved by using the asymmetric transverse filter (see the full curve in Fig. 8.25(a)). For $n = 10$ the system efficiency approximates to its maximum value. However, it must also be considered that the cost of realization of the DFE increases with increasing n (the number of postcursors to be compensated), and that inaccuracies in the feedback network become much more apparent (see Fig. 8.24(b) and Section 8.7).

The rate of increase in the system efficiency with increasing n is slower for a quaternary system ($M = 4$) than for a binary system ($M = 2$). This means that, even

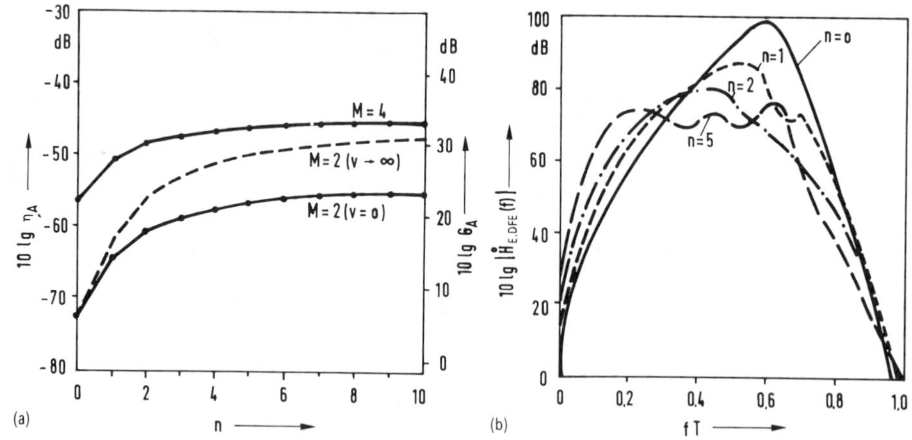

Fig. 8.25 Optimum Nyquist equalizer for a coaxial pair system with DFE ($a_* = 80\,\text{dB}$): (a) system efficiency $10\lg\eta_A$ where $F(f) = 1$; (b) amplitude response $|\mathring{H}_{E,DFE}(f)|$ where $M = 2$.

with Nyquist equalization, DFE is most effective for a binary system. Despite this, with complete DFE ($n \to \infty$) multilevel systems are still superior to the binary system.

Figure 8.25(b) shows the absolute value of the frequency response

$$\mathring{H}_{E,DFE}(f) = H_{MF}(f)H_{TF}(f)H_{TF_2}(f) \tag{8.88}$$

for various values of n, where a level number $M = 2$ is assumed. It is assumed that $v = 0$ and $c_0 = 1$. In a system without DFE ($n = 0$) the equalizer frequency response $\mathring{H}_E(f)$ exhibits a distinct maximum in the region of the Nyquist frequency $f_N = 1/2T$. The higher-frequency spectral components are increasingly attenuated by the additional transverse filter $H_{TF_2}(f)$, and hence the (normalized) noise power becomes ever smaller. In the limiting case $n \to \infty$ the amplitude response $|\mathring{H}_{E,DFE}(f)|$ is approximately constant over a wide frequency range. This means that in this case the noise power spectrum $L_{\overset{\times}{d}}(f)$ at the detector is approximately frequency independent, so that the individual noise sample values $\overset{\times}{d}(vT)$ are mutually uncorrelated.

Generalization to systems with intersymbol interference So far it has been assumed that the asymmetric transverse filter $H_{TF_2}(f)$ only produces postcursors, which can be compensated by using DFE ($v = 0$). Hence the corrected basic detection pulse is a Nyquist pulse. If the filter also has negative delay elements (i.e. $v > 0$), the basic detection pulse has precursors $g_d(-T) \ldots g_d(-vT)$ in addition to the n postcursors, which cannot be eliminated for reasons of causality and thus lead to intersymbol interference.

For such an equalizer (see Fig. 8.24(a)) the optimization condition is (see eqns (8.55) and (8.79))

$$\eta_A = \log_2 M \frac{\{c_0/(M-1) - \sum_{v=-v}^{-1} |c_v|\}^2}{\sum_{v=-v}^{n} \sum_{\mu=-v}^{n} k_{\mu-v} c_v c_\mu} \overset{!}{=} \text{maximum} \tag{8.89}$$

This is based on nonredundant systems with ideal DFE. If a filter coefficient c_ν ($\nu < 0$) has a value other than zero, the eye opening becomes smaller. However, the noise power can also be reduced in this way.

In principle we can use analytical procedures to determine the optimum filter coefficients $\mathring{c}_{-v}, \ldots, \mathring{c}_n$, and hence we obtain a nonlinear set of equations of degree $n + v$. However, an iterative numerical optimization of the coefficients on a computer is preferable for large values of n and v, in which case the coefficients $c_1, \ldots, c_n$ according to eqn (8.86) and $c_{-1}, \ldots, c_{-v} = 0$ are suitable start values.

The system efficiency numerically determined from eqn (8.89) for $M = 2$ is plotted in Fig. 8.25(a) for a case in which v is very large ($v \to \infty$, broken curve). It shows that, for a coaxial cable system ($a_* = 80$ dB), a gain in the signal-to-noise ratio of about 8 dB can be achieved relative to the optimum binary Nyquist system ($M = 2$, $v = 0$). The equalizer in Fig. 8.24(a) with $n \to \infty$ and $v \to \infty$ and with optimum filter coefficient $\mathring{c}_\nu$ is the optimum equalizer for a digital system with DFE [8.13].

Optimization of multilevel systems ($M > 2$) with DFE results in the filter coefficients $\mathring{c}_{-1} = \ldots = \mathring{c}_{-v} = 0$. The other coefficients are as defined by eqn (8.86). This result states that in a multilevel system with DFE the optimum equalizer is a Nyquist equalizer without precursors ($v = 0$) corresponding to eqn (8.88).

It can be shown in an analogous way that in a system without DFE the optimum Nyquist equalizer defined in eqn (8.62) is always (i.e. for all M) the optimum equalizer. To prove this result we assume as a basis the general equalizer from Fig. 8.24(a). Instead of eqn (8.89) the optimization condition for a system without DFE is

$$\eta_A = \log_2 M \frac{\{c_0/(M-1) - \sum_{\nu=-v}^{-1} |c_\nu| - \sum_{\nu=1}^{n} |c_\nu|\}^2}{\sum_{\nu=-v}^{n} \sum_{\nu=-v}^{n} k_{\mu-\nu} c_\nu c_\mu} \overset{!}{=} \text{maximum} \qquad (8.90)$$

Under certain limitations, which are always satisfied by channels used in practice, the (numerical) optimization of the filter coefficients produces $\mathring{c}_\nu = 0$ for $\nu \neq 0$ and all level numbers M.

8.4 Joint optimization of transmitter and equalizer

In Sections 8.1, 8.2 and 8.3 the characteristics of either the receiver (with a given transmitter) or the transmitter (for a given receiver) were optimized in isolation. By means of common optimization of transmitter and equalizer we obtain the optimum system for the channel under consideration; symbol-by-symbol detection is assumed. A further improvement is possible only by means of more complicated receiver strategies, e.g. the Viterbi receiver (see Chapter 6).

In this section the joint optimization of the transmitter and equalizer of a Nyquist system is described; a nonredundant transmitter signal and a detector without DFE are assumed. The difference between power-limited and peak-value-limited systems is taken into account (see eqns (7.1) and (7.2)).

8.4.1 Joint optimization under power limitation

According to Chapter 7, in the joint optimization of transmitter and receiver under power limitation the frequency responses $H_S(f)$ and $H_E(f)$ and the level number M have to be determined such that the system efficiency η_L is a maximum. Therefore, from eqns (8.70) and (8.77), the following must be true for a nonredundant M-level Nyquist system:

$$\eta_L = \frac{3\log_2 M}{M^2-1}\left\{T\int_{-\infty}^{+\infty}|H_S(f)|^2\,df\,T\int_{-\infty}^{+\infty}\frac{|H_N(f)|^2}{|H_S(f)|^2|H_K'(f)|^2}\,df\right\}^{-1} \overset{!}{=} \text{maximum} \tag{8.91}$$

It has been implicitly taken into consideration that the overall frequency response should be a Nyquist frequency response (see eqn (8.41)): $H_S(f)H_K(f)H_E(f) = H_N(f)$. Colored noise ($F(f) \neq$ constant) is also taken into account by the modified channel frequency response $H_K'(f)$ from eqn (8.61).

If a fixed level number M is assumed, the optimization problem can be reduced to minimization of the term in braces in eqn (8.91). This optimization problem can be solved in two steps. First the optimum transmitter frequency response $\mathring{H}_S(f)$ is calculated as a function of the Nyquist frequency response $H_N(f)$. Then $H_N(f)$ is optimized [1.15]. The Schwartz inequality [1.4]

$$\int_{-\infty}^{+\infty}|H_1(x)|^2\,dx\int_{-\infty}^{+\infty}|H_2(x)|^2\,dx \geqslant \left\{\int_{-\infty}^{+\infty}|H_1(x)H_2(x)|\,dx\right\}^2 \tag{8.92}$$

can be used to rewrite the term in braces in eqn (8.91):

$$\begin{matrix}\text{term}\\ \text{in braces}\end{matrix} = T\int_{-\infty}^{+\infty}|H_S(f)|^2\,df\,T\int_{-\infty}^{+\infty}\frac{|H_N(f)|^2}{|H_S(f)|^2|H_K'(f)|^2}\,df \geqslant \left\{T\int_{-\infty}^{+\infty}\frac{H_N(f)}{H_K'(f)}\,df\right\}^2 \tag{8.93}$$

When the transmitter amplitude response

$$|H_S(f)| = |\mathring{H}_S(f)| = \frac{1}{\gamma}\left\{\frac{|H_N(f)|}{|H_K'(f)|}\right\}^{1/2} = \frac{1}{\gamma}\{F(f)\}^{1/4}\left\{\frac{|H_N(f)|}{|H_K(f)|}\right\}^{1/2} \tag{8.94}$$

is inserted in this inequality, the equality is valid and consequently the smallest possible value is found for the denominator which is being minimized. However, this means that the desired optimum transmitter frequency response under power limitation is given by eqn (8.94). γ is needed for normalization reasons only and it can be determined from the condition $T\int \mathring{H}_S(f)\,df = 1$ (see eqn (2.16)).

From eqns (8.61), (8.91) and (8.93) we obtain the system efficiency η_L with optimum transmitter frequency response $\mathring{H}_S(f)$:

$$\eta_L = \frac{3\log_2 M}{M^2-1}\left[T\int_{-\infty}^{+\infty}\{F(f)\}^{1/2}\frac{|H_N(f)|}{|H_K(f)|}\,df\right]^{-2} \tag{8.95}$$

The second step of the optimization process now consists of determining the Nyquist frequency response $H_N(f)$ so that the integral has the smallest possible value.

The optimum Nyquist frequency response $\mathring{H}_N(f)$ depends only on the spectral noise factor $F(f)$ and the channel frequency response $H_K(f)$ and cannot, in general, be derived in closed form. However, as shown in Fig. 8.26 this optimization problem can be solved graphically. The optimum Nyquist frequency response $\mathring{H}_N(f)$ in Fig. 8.26(a) is built up from rectangular components. With white noise these components are situated where the channel frequency response $H_K(f)$ exhibits its largest values [1.15]. In addition it should be noted that the first Nyquist criterion (eqn (8.42)) must be satisfied. This means that the selected partial frequency responses shifted by multiples of $2f_N = 1/T$ must produce a rectangular (brick-wall) low-pass filter of bandwidth $\pm f_N$.

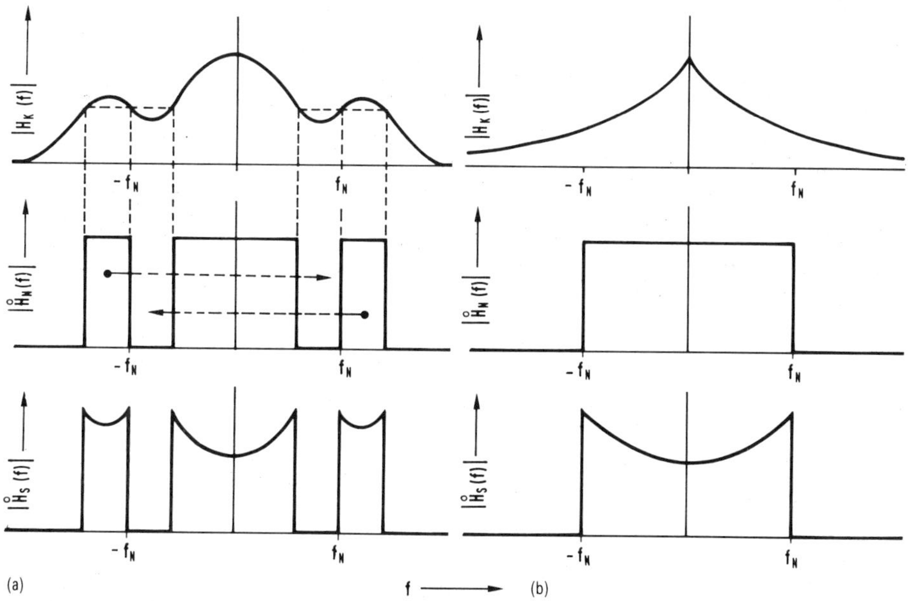

Fig. 8.26 Joint optimization of transmitter and equalizer frequency responses with white noise under power limitation: (a) arbitrary $|H_K(f)|$; (b) monotonically decaying $|H_K(f)|$.

The absolute value of the optimum transmitter frequency response $\mathring{H}_S(f)$ can be determined from eqn (8.94). However, no statement concerning the phase of $\mathring{H}_S(f)$ is possible.

We obtain from eqn (8.41) the optimum equalizer frequency response under power limitation and with an optimum transmitter:

$$|\mathring{H}_E(f)| = \frac{\mathring{H}_N(f)}{\mathring{H}_S(f) H_K(f)} = \frac{\gamma}{\{F(f)\}^{1/4}} \left\{ \frac{|\mathring{H}_N(f)|}{|H_K(f)|} \right\}^{1/2} \tag{8.96}$$

If $|H_K(f)|$ decreases monotonically, we obtain for the optimum Nyquist frequency response a rectangular low-pass filter with a (one-sided) bandwidth $f_N = 1/2T$ (see Fig. 8.26(b)). Hence with white noise the optimum frequency responses of the transmitter and the equalizer in the frequency range $-f_N$ to $+f_N$ are given by

$$|\mathring{H}_S(f)| = \frac{1}{\gamma}\left\{\frac{1}{|H_K(f)|}\right\}^{1/2} \tag{8.97a}$$

$$|\mathring{H}_E(f)| = \gamma\left\{\frac{1}{|H_K(f)|}\right\}^{1/2} \tag{8.97b}$$

Outside the Nyquist band $\mathring{H}_S(f) = \mathring{H}_E(f) = 0$. This equation states that, if power limitation and Nyquist equalization are assumed, the optimum amplitude responses of transmitter and equalizer have the same shape and hence each contributes half the equalization of the channel frequency response $H_K(f)$. This type of equalization is referred to as the **root–root characteristic** in what follows. No other combination of transmitter and equalizer frequency responses results in a higher system efficiency η_L; therefore under the co-condition of power limitation this system is the optimum.

However, it should be noted that the basic detection pulse relating to this optimum system is the si pulse and that this has only a vanishingly small temporal eye opening. If a Nyquist frequency response with edges of finite slope (e.g. a cosine roll-off low-pass filter) is used, the temporal eye opening can be matched to the practical requirements (see Table 8.2). As the loss in signal-to-noise ratio associated with this is insignificant, the following example is simply based on the theoretical optimum.

Example For a nonredundant M-level coaxial cable system (characteristic attenuation value a_* and noise factor F) the following is true for optimum equalization corresponding to the root–root characteristic (see eqns (2.55) and (8.95)):

$$\eta_L = \frac{3\log_2 M}{(M^2-1)F}\left[T\int_{-1/2T}^{1/2T}\exp\left\{a_*\left(\frac{2|f|}{R}\right)^{1/2}\right\}\mathrm{d}f\right]^{-2} \approx \frac{3a_*^{\,2}}{4(M^2-1)F}\exp\left\{-\frac{2a_*}{(\log_2 M)^{1/2}}\right\} \tag{8.98}$$

The right-hand approximation is true for $a_* \gg 1$ (a_* in nepers). For $a_* = 10\,\mathrm{Np}$ the error is less than 1 dB.

By differentiation of the right-hand approximation we obtain for the optimum level number

$$\mathring{M} \approx \exp\left[\left\{\frac{(\ln 2)^{1/2}a_*}{2}\right\}^{2/3}\right] \tag{8.99}$$

where $\mathring{M}$ is an integer. For example when $a_* = 80\,\mathrm{dB}$, $\mathring{M} \approx 12$.

Figure 8.27 shows the system efficiency $10\lg\eta_L$ as a function of the characteristic attenuation value a_* for the binary system ($M = 2$) and for the optimum level number ($M = \mathring{M}$). The broken curves are for the optimum basic transmitter pulse

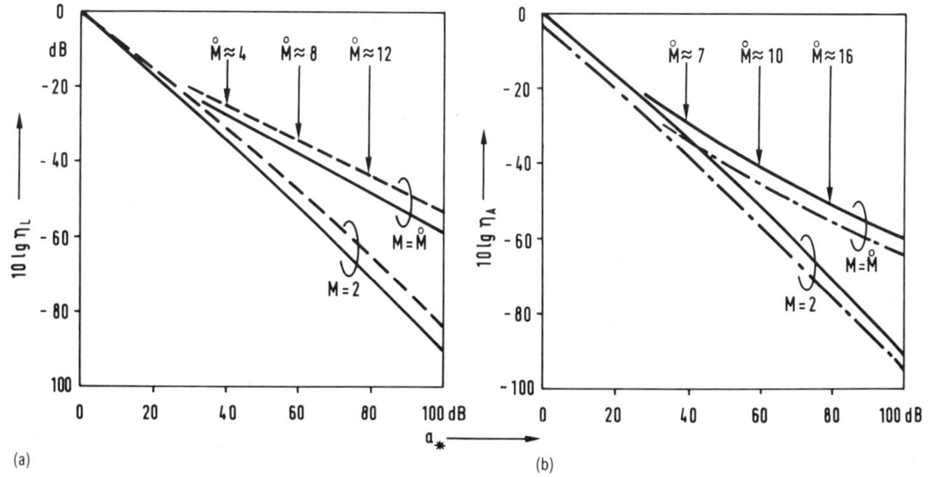

Fig. 8.27 Maximum system efficiency for a coaxial cable system with optimum Nyquist equalization ($F = 1$): (a) power limitation (—, NRZ rectangular transmitter pulse;---, optimum transmitter pulse under power limitation, eqn (8.97)); (b) peak-value limitation (—, NRZ rectangular transmitter pulse, $T_s = T$;—·—, RZ rectangular transmitter pulse, $T_s = T/2$).

under power limitation (see eqn (8.97a)). The curves for systems with rectangular NRZ transmitter pulses and optimum Nyquist equalization are also drawn (full curves) for comparison. This figure makes it plain that a gain of several decibels relative to a Nyquist system with NRZ rectangular transmitter pulses can be achieved by using root–root equalization if power limitation is assumed. This gain increases with increasing characteristic attenuation value. For example, with a_* = 80 dB and $M = 2$ it is about 6 dB.

The optimum transmitter frequency response $\mathring{H}_S(f)$ under power limitation is plotted in Fig. 8.28(a) for the parameter values $a_* = 80$ dB and $M = 2$ (see eqn (8.97a)). The optimum equalizer frequency response $\mathring{H}_E(f)$ in the case of white noise ($F(f)$ = constant) has the same shape as $\mathring{H}_S(f)$ (see eqn (8.97b)).

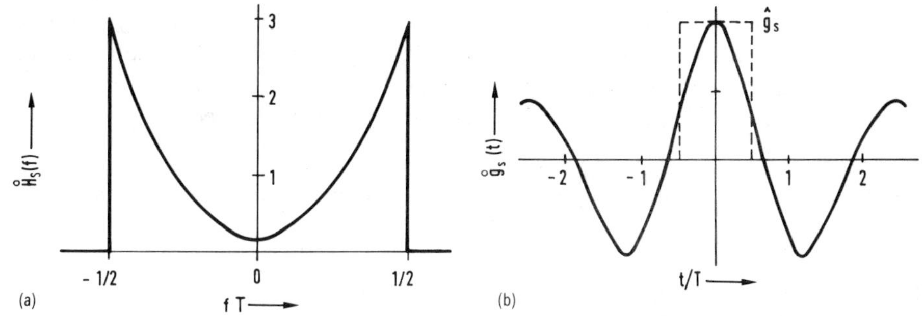

Fig. 8.28 Transmitter optimization of a binary coaxial cable system under power limitation ($a_* = 80$ dB, $F(f) = 1$): (a) optimum transmitter frequency response; (b) optimum basic transmitter pulse (with NRZ rectangular pulse for comparison).

This figure shows that for the systems considered here $\mathring{H}_S(f)$ increases very sharply at the frequencies $f = \pm f_N$. Owing to this marked "pre-emphasis" the oscillation of the optimum basic transmitter pulse $\mathring{g}_s(t)$ shown in Fig. 8.28(b) decreases more slowly than in proportion to $1/t$. Hence the peak value s_{max} of the transmitter signal is very high, and so this system is also likely to be unsuitable where (at least slight) limitation of the peak value is present in addition to the power limitation.

8.4.2 Joint optimization under peak-value limitation

The common optimization of transmitter and equalizer under peak-value limitation means that the frequency responses $H_S(f)$ and $H_E(f)$ both have to be determined so that the system efficiency η_A is a maximum. The optimum equalizer frequency response $\mathring{H}_E(f)$ for a given transmitter is defined by eqn (8.62) (system without DFE) or eqn (8.88) (system with DFE). The mutual optimization of $H_S(f)$ and $H_E(f)$ can thus be performed by optimization of the transmitter frequency response $H_S(f)$ as a function of $\mathring{H}_E(f)$. For a given level number M and a detector without DFE the optimization rule is (see eqns (8.70), (8.76) and (7.40))

$$\frac{s_{max}{}^2}{\hat{g}_s{}^2} T \int_{-f_N}^{f_N} \frac{df}{\sum_{\kappa=-\infty}^{+\infty} |H_S(f-\kappa/T)|^2 |H_K'(f-\kappa/T)|^2} \stackrel{!}{=} \text{minimum} \qquad (8.100)$$

where $f_N = 1/2T$ is the Nyquist frequency.

We stipulate for the following that the basic transmitter pulse is limited in time to only the symbol duration T, i.e. $g_s(t) = 0$ for $|t| > T/2$. With this assumption $s_{max} = \hat{g}_s$ and so the optimization condition in eqn (8.100) is simplified. The optimum transmitter frequency response $\mathring{H}_S(f)$ is to be determined as a function of the channel frequency response $H_K'(f)$ related to the noise factor $F(f)$ (see eqn (8.61)) so that the integral is a minimum. If it can be shown that for a particular $H_S(f)$ the integrand itself has the smallest possible value, in the whole integration range from $-f_N$ to $+f_N$, then this $H_S(f)$ is the desired optimum transmitter frequency response $\mathring{H}_S(f)$. It is considered here that the denominator of the integrand can only be positive because of the summation of absolute values. From this a further simplified sufficient (but not necessary) optimization condition can be derived:

$$\text{denominator} = \sum_{\kappa=-\infty}^{+\infty} \left| H_S\left(f - \frac{\kappa}{T}\right)\right|^2 \left| H_K'\left(f - \frac{\kappa}{T}\right)\right|^2 \stackrel{!}{=} \text{maximum} \quad \text{for} \quad -f_N \leqslant f \leqslant f_N \qquad (8.101)$$

The co-condition of peak-value limitation (eqn (7.2)) is very difficult to formulate in the frequency domain. Therefore we start from the time domain and approximate the basic transmitter pulse by the step function shown in Fig. 8.29(a). In this way the basic transmitter pulse $g_s(t)$ can be represented by the sum of $2L+1$ weighted

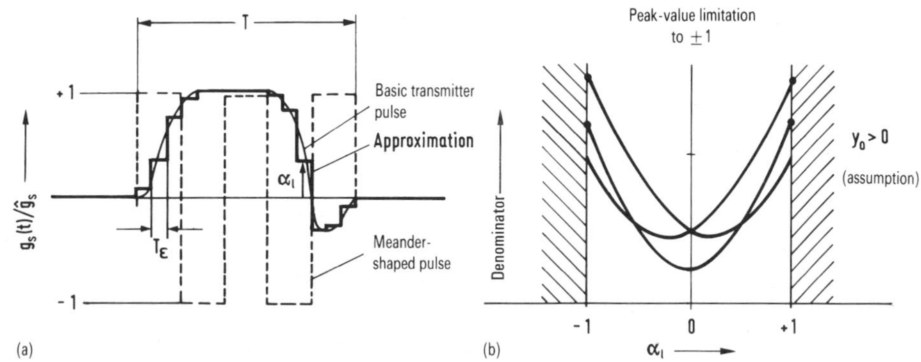

Fig. 8.29 Optimization of the transmitter pulse under peak-value limitation and optimum Nyquist equalization: (a) approximation of the basic transmitter pulse; (b) the denominator in eqn (8.105) as a function of a single transmitter pulse coefficient α_l ($\alpha_{i \neq l}$ = constant).

elementary rectangular pulses of duration $T_\varepsilon = T/(2L+1)$:

$$g_s(t) = \sum_{l=-L}^{+L} \alpha_l \hat{g}_s \operatorname{rec}\left(\frac{t - lT_\varepsilon}{T_\varepsilon}\right) \tag{8.102}$$

The α_l ($l = -L, \ldots, L$) are the **transmitter pulse coefficients** which, because of the peak-value limitation, must lie between -1 and $+1$. In a trend to the limit $L \to \infty$ each arbitrary continuous basic transmitter pulse can be sufficiently accurately described if it is time as well as peak-value limited.

With this approximation we have for the transmitter frequency response according to eqn (2.15)

$$H_S(f) = \sum_{l=-L}^{+L} \alpha_l \frac{T_\varepsilon}{T} \operatorname{si}(\pi f T_\varepsilon) \exp(-j2\pi f l T_\varepsilon) \tag{8.103}$$

After some trigonometrical rearrangement we obtain from this the square of the magnitude of the transmitter frequency response:

$$|H_S(f)|^2 = H_S(f) H_S{}^*(f) = \frac{T_\varepsilon^2}{T^2} \operatorname{si}^2(\pi f T_\varepsilon) \sum_{k=-L}^{+L} \sum_{l=-L}^{+L} \alpha_k \alpha_l \cos\{2\pi(l-k) f T_\varepsilon\} \tag{8.104}$$

Therefore the optimization condition of eqn (8.101) becomes

$$\text{denominator} = \sum_{k=-L}^{L} \sum_{l=-L}^{L} \alpha_k \alpha_l y_{l-k} \stackrel{!}{=} \text{maximum for } -f_N \leqslant f \leqslant f_N \tag{8.105}$$

The $2L+1$ different coefficients ($i = 0, \ldots, 2L$)

$$y_i = \frac{T_\varepsilon^2}{T^2} \sum_{\kappa=-\infty}^{+\infty} \left| H_K{}'\left(f - \frac{\kappa}{T}\right)\right|^2 \operatorname{si}^2\left\{\pi\left(f - \frac{\kappa}{T}\right) T_\varepsilon\right\} \cos\left\{2\pi i\left(f - \frac{\kappa}{T}\right) T_\varepsilon\right\} \tag{8.106}$$

depend (apart from T_ε) only on the related channel frequency response $H_K{}'(f)$, and they are referred to in the following as the **channel coefficients**.

The transmitter pulse coefficients α_l ($l = -L,...,L$) now have to be determined so that the denominator of eqn (8.105) is a maximum at every frequency in the range from $-f_N$ to $+f_N$. This optimization procedure is described in detail in refs 3.12 and 8.33 and so will only be outlined briefly here.

If, in each case, we consider the denominator as a function of a particular transmitter pulse coefficient α_l ($l = -L,...,L$) while the other $2L$ coefficients ($\alpha_{i\neq l}$) remain constant, we obtain parabolic curves (see Fig. 8.29(b)). The shape of the individual curves depends on the transmitter pulse coefficients $\alpha_{i\neq l}$ and the channel coefficients y_i. As according to eqn (8.106) the coefficient y_0 is always greater than zero, all these parabolae are concave upwards. It thus follows that, because of the peak-value limitation (i.e. $-1 \leqslant \alpha_l \leqslant +1$), the optimum value for the transmitter pulse coefficient α_l is ± 1 and is independent of the other coefficients $\alpha_{i\neq l}$. This statement is true for all l ($l = -L,...,L$) and all frequencies from $-f_N$ to $+f_N$. Therefore in the frequency range of interest the denominator is an exact minimum if all the transmitter pulse coefficients $\alpha_{-L} \ldots \alpha_L$ are either $+1$ or -1. This means that the basic transmitter pulse under peak-value limitation exhibits a meander shape, i.e. the pulse has the amplitude $+\hat{g}_s$ or $-\hat{g}_s$ from $-T/2$ to $+T/2$ and outside this time range is zero. Such a pulse is shown by broken lines in Fig. 8.29(a). All meander-shaped pulses have in common the fact that their transmitter energy $\hat{g}_s{}^2 T$ is a maximum under peak-value limitation. The NRZ rectangular transmitter pulse is a special case of the meander-shaped pulse.

If we take this result into consideration the $2L+1$ continuous-value optimization parameters ($-1 \leqslant \alpha_l \leqslant +1$) can be converted into a similar number of dual-value parameters ($\alpha_l = \pm 1$), so that the maximization of eqn (8.105) is significantly simplified and it can be solved for any arbitrary transmission channel using a computer.

For many transmission channels it is possible to perform the optimization of the transmitter pulse coefficients α_l implicitly as an analysis of the channel coefficients y_i. If, for example, the frequency response in a radio channel can be set constant ($H_K'(f) = K$), then for $i \neq 0$ the channel coefficients are $y_i = 0$ (see eqn (8.106) and ref. 8.33). However, $y_0 = K^2 T_\varepsilon/T = K^2/(2L+1)$ and hence the denominator of eqn (8.105) is independent of the transmitter pulse coefficients $\alpha_l = \pm 1$ and has the value

$$\text{denominator} = \sum_{l=-L}^{L} \alpha_l{}^2 y_0 = K^2 \tag{8.107}$$

from which the system efficiency $\eta_A = K^2(\log_2 M)/(M-1)^2$ can be calculated (see eqns (8.76) and (8.100)). Hence it is seen that for a frequency-independent channel with white noise every meander-shaped transmitter pulse time limited to T is the optimum as long as it has the maximum energy $\hat{g}_s{}^2 T$. With these assumptions the NRZ rectangular pulse is also an optimum as a special case of a meander-shaped pulse.

This result can be understood if the spectra of a "high frequency" meander-shaped pulse and an NRZ rectangular pulse are compared. In contrast with the

NRZ rectangular pulse the meander-shaped pulse exhibits fewer spectral components at low frequencies and hence more components at higher frequencies. However, as all frequencies are transmitted across the channel equally well it is immaterial whether the basic transmitter pulse has a majority of high- or low-frequency spectral components.

It is further shown in ref. 3.12 that under peak-value limitation the rectangular NRZ pulse is the only optimum basic transmitter pulse if the channel coefficients for all frequencies $-f_N \leqslant f \leqslant f_N$ satisfy the following conditions:

$$y_i > 0 \qquad \text{for} \qquad 1 \leqslant i \leqslant L \tag{8.108a}$$

$$y_i \leqslant y_{i-1} \qquad \text{for} \qquad 1 \leqslant i \leqslant L \tag{8.108b}$$

$$y_i \geqslant -y_{2L+1-i} \qquad \text{for} \qquad 1 \leqslant i \leqslant L \tag{8.108c}$$

All the channels with monotonically increasing attenuation examined here (brick-wall channel, Gaussian channel, coaxial pair and symmetric pair) satisfy these conditions. Therefore, the NRZ rectangular transmitter pulse is optimum for these channels if peak-value limitation is assumed. Figure 8.27(b) shows, for example, that for a binary coaxial pair system a loss in signal-to-noise ratio of about 4 dB has to be tolerated for a (rectangular) RZ transmitter pulse with a duty factor $T_s/T = 0.5$ (broken curve) compared with the NRZ rectangular pulse (full curve). A qualitatively similar result is obtained for the optimum level number ($M = \mathring{M}$). It should be noted that for a given characteristic attenuation value a_* the optimum level number $\mathring{M}$ in the case of peak-value limitation is somewhat higher than under power limitation (see Fig. 8.27(a) and eqn (8.99)).

8.5 Coding optimization

In previous sections it has generally been assumed that the level number M is the same at the transmitter and the receiver. The only exception to this was in Section 8.1.3 in which partial response coding at the receiver, from which quite beneficial results were gained, was described. In that description, however, some system parameters that could be freely chosen were assumed to be fixed (e.g. a nonredundant transmitter signal and a linear coding network at the receiver) and these limitations are not really necessary.

An optimization algorithm which is applicable to all digital systems and which is based on the most general block diagram in Fig. 8.30 is described in ref. 8.34. The transmitter contains an arbitrary coding device which maps every N_q binary source symbols onto a block of N code symbols. Thus the following is true for the time period which is available for transmission of one code symbol or one transmitter pulse:

$$T = \frac{N_q}{N} T_q \tag{8.109}$$

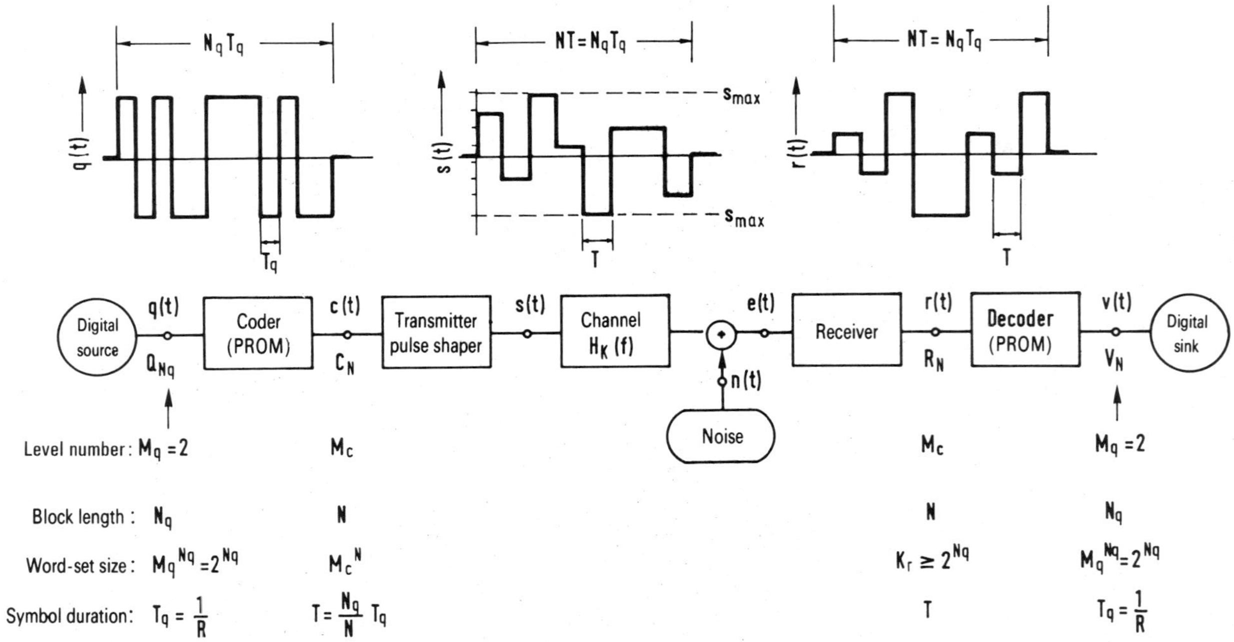

Fig. 8.30 General block diagram of a digital baseband transmission system with (nonlinear) coding, and the corresponding signals for system parameters $M_q = 2$, $M_c = 8$, $M_r = 4$ and $T = 1.5T_q$.

A further coding parameter is the level number M_c, which is generally very large and in the limiting case can approach infinity, from which an approximately continuous-value transmitter signal results. The coder could emit a total of up to M_c^N different sequences within the block duration $NT = N_q T_q$. However, as only 2^{N_q} of these code symbol sequences are assigned to a source symbol sequence by the coding rule, the relative code redundancy defined by eqn (4.4) is given by

$$r_c = 1 - \frac{N_q}{N \log_2 M_c} \tag{8.110}$$

By analogy with the notation used in Chapter 6, the source symbol sequence consisting of N_q symbols will be abbreviated in this section to Q_{N_q}. Similarly C_N denotes the coded symbol sequence consisting of N symbols.

Various receiver designs are possible, e.g. the optimum receiver (correlation or Viterbi receiver) as described in Chapter 6 or a conventional receiver with a threshold detector. It is merely stipulated that the receiver can distinguish within a block precisely K_r different regenerated symbol sequences $R_N^{(k)}$ $(k = 1, \ldots, K_r)$. M_r is the level number of the regenerated signal $r(t)$, which can, in principle, differ from the level number M_c of the coded signal $c(t)$.

By means of the high code level number M_c and the high redundancy associated with it, transmitter signals can be produced which the receiver can distinguish as well as possible. For example, the permitted transmitter signals can be defined so that the intersymbol interference which occurs at the receiver in conventional systems is already compensated for at the transmitter.

To describe the optimization method defined in ref. 8.34 it is first assumed that the other optimizable system parameters (e.g. the transmitter pulse shaper and the "structural design" of the receiver) are fixed for the optimization. Thus the purpose of the optimization is to determine the coding parameters (e.g. the level number M_c and the symbol duration T) and particularly the coding rule (i.e. the relationship between the source and the code symbol sequences) so that the mean bit error probability $p_B = \overline{P(v_\nu \neq q_\nu)}$ of the whole transmission system is a minimum.

If we denote by p_k the (conditional) mean bit error probability under the assumption that the symbol sequence $C_N^{(k)}$ was transmitted, we obtain from eqn (6.101) an approximation for the mean bit error probability to be minimized:

$$p_B = 2^{-N_q} \sum_{k=1}^{2^{N_q}} p_k \tag{8.111}$$

This calculation corresponds to an ensemble averaging over all possible source symbol sequences $Q_{N_q}^{(k)}$ or over all allowed code symbol sequences $C_N^{(k)}$.

The basis of the coding optimization described here is that from the many M_c^N possible code symbol sequences $C_N^{(i)}$ those 2^{N_q} sequences $C_N^{(k)}$ can be selected which have the most beneficial transmission characteristics within the present boundary conditions, i.e. those which collectively lead to the minimum error probability. In particular the code optimization can be split into the following steps.

(a) Fix the suitable start values for the coding parameters, i.e. for the symbol duration T and the level number M_c of the coder. (The block length N is defined by T.)

(b) Calculate the basic receiver pulse or the basic detection pulse and the noise power of this signal.

(c) Select the most beneficial code symbol sequences, i.e. those sequences which collectively lead to the minimum error probability.

(d) Vary the code parameters; continue from (b).

Note If the characteristics of the filters of the transmitter and the receiver are also varied, this algorithm illustrates the most general concept for determining the optimum digital system. However, with high level numbers and large block lengths the computing cost can increase rapidly.

8.5.1 Optimum coding for an optimum receiver (Viterbi detector)

The difficulty of selecting the 2^{N_q} most beneficial sequences $C_N^{(k)}$ lies in the fact that the (conditional) error probabilities p_k also generally depend on the other $2^{N_q}-1$ selected code symbol sequences $C_N^{(j)}$ ($j \neq k$). This will be made clear by the following example.

For an optimum receiver and Gaussian white noise with noise power density L_0 the approximation defined in Section 6.4.2 is valid:

$$p_k = \min_{j \neq k} Q\left\{\left(\frac{\Delta E_{kj}}{4L_0}\right)^{1/2}\right\} \qquad \begin{matrix} k = 1 \ldots 2^{N_q} \\ j = 1 \ldots 2^{N_q} \end{matrix} \tag{8.112}$$

where ΔE_{kj} is the energy distance between the kth and jth useful received signals (see eqn (6.97)). It is now appropriate to derive from this equation the effective signal-to-noise ratio of the kth code symbol sequence by analogy with eqn (7.4):

$$\rho_k = \{Q^{-1}(p_k)\}^2 = \min_{j \neq k} \left(\frac{\Delta E_{kj}}{4L_0}\right) \tag{8.113}$$

where $Q^{-1}(x)$ is the inverse function of the complementary Gaussian error integral shown in the Appendix, Tables A1 and A2.

For optimization the 2^{N_q} sequences $C_N^{(k)}$ with maximum ρ_k must be chosen from the M_c^N possible code symbol sequences $C_N^{(i)}$. This can be achieved by, for example, determining all M_c^{2N} possible effective signal-to-noise ratios [8.34]

$$\rho_{ij} = \frac{\Delta E_{ij}}{4L_0} \qquad i = 1 \ldots M_c^N \qquad j = 1 \ldots M_c^N \tag{8.114}$$

If we arrange these ρ_{ij} into an $M_c^N \times M_c^N$ matrix, the symbol sequence pair (i^*, j^*) with the smallest effective signal-to-noise ratio can be determined. To ensure that this minimum signal-to-noise ratio ρ_{i*j*} no longer occurs at the next iteration step, either the i^*th or the j^*th symbol sequence must be eliminated. If, for example, the i^*th code symbol sequence has a lower effective signal-to-noise ratio relative to the remaining sequences $C_N^{(j)}$ $(j \neq i^*, j^*)$, i.e. if

$$\min_{j \neq i^*, j^*} \rho_{i^*j} < \min_{i \neq i^*, j^*} \rho_{ij^*} \tag{8.115}$$

the i^*th sequence must be eliminated; otherwise, the j^*th sequence must be eliminated. Elimination of the i^*th code symbol sequence means removal of the i^*th row and the i^*th column from the matrix (ρ_{ij}).

The matrix (ρ_{ij}) is successively reduced by repetition of this elimination algorithm. When the number of rows and columns in this matrix is only 2^{N_q}, the most effective code symbol sequences $C_N^{(k)}$ have been determined.

8.5.2 Optimum coding for a threshold detector

An M_r-level threshold device can distinguish precisely $K_r = M_r^N$ different symbol sequences within a block duration $NT = N_q T_q$. However, as up to 2^{N_q} different source symbol sequences can be output within this block duration and the same number of different code symbol sequences $C_N^{(k)}$ $(k = 1, \ldots, 2^{N_q})$ is possible, K_r must be greater than or equal to 2^{N_q}. If M_c is greater than M_r, the number M_c^N of possible (*not* permissible) code symbol sequences is also greater than the number M_r^N of sequences which can be distinguished at the receiver. This means that, as a rule, several possible code symbol sequences are interpreted by the receiver as the same regenerated symbol sequence.

This is illustrated in Fig. 8.31 by the example of a quaternary threshold detector ($M_r = 4$) with three decision values E_1, E_2 and E_3. Both the useful detection signals $\tilde{d}_j(t)$ and $\tilde{d}_k(t)$ shown here are, at least in the absence of noise, interpreted by the

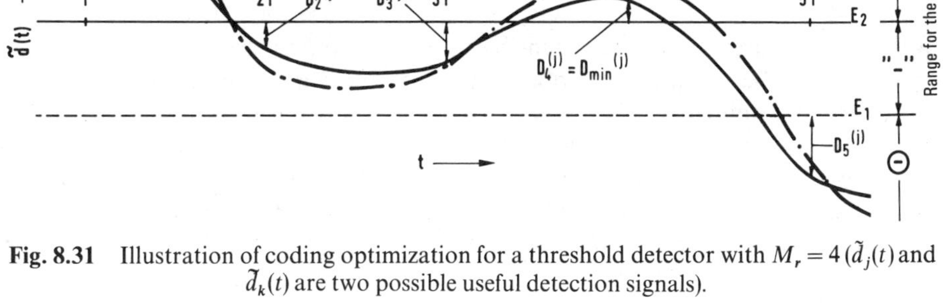

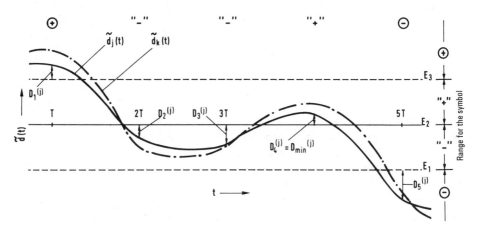

Fig. 8.31 Illustration of coding optimization for a threshold detector with $M_r = 4$ ($\tilde{d}_j(t)$ and $\tilde{d}_k(t)$ are two possible useful detection signals).

detector as the same regenerated sequence $\langle r_\nu \rangle = \ldots \oplus - - + \ominus \ldots$ because the sample values $\tilde{d}_j(\nu T)$ and $\tilde{d}_k(\nu T)$ at the detection times $T, \ldots, 5T$ each lie between the same decision values.

That sequence $C_N^{(k)}$ which has the lowest mean bit error probability

$$p_k = \min_j p_j \tag{8.116}$$

must now be selected from the possible code symbol sequences $C_N^{(j)}$ which lead to a quite distinct regenerated sequence $R_N^{(k)}$. The minimum is then derived for all possible code symbol sequences $C_N^{(j)}$ which are seen by the detector as the same regenerated symbol sequence $R_N^{(k)}$. We then have from eqn (3.9) the error probability of the jth sequence:

$$p_j = \frac{1}{N} \sum_{\nu=1}^{N} Q\left(\frac{D_\nu^{(j)}}{N_d^{1/2}}\right) \tag{8.117}$$

where N_d is the detection noise power. The detection distance $D_\nu^{(j)}$ denotes the distance between the useful detection signal $\tilde{d}_j(t)$ at the detection time νT and the next threshold (see eqn (3.6) and Section 3.4.3):

$$D_\nu^{(j)} = \min_{\mu = 1, \ldots, M-1} |\tilde{d}_j(\nu T) - E_\mu| \tag{8.118}$$

We then have, approximately, from the results in Section 3.5

$$p_j \approx Q\left(\frac{D_{\min}^{(j)}}{N_d^{1/2}}\right) \quad \text{where} \quad D_{\min}^{(j)} = \min_{\nu = 1, \ldots, N} D_\nu^{(j)} \tag{8.119}$$

In the example of Fig. 8.31 the error probability p_k of the sequence $C_N^{(k)}$ corresponding to the useful detection signal $\tilde{d}_k(t)$ is smaller than p_j, as the minimum detection distance $D_{\min}^{(k)}$ is larger than $D_{\min}^{(j)}$.

Hence a total of M_r^N sequences $C_N^{(k)}$ are preselected from the M_c^N possible code symbol sequences $C_N^{(i)}$, each of which relates to a regenerated sequence $R_N^{(k)}$. The 2^{N_q} sequences with the lowest error probabilities can then be determined from this. These are to be used for the transmission. For this we can, for example, redesignate the M_r^N different error probabilities

$$p_k \approx Q\left(\frac{D_{\min}^{(k)}}{N_d^{1/2}}\right) \quad k = 1, \ldots, M_r^N \tag{8.120}$$

so that p_k increases with k. It is sensible to select the first 2^{N_q} sequences for the transmission. The mean error probability p_M is then approximately equal to the 2^{N_q}th error probability p_k.

If we assume a threshold detector without DFE and a given pulse-shaper frequency response $H_I(f)$, the optimization algorithm we describe here results in coding which exhibits almost no intersymbol interference at the receiver. However, an assumption for this is that $M_c \gg M_r$ [8.15].

Figure 8.32 shows, for example, the transmitter signals and the corresponding

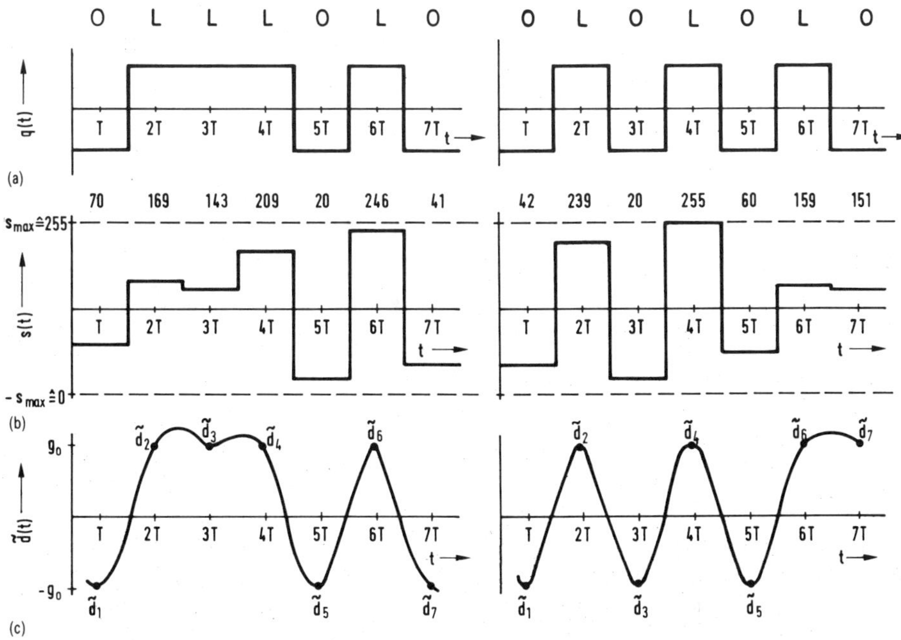

Fig. 8.32 Generation of Nyquist systems by coding (predistortion): (a) source signal; (b) transmitter signal; (c) useful detection signal.

useful detection signals for two selected symbol sequences. The example is based on a Gaussian pulse shaper with cut-off frequency $f_I = 0.3R$ and a binary threshold detector ($M_r = 2$). The level number of the coder (and hence of the transmitter signal) is chosen to be very large: $M_c = 256$. The possible code symbols c_μ are denoted by 1 to 256.

Although the pulse-shaper frequency response considered here would lead to high intersymbol interference with a nonredundant binary signal (see Fig. 8.1(a)), a suitable choice of 2^N code symbol sequences from all 256^N possible sequences ensures that all useful detection signals $\tilde{d}_k(t)$ occurring at the detection times $T_D + \nu T$ have, to a good approximation, one of the values $\pm g_0$. In this way intersymbol-interference-free detection is made possible by the use of coding.

The coding method proposed here can therefore be interpreted as the "digital realization" of a Nyquist system which also provides intersymbol-interference-free detection if the overall frequency response

$$H_S(f)H_K(f)H_E(f) \neq H_N(f) \tag{8.121}$$

deviates widely from a Nyquist frequency response (eqn (8.41)). This method offers the advantage that it can be relatively simply realized with a programmable read-only memory (PROM) at the transmitter and hence it is less sensitive to parameter fluctuations than is the analog realization of a Nyquist system (the Nyquist equalizer in Section 8.3).

Figure 8.33(a) shows the effect of such a predistortion for a coaxial cable system with characteristic attenuation value $a_* = 80\,\text{dB}$. The curve labeled "Nyquist by predistortion" shows that, by using this coding, a gain in signal-to-noise ratio can be achieved which almost equals the gain given by (linear) Nyquist equalization. The second curve from the top is for a system with a multilevel transmitter signal ($M_c = 256$) and a Gaussian pulse shaper with optimized cut-off frequency. The level number of the threshold device is plotted along the abscissa ($M = M_r$). It can be seen that this type of coding produces a considerable improvement relative to the nonredundant system with the same pulse shaper (Gaussian low pass). However, the curve is always 1–2 dB below that of the nonredundant Nyquist system.

8.6 Comparison of the optimized systems

In this section the optimized systems of the previous sections are compared on the basis of the use of a normal coaxial pair as the transmission channel. In general the comparison is performed for nonredundant M-level codes and for rectangular NRZ transmitter pulses with the co-condition of peak-value limitation. Similar curves can be derived for the power-limitation case [3.12]. However, it should be remembered in the interpretation of Figs 8.33 and 8.34 that practical inaccuracies

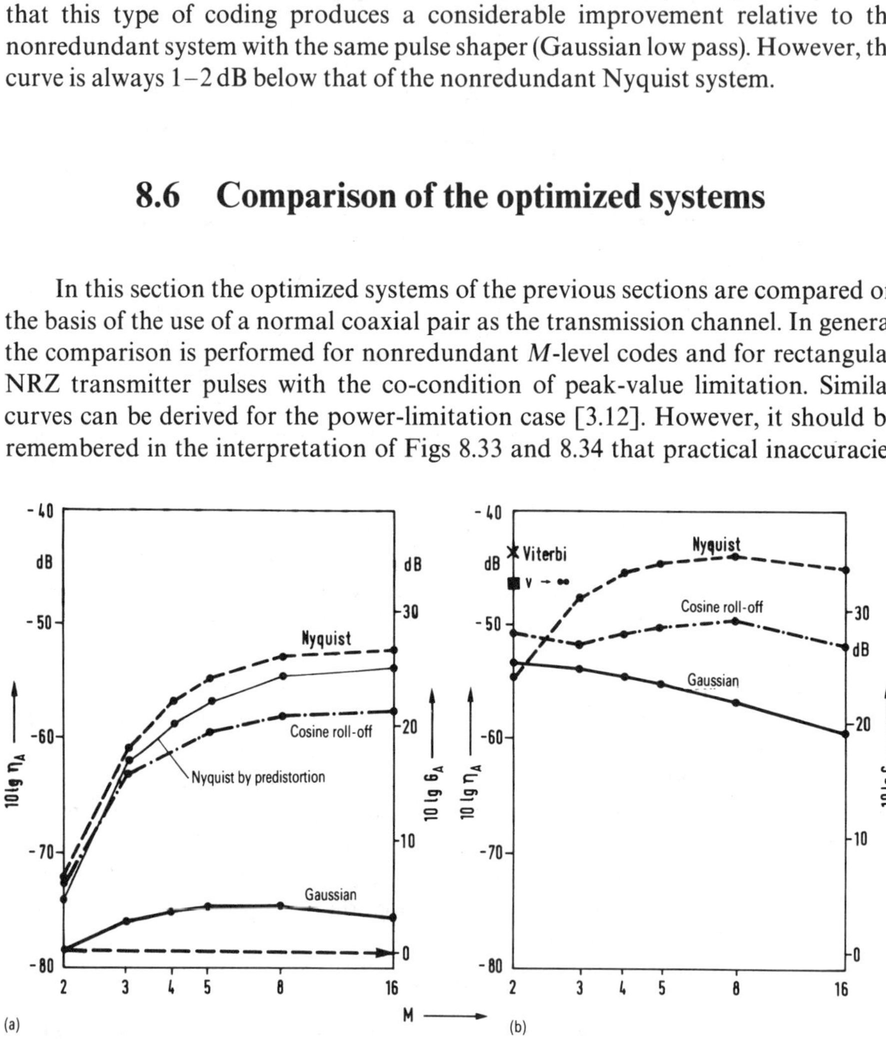

Fig. 8.33 Efficiency $10\lg\eta_A$ and gain $10\lg G_A$ in the signal-to-noise ratio of a system with peak-value limitation as a function of the level number M ($a_* = 80\,\text{dB}$): (a) without DFE; (b) with ideal DFE. The curves labeled "Nyquist by predistortion" are for the system in Fig. 8.32, where $M = M_r$ represents the level number of the regenerated signal. The level number M_c of the transmitter signal is 256. The pulse shaper is Gaussian.

are not taken into account here, so that the curves only represent the theoretically achievable limits. The effect of tolerances is examined in Section 8.7.

Figure 8.33(a) shows the system efficiency as a function of the level number M for systems without DFE. This figure is for a characteristic attenuation value $a_* = 80\,\text{dB}$ and white noise ($F = 1$). For a larger noise factor ($F > 1$), η_A is smaller by the factor F.

The system efficiency for the binary system ($M = 2$) with a Gaussian pulse shaper is about $-78.6\,\text{dB}$. The gain G_A in the signal-to-noise ratio can be read off the right-hand scale of Fig. 8.33 and by definition is equal to 0 dB for this reference system.

If the characteristic attenuation value a_* is sufficiently large, a significant gain in the signal-to-noise ratio of a system without DFE can be achieved by means of a multilevel transmission. This improvement is due to the fact that by raising the level number to $M > 2$ the symbol rate is reduced by a factor $\log_2 M$. Thus the transmitter power spectrum is shifted to lower frequencies at which the coaxial pair has lower attenuation. For example, in a system with a Gaussian pulse shaper a signal-to-noise ratio improvement of about 4 dB can be achieved by changing from $M = 2$ to $\mathring{M} = 8$ (see Section 8.1.1).

A further marked gain in the signal-to-noise ratio results from using a pulse shaper with steeper edges than is the case for the Gaussian low-pass filter, e.g. a cosine roll-off low-pass filter as shown in Section 8.1.5. Optimization of the roll-off factor here leads to relatively small values, which has a beneficial effect on the noise power. The intersymbol interference caused by this pulse shaper is relatively low, which is particularly advantageous in multilevel systems. For $M \geqslant 8$ the gain relative to the reference system is more than 20 dB.

It can also be seen from Fig. 8.33(a) that for receivers without DFE a system with optimum Nyquist equalization results in the maximum system efficiency for all values of level number (see Section 8.3.2). For $a_* = 80\,\text{dB}$ we obtain a gain of about 26.5 dB in the signal-to-noise ratio for the optimum level number $\mathring{M} \approx 16$.

For reasons of completeness the system described in Section 8.5.2 is also plotted in this figure. In this system the Nyquist characteristics are achieved by means of coding. The level number $M = M_r$ defines the level number of the detector, and the transmitted signal has a very much larger level number ($M_c = 256$). It can be seen from this representation that a large gain in the signal-to-noise ratio is also possible with a Gaussian pulse shaper if the coding is incorporated into the optimization.

Similarly, quite substantial gains can be achieved by using (ideal) DFE (see Fig. 8.33(b)). DFE is most effective in binary systems with Gaussian pulse shapers. Here the maximum gain obtained by using ideal DFE is more than 25 dB. This means that a binary system with a receiver with DFE and Gaussian pulse shaping is superior to multilevel systems. However, if greater effort is given to linear signal equalization, e.g. by means of optimum Nyquist equalization as shown in Section 8.3.3, multilevel transmission produces more satisfactory results than binary transmission even for a receiver with DFE. The optimum level number $\mathring{M} = 8$, however, is less than for a system without DFE.

Table 8.4 Constants K_A and K_* for a number of systems (with the assumptions of eqn (7.71))

Receiver		*Nonredundant coding*					*Redundant coding*			
		$M=2$	$M=3$	$M=4$	$M=8$	$M=16$	*AMI code*	*Duo-binary*	*Mod duo-binary*	*4B3T codes*
Gaussian pulse shaper without DFE	K_A (dB)	9.4	6.5	4.4	−1.3	−6.5	8.7	3.4	3.5	9.9
	K_*	1.10	1.03	0.99	0.91	0.86	1.27	0.98	1.11	1.17
Gaussian pulse shaper with ideal DFE	K_A (dB)	−8.0	−8.1	−10.0	−13.5	−17.6	−16.3	−21.9	−16.2	−10.0
	K_*	0.57	0.57	0.56	0.54	0.52	0.61	0.49	0.62	0.61
Optimum Nyquist equalizer without DFE	K_A (dB)	4.5	−0.3	−3.1	−9.3	−15.0	−1.5	−1.5	−1.5	2.2
	K_*	0.96	0.76	0.67	0.54	0.46	0.96	0.96	0.96	0.85
Optimum Nyquist equalizer with ideal DFE	K_A (dB)	−0.8	−4.9	−7.4	−13.0	−18.3	−6.8	−6.8	−6.8	−4.7
	K_*	0.67	0.53	0.47	0.39	0.34	0.67	0.67	0.67	0.58

The optimum binary system with DFE (see Fig. 8.24(a), $v \to \infty$) is indicated in Fig. 8.33(b) by a solid square. In this system the postcursors of the pulses are eliminated by DFE, but the precursors cause intersymbol interference. The use of this extremely-narrow-band equalizer under the present assumptions results in a gain of about 8 dB in the signal-to-noise ratio of a binary system with DFE relative to that which can be achieved using Nyquist equalization ($v = 0$). This means that a very-narrow-band system with intersymbol interference is the optimum for the case of binary transmission and DFE.

The binary Viterbi receiver which was described in Chapter 6 is indicated by a cross in Fig. 8.33(b). It can be seen that this system, which eliminates the effect of all precursors and postcursors, yields a system efficiency which is about 2.5 dB greater than that of the optimum binary system ($v \to \infty$) with symbol-by-symbol threshold decision and ideal DFE.

Further improvement in the signal-to-noise ratio can be obtained by using a multilevel Viterbi receiver [6.20]. This system is already very close to the absolute upper limit which can be derived from the channel capacity (see Section 7.4).

The maximum transmittable bit rate $R_{\max}$ (for a given cable length l) and the maximum regenerator-section length $l_{\max}$ (at a given bit rate R) can be determined from eqns (7.72)–(7.77). For this the constants K_{A} and K_{*} are needed. These describe the (approximately linear) relationship between the system efficiency $10 \lg \eta_{\mathrm{A}}$ and the characteristic cable attenuation value a_{*} of the coaxial pair and are given in Table 8.4 for the systems optimized above.

Figure 8.34 shows the maximum regenerator-section length $l_{\max}$ for a normal coaxial pair (2.6 mm/9.5 mm). If $l_{\max}$ had been plotted logarithmically, straight lines would have been obtained (see Fig. 7.7). Figure 8.34(a) is for binary systems. It can be seen that in the case of a threshold detector without DFE the regenerator-section length can only be about half that obtained with an optimum (Viterbi) receiver. However, the maximum regenerator-section length for an optimum binary system with DFE is only about 10%–15% lower than that for a binary Viterbi receiver if the error propagation effect is not taken into consideration. Figure 8.34(b) is based on the optimum level number $M = \mathring{M}$. In this case the improvement due to DFE is lower than that for binary transmission. The regenerator-section length allowed by a multilevel system ($\mathring{M} \approx 8$) with Nyquist equalization ($v = 0, n \to \infty$) and ideal DFE is approximately the same as that obtained using a binary Viterbi receiver.

The essential points of Figs 8.33 and 8.34 can be summarized as follows. Attainment of a significant improvement over the simple reference system (binary signal, Gaussian pulse shaper, receiver without DFE) can be achieved using two fundamentally different approaches: (a) realize a multilevel system with very good signal equalization, preferably a Nyquist system; (b) realize a very-narrow-band binary system with DFE or with a Viterbi receiver. In the first case there is the large expense of the (Nyquist) equalizer and the (multilevel) detector. However, in the second case the DFE or the Viterbi detector must be produced to very narrow tolerances.

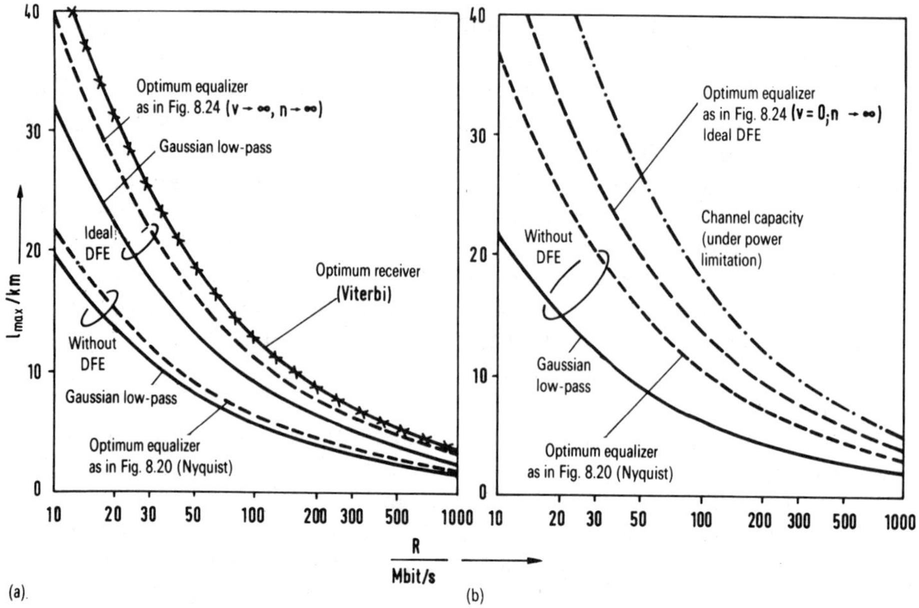

Fig. 8.34 Maximum regenerator-section length l_{max} as a function of the bit rate R for (a) binary systems ($M = 2$) and (b) optimum level number ($M = \mathring{M}$). The following assumptions are made: NRZ rectangular pulses ($s_{max} = 3$ V), white noise ($F = 6$) and a normal coaxial pair (2.6 mm/9.5 mm) ($\alpha_2 = 2.36$ dB km^{-1} MHz$^{-1/2}$).

8.7 System optimization with consideration of tolerances

So far all system components have been assumed to be ideal. However, in the realization of a transmission system unavoidable tolerances occur so that we must depart from the assumption of ideal behavior. If the inaccuracies in realization are taken into account in system optimization, different values are obtained for the optimum system parameters than in the tolerance-free case.

The most important effects of tolerances are summarized in refs 8.7 and 8.9, and are as follows:

(a) deviation of the actual channel frequency response from that assumed, e.g. because of deviation from the nominal length of the transmission medium or because of seasonal temperature variations;

(b) deviation from the optimum basic transmitter pulse, e.g. because of the finite slope of the leading and trailing edges of the pulse or because of asymmetry;

(c) deviation from the calculated optimum equalizer frequency response caused by reflection (mismatching) and temperature deviations as well as by tolerances and aging of the components;

(d) increase in the noise power density owing to impedance mismatching [5.17], pulse disturbances [2.2], near-end and far-end cross-talk [2.26] etc.

(e) deviation from the optimum decision values ("threshold drift") owing to shift of the operating point for example [8.7];

(f) deviation from the optimum detection time caused by a jittery clock signal or a sample pulse of finite width [2.20];

(g) inaccuracies in the decision feedback network so that the interfering postcursors of the detection pulse cannot be completely eliminated [8.9].

Whereas the worst-case signal-to-noise ratio under tolerance-free conditions was (eqn (7.49))

$$\rho_{\mathrm{U}} = \frac{\{o(T_{\mathrm{D}})/2\}^2}{N_d}$$

when tolerances are taken into account this becomes

$$\rho_{\mathrm{U,tol}} = \frac{\{o(T_{\mathrm{D}})/2 - \Delta A\}^2}{N_d + \Delta N_d} \tag{8.122}$$

ΔA includes all tolerance effects which contribute to a reduction in the half eye opening ("additive tolerances"), i.e. (a)–(c) and (e)–(g). The deviation ΔN_d of the noise power from its theoretical value is due to effects (c) and (d) only. It is already apparent from this that the equalizer frequency response $H_{\mathrm{E}}(f)$ affects both ΔA and ΔN_d, so that in general ΔA and ΔN_d cannot be treated as completely independent tolerance parameters.

The effects specified above result in a signal-to-noise ratio which is smaller than that of the tolerance-free system by an amount defined as the **loss factor (due to tolerances)**:

$$V_{\mathrm{tol}} = \frac{\rho_{\mathrm{U}}}{\rho_{\mathrm{U,tol}}} = \frac{1 + \Delta N_d/N_d}{\{1 - 2|A|/o(T_{\mathrm{D}})\}^2} \tag{8.123}$$

Therefore this also results in a lower system efficiency. In this definition of V_{tol}, N_d and $o(T_{\mathrm{D}})$ are the noise power and the vertical eye opening in the tolerance-free case. The loss 10 lg V_{tol} in signal-to-noise ratio is plotted in Fig. 8.35 as a function of the tolerance parameters ΔA and ΔN_d. If $|\Delta A| \ll o(T_{\mathrm{D}})/2$ and $\Delta N_d \ll N_d$,

$$V_{\mathrm{tol}} \approx 1 + \frac{\Delta N_d}{N_d} + \frac{4|A|}{o(T_{\mathrm{D}})} \tag{8.124}$$

It can be seen from this equation that the additive tolerances ΔA, which lead to a reduction in the vertical eye opening, become more marked the smaller is the eye opening $o(T_{\mathrm{D}})$ existing in the tolerance-free case. For this reason, the effect of a threshold drift for a system with a Gaussian pulse shaper is very much greater than that for a Nyquist system. In the same way the effect on a multilevel system is greater than that on a binary system [3.12].

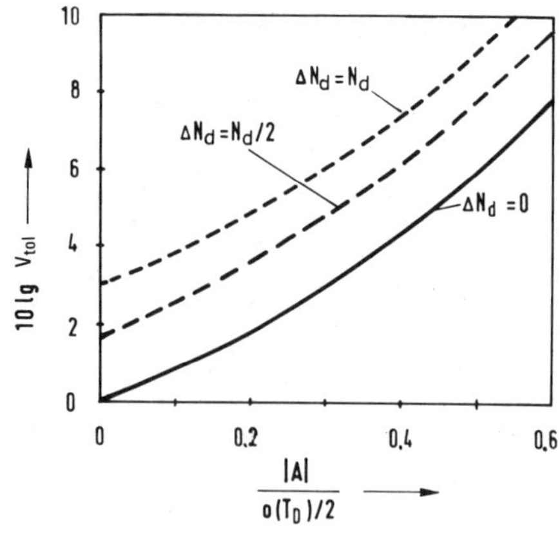

Fig. 8.35 Loss in the signal-to-noise ratio owing to tolerances.

If we consider the additive tolerances existing at the time of system optimization, i.e. if we use the signal-to-noise ratio $\rho_{U,tol}$ given by eqn (8.123) as the optimization criterion instead of ρ_U, we obtain another optimum cut-off frequency $\mathring{f}_I$ which increases with increasing additive tolerances ΔA in the system optimization. In the tolerance-free case ($\Delta A = 0$) this leads to a worsening of the signal-to-noise ratio. However, if the assumed tolerances are present, the system with the higher cut-off frequency (and hence the larger eye opening) is superior to the system optimized for the tolerance-free case.

Tolerances (a), (b) and (c) affect the basic detection pulse, in which case the eye opening and hence the achievable signal-to-noise ratio can deteriorate. These tolerance effects must also be taken into consideration in the tolerance parameter ΔA. If the worst-case tolerances are combined, a tolerance envelope can be produced for the basic detection pulse $g_d(t)$ (see Fig. 8.36). In the calculation of the vertical eye opening using eqn (3.51) it must now be remembered that the "worst case" occurs whenever the main value $g_d(T_D)$ has the smallest value and the precursors and postcursors $g_d(T_D + \nu T)$, $\nu \neq 0$, approach the largest values. Therefore examination of tolerance effects must be performed using the "worst-case basic detection pulse" drawn in Fig. 8.36.

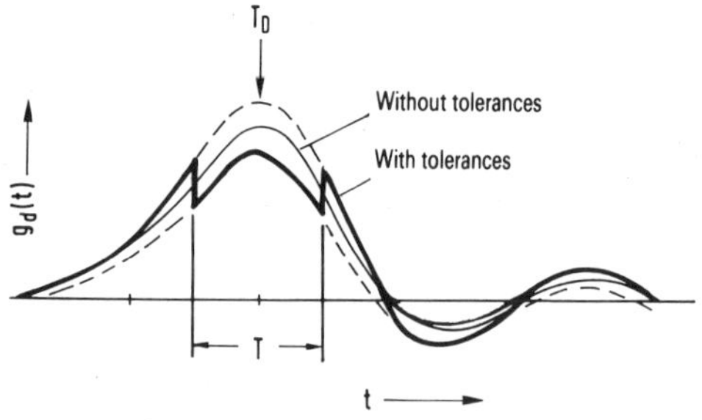

Fig. 8.36 Effect of tolerances on the basic detection pulse $g_d(t)$.

The detection noise power N_d given by eqn (2.70) also changes owing to equalizer tolerances. It can be either increased or reduced relative to the original value N_d. Like the additional disturbances (d), this effect can be included in worst-case analysis by a suitable increase in the noise factor $F(f)$.

Finally, the effect of nonideal DFE will be estimated. The optimum Nyquist system with DFE, which was derived in Section 8.3, will be considered for this purpose. Systems with DFE which have intersymbol interference exhibit similar behavior. In this system the basic detection pulse has no precursors ($v = 0$) but n postcursors. These are eliminated by means of DFE. The optimum basic detection pulse $\mathring{g}_d(t)$ is illustrated in Fig. 8.37(a), assuming a binary system ($M = 2$), for various values of n. As n increases (i.e. as the number of postcursors that can be compensated by the DFE increases) the pulse spectrum becomes more concentrated at low frequencies and the gain in the signal-to-noise ratio relative to the optimum system without DFE ($n = 0$) becomes larger. This fact has already been shown by Fig. 8.25(a), although ideal DFE was assumed in that case.

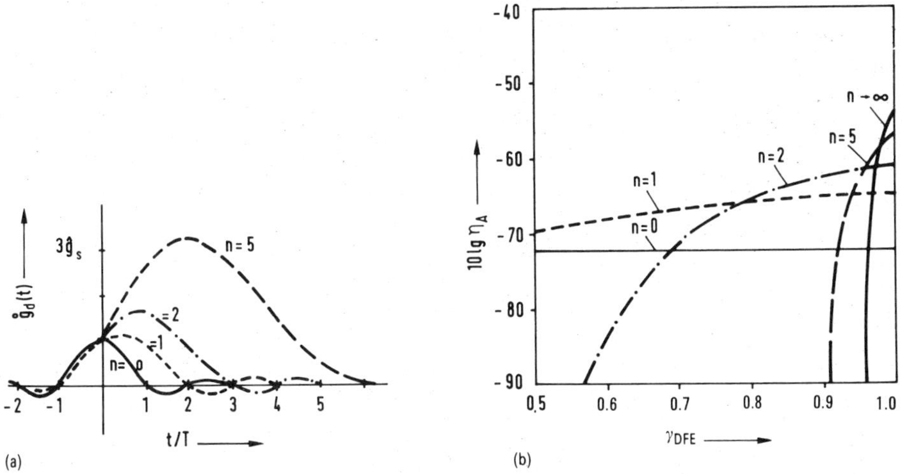

Fig. 8.37 Effect of non-ideal DFE in a Nyquist system (Section 8.3.3): (a) optimum basic detection pulse; (b) system efficiency with respect to γ_{DFE}. The following assumptions are made: binary NRZ rectangular pulses, a coaxial pair ($a_* = 80$ dB), white noise ($F = 1$) and no precursors ($v = 0$).

If we now assume that DFE does not completely compensate for all n postcursors but only compensates to a certain degree, instead of eqn (5.12) we obtain the following for the vertical half-eye opening of a bipolar M-level nonredundant system:

$$\frac{o(T_D)}{2} = \frac{g_d(T_D)}{M-1} - \sum_{\nu=1}^{v} |g_d(T_D - \nu T)| - (1 - \gamma_{DFE}) \sum_{\nu=1}^{n} |g_d(T_D + \nu T)| \qquad (8.125)$$

Here the factor γ_{DFE} is used as a measure of the accuracy of realization of the DFE.

This factor defines what proportion of the postcursors in the transmission of the worst-case symbol sequences are actually compensated for by DFE: $\gamma_{\mathrm{DFE}} = 0$ corresponds to a system without DFE, while in the case of ideal DFE $\gamma_{\mathrm{DFE}} = 1$.

The system efficiency η_{A} (eqn (7.17)) is plotted against the parameter γ_{DFE} in Fig. 8.37(b). The curve labeled $n = 0$ is for the optimum Nyquist system without DFE (Section 8.3.2). As the basic detection pulse has no postcursors at times $T_{\mathrm{D}} + vT$, no improvement can be gained from DFE. The system efficiency is therefore independent of γ_{DFE}. However, the curve labeled $n = 1$ was optimized under the assumption that the first postcursor is completely eliminated by DFE. With ideal DFE ($\gamma_{\mathrm{DFE}} = 1$) a gain of about 8 dB is achieved relative to the optimum system without DFE ($n = 0$), but if the DFE is not ideal ($\gamma_{\mathrm{DFE}} < 1$) this gain is reduced. For $\gamma_{\mathrm{DFE}} = 0.5$ (i.e. the DFE compensates only 50% of the postcursors) this system is only 2.5 dB better than the system without DFE.

The larger we choose the value of n, the better the system assuming ideal DFE ($\gamma_{\mathrm{DFE}} = 1$) becomes. However, Fig. 8.37(b) makes it clear that with increasing n the DFE must be realized increasingly accurately. If, for example, five postcursors are completely eliminated by means of the DFE ($n = 5$, $\gamma_{\mathrm{DFE}} = 1$), the gain in the signal-to-noise ratio relative to the system without DFE is more than 15 dB. If only 92% of the postcursors are compensated, this system has the same error probability as the system without DFE. Even when $\gamma_{\mathrm{DFE}} = 0.9$ the eye is closed and the error probability is about 0.5.

This figure should show in particular that the curves in Figs 8.33 and 8.34 define only theoretically achievable limits and should therefore not be used as a practical basis. If we are satisfied in this example with the parameters $n = 2$ and $\gamma_{\mathrm{DFE}} = 0.9$, the gain in the signal-to-noise ratio relative to the optimum system without DFE is still 10 dB. This system is about 8 dB worse than the optimum system with ideal DFE ($n \to \infty, \gamma_{\mathrm{DFE}} = 1$).

Chapter 9

Optical Transmission Systems

Contents The characteristics of optical digital systems are discussed in this chapter. In Section 9.1 the individual components (optical transmitter, optical channel and optical receiver) are described using systems theory and models for their signal transmission behavior are defined. The error probability of an optical system is calculated in Section 9.2. In Sections 9.3 and 9.4 it is shown how the optimization of optical transmitters and receivers differs from that of the electrical systems described in Chapters 7 and 8.

Assumptions Although multilevel or coded transmission is possible in principle with an optical digital system, for the purposes of this presentation we limit ourselves to the case of nonredundant binary transmission with threshold detection. The significant difference from Chapters 2–8 is that, because of the presence of signal-dependent (and hence time-dependent) shot noise, perturbations can no longer be assumed to be stationary. This affects the calculation of the error probability and consequently the system optimization.

9.1 Components of an optical digital system

Figure 9.1 shows the principle of signal transmission via an optical waveguide. A suitable semiconductor device, e.g. a laser diode or a light-emitting diode, is stimulated by an electric current to emit light which is transmitted along an optical waveguide (optical fiber) to the receiving device. In the receiver the incoming optical signal is reconverted to an electrical current by a photodiode, e.g. an avalanche photodiode.

The advantages of optical fibers over conventional electrical conductor pairs are many, and only a few are listed here: large bandwidth, low attenuation value, small diameter, low weight and insensitivity to electrical disturbances (e.g. cross-talk). The application of optical fibers to the transmission of digital signals is dealt with here. The physical and technological characteristics of the optical system are not described in detail as they have been discussed extensively elsewhere [1.9, 1.30, 9.8, 9.10].

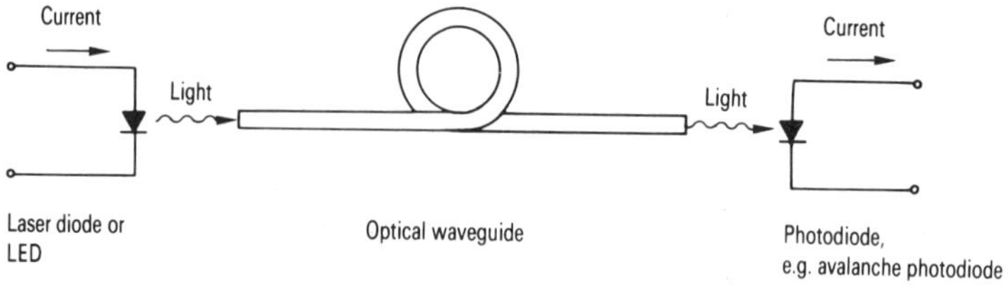

Fig. 9.1 Principle of signal transmission via an optical waveguide.

9.1.1 Optical transmitter

Comparison of the block diagrams shown in Figs 2.1 and 9.2 shows the similarities and differences between electrical and optical digital systems. In the optical system the conversion of the electrical source signal $q(t)$ into the optical transmitter signal $s(t)$ is achieved by means of a **laser diode** (from light amplification by stimulated emission of radiation) or a **light-emitting diode** (LED). In both cases the transmitted signal can be represented in the same way as for an electrical digital system (see eqn (2.11) and eqn (9.1) in Fig. 9.2) if the "pattern effect" [9.8] which occurs to some degree with the laser diode is neglected. Here, however, $s(t)$ is an (optical) power, so that the condition $s(t) \geqslant 0$ must always be satisfied. Therefore it is necessary to consider the optical transmission system as unipolar. In a binary system this means that the amplitude coefficients a_v are either zero or unity.

The basic transmitter pulse $g_s(t)$ of an optical system is an **optical power pulse** with units of milliwatts. When a laser diode is used $g_s(t)$ can be assumed to be Gaussian to a good approximation [9.10]. However, if an LED is used, $g_s(t)$ is similar to the response of a first-order low-pass filter to a rectangular pulse (see the Appendix, Table A3).

The **optical base component** s_0 is particularly necessary in high-rate systems to reduce the response time of the diode, and it corresponds to the transmitted signal in the transmission of the long O sequence. Generally s_0 is about 10%–30% of the amplitude $\hat{g}_s$ of the optical power pulse.

Another fundamental characteristic of the optical transmitter is the **optical wavelength** λ_L which depends on the semiconductor material used. Wavelengths of 850, 1300 and 1500 nm are frequently used. As optical fibers exhibit minimum dispersion at the wavelength $\lambda_L \approx 1300$ nm, this wavelength is particularly suitable for multimodal transmission. However, in single-mode fibers a wavelength of 1500 nm has advantages because of its extremely low attenuation.

An LED emits noncoherent light with a typical linewidth $\Delta\lambda_L$ of 30–40 nm; however, because of the coherent emission the linewidth of the light emitted by a laser diode is only a few nanometers. In addition, the light emitted by a laser diode is more sharply focused than that emitted by an LED so that its coupling efficiency (30%–40%) is very much higher than that of an LED (about 1%). A further

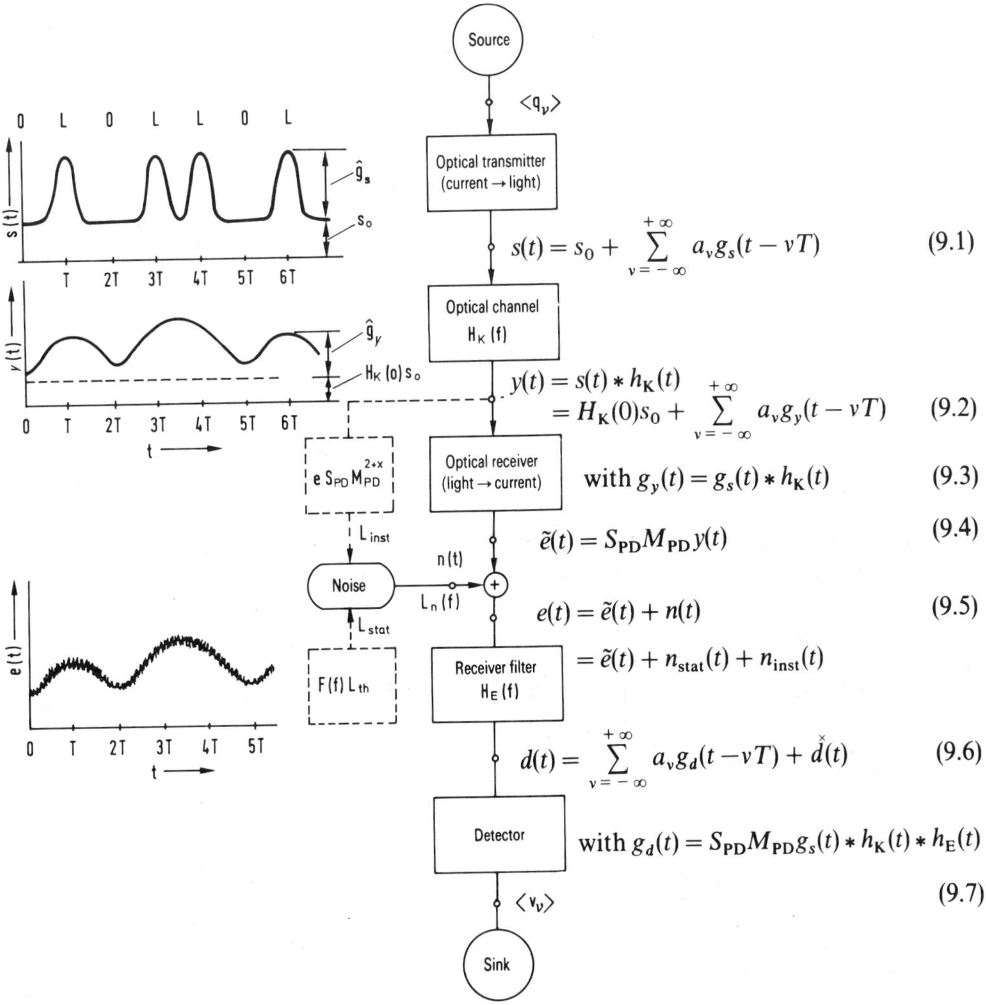

$$\langle q_\nu \rangle$$

$$s(t) = s_0 + \sum_{\nu=-\infty}^{+\infty} a_\nu g_s(t - \nu T) \tag{9.1}$$

$$y(t) = s(t) * h_K(t) = H_K(0) s_0 + \sum_{\nu=-\infty}^{+\infty} a_\nu g_y(t - \nu T) \tag{9.2}$$

$$\text{with } g_y(t) = g_s(t) * h_K(t) \tag{9.3}$$

$$\tilde{e}(t) = S_{PD} M_{PD} y(t) \tag{9.4}$$

$$e(t) = \tilde{e}(t) + n(t) = \tilde{e}(t) + n_{stat}(t) + n_{inst}(t) \tag{9.5}$$

$$d(t) = \sum_{\nu=-\infty}^{+\infty} a_\nu g_d(t - \nu T) + \check{d}(t) \tag{9.6}$$

$$\text{with } g_d(t) = S_{PD} M_{PD} g_s(t) * h_K(t) * h_E(t) \tag{9.7}$$

$$\langle v_\nu \rangle$$

Fig. 9.2 Block diagram of an optical digital system.

advantage of the laser diode is the fact that it can be modulated with very much higher frequencies than is possible for an LED.

9.1.2 Optical transmission channel

The transmission medium of an optical communications system is an **optical waveguide** (optical fiber). This comprises a cylindrical core consisting mostly of fused silica which is sheathed in a material with a somewhat lower refractive index. If an optical signal $s(t)$ is coupled to the fiber, light rays (modes) are propagated along the optical waveguide. Optical losses occur because of absorption by impurities in the fiber and scattering by inhomogeneities in the glass, so that the optical power $y(t)$ at the end of the fiber is lower than the coupled optical power. These losses are included in the **fiber attenuation** which also takes account of the optical power reduction due to coupling and decoupling.

The regenerator-section length l of an optical transmission system is limited by dispersion, which causes pulse spreading and hence intersymbol interference, as well as by attenuation. There are two reasons for the dispersion.

(a) In general the refractive index of the material is dependent on the wavelength λ_L of light (**material dispersion**). This effect is particularly strong if an LED is used as the optical transmitter. Because of the large range of wavelengths ($\Delta\lambda_L = 40$ nm) differences in delay arise which cause a broadening of the transmitted optical pulse.

(b) If the light is not coupled exactly parallel to the fiber axis, several optical waves (modes) are capable of propagation by virtue of total reflection from the cladding. There is a difference in delay between a mode which is coupled to the fiber with the angle $\phi = 0$ and a mode with $\phi \neq 0$ which leads to pulse broadening or **modal dispersion**.

Modal dispersion plays a particularly important role in **step index fibers** in which very many modes can propagate. It can be reduced by about three orders of magnitude if a **graded-index fiber** is used. In this the refractive index decreases from a maximum at the center of the fiber to the refractive index of the cladding in an almost parabolic fashion. The differences in delay between the individual modes are significantly lower because of this continuous variation in refractive index. However, it is not possible to eliminate modal dispersion completely by using a graded-index fiber. If the core diameter of the fiber is sufficiently small (about 1–4 μm), only a single mode can be propagated and we refer to a **single-mode fiber**. Modal dispersion vanishes with such a fiber so that very large bandwidths can be achieved.

According to systems theory, by analogy with electrical systems, both the attenuation (pulse power reduction) and the dispersion (pulse broadening) can be taken into account by the channel frequency response $H_K(f) \bullet\!\!-\!\!\circ\, h_K(t)$, so that the optical received signal $y(t)$ can be calculated from eqn (2.43). The first term in eqn (9.2) (see Fig. 9.2) is due to the optical base component s_0, and the **basic optical receiver pulse** $g_y(t)$ is due to the basic transmitter pulse (see eqns (2.42) and (9.3)).

With the spectra $S(f) \bullet\!\!-\!\!\circ\, s(t)$ and $Y(f) \bullet\!\!-\!\!\circ\, y(t)$ the attenuation value of an optical waveguide is

$$a_K(f) = \frac{1}{2}\ln|H_K(f)| = \ln\left\{\frac{|S(f)|}{|Y(f)|}\right\}^{1/2} \tag{9.8}$$

The attenuation constant differs by a factor of 2 relative to an electrical system as $s(t)$ and $y(t)$ are power signals here (see eqn (2.35)).

The precise calculation of the frequency response of an optical waveguide is generally very complicated. It was shown in ref 9.10 that $H_K(f)$ can be represented by the sum of a set of weighted Gaussian functions. However, in many cases the channel frequency response can be approximated with sufficient accuracy by a single Gaussian low-pass filter:

$$H_K(f) \approx \exp(-2a_0)\exp\{-\pi(f\Delta t_K)^2\} \tag{9.9}$$

a_0 (in nepers) is the direct signal attenuation value which is made up of a constant and a component which is a function of length:

$$a_0 = a_{\text{const}} + \alpha_0 l \tag{9.10}$$

The constant component a_{const} accounts for the losses due to coupling and decoupling as well as the light lost at the splice points. α_0 is the kilometric attenuation coefficient which is of the order of 1 dB km^{-1}.

The **dispersion constant** Δt_K is the reciprocal value of the systems theory bandwidth Δf_K given by eqn (2.37). It depends on both the material dispersion (Δt_{ma}) and the modal dispersion (Δt_{mo}):

$$\Delta t_K{}^2 = \Delta t_{ma}{}^2 + \Delta t_{mo}{}^2 \tag{9.11}$$

The component attributable to material dispersion is proportional to the length l, i.e. $\Delta t_{ma} = K_{ma} l$. For fibers in common use the constant K_{ma} of proportionality lies between 0.5 and 10 ns km^{-1} for LEDs and between 0.01 and 0.2 ns km^{-1} for laser diodes [9.16]. The component due to modal dispersion is given by

$$\Delta t_{mo} = K_{mo}\left[l l_K + \frac{l_K{}^2}{2}\left\{\exp\left(-\frac{2l}{l_K}\right) - 1\right\}\right]^{1/2} \tag{9.12}$$

where K_{mo} is a constant specific to the fiber (of the order of 1 ns km^{-1}) and l_K is the coupling length. In practice, values for l_K lie in the range 2–20 km. For short fiber lengths ($l < l_K$), $\Delta t_{mo} \approx K_{mo} l$ increases approximately in proportion to l. However, the following approximation is valid for $l > l_K$:

$$\Delta t_{mo} \approx K_{mo} l_K \left(\frac{l}{l_K} - \frac{1}{2}\right)^{1/2} \tag{9.13}$$

9.1.3 Optical receiver

The conversion of the received optical signal $y(t)$ back into the electrical signal $e(t)$ is achieved by means of a **photodiode**. This is a semiconductor device operated under reverse bias which in the model can be envisaged as a controlled current source with a useful current output $\tilde{e}(t)$. The effect of noise will be ignored at first.

As photodiodes can generally be assumed to be very wideband relative to the subsequent receiver filter $H_E(f)$, $\tilde{e}(t)$ has the same shape as the optical receiver signal $y(t)$ (see eqn (9.4) in Fig. 9.2). The **dark current** (i.e. the output current with no light falling onto the photodiode) is negligible here. The **responsivity** S_{PD} (units, A W^{-1}) of the photodiode, which can be calculated from the **quantum efficiency** η_Q and the wavelength λ_L, is defined as follows:

$$S_{PD} = \eta_Q \frac{\lambda_L e}{hc} = \eta_Q \frac{e}{E_{ph}} \tag{9.14}$$

In this equation $e = 1.6 \times 10^{-19}$ A s is the elementary charge, $h = 6.62 \times 10^{-34}$ W s^2 is Planck's constant and $c = 3 \times 10^8$ m s^{-1} is the velocity of light in a vacuum. $E_{ph} = hc/\lambda_L$ is the energy of a photon at wavelength λ_L.

In an **avalanche photodiode** (APD) every primary electron resulting from a light impact produces a (statistically variable) number Z_{PD} of secondary electrons by virtue of an "avalanche effect". This means that the useful output current $\tilde{e}(t)$ relative to a simple photodiode is greater by the **mean avalanche gain** $M_{PD} = \overline{Z_{PD}}$; p–i–n diodes have a gain M_{PD} of unity.

In Fig. 9.2 all disturbances are included in the noise current $n(t)$ which is superimposed on the useful current $\tilde{e}(t)$ (see eqn (9.5)). In an optical system the total noise $n(t)$ is generally composed of two components with a Gaussian distribution: (a) the thermal noise $n_{stat}(t)$ and (b) the photodiode shot noise $n_{inst}(t)$.

The statistical parameters (e.g. the probability density function and the power spectrum) of the additive thermal noise current $n_{stat}(t)$ are independent of signal and time. This component is defined by the **stationary noise power spectrum** (see eqn 2.50))

$$L_{stat}(f) = F(f)L_{th} = 2F(f)k_B\theta G_{th} \tag{9.15}$$

in units of A^2 Hz^{-1}. $G_{th} = 1/R_{th}$ is the equivalent noise conductance. It should be noted that the thermal noise power density L_{th} in this case is greater than in an electrical system as no resistance matching is present (see eqn (2.48)).

In contrast with the thermal noise the photodiode shot noise depends on the instantaneous value $y(t)$ of the optical power, i.e. $n_{inst}(t)$ is signal dependent. Hence the moments of this noise component are also time dependent (see Fig. 9.2). The **shot noise** is caused by the statistically varying number of charge carriers which contribute to the diode current $e(t)$. The number of charge carriers per unit time can be described by a Poisson distribution.

If all spectral components of the received optical signal $y(t)$ have frequencies which are very low compared with the frequency of the light, the shot noise, which is generally frequency and time dependent, can be replaced by a time-dependent white noise for which the noise power spectrum is [9.8]

$$L_{inst}(f, t) \approx L_{inst}(0, t) = eS_{PD}M_{PD}{}^{2+x}y(t) \tag{9.16}$$

The empirically determined quantity $M_{PD}{}^x$ is often referred to as the **excess noise factor**. The **excess noise exponent** x describes the increase in noise current owing to avalanche multiplication: x lies between 0.3 and 0.5 for silicon APDs ($\lambda_L = 850$ nm), and $x \approx 0.5$ or $x \approx 1$ for germanium or InGaAs APDs with a wavelength of 1300 nm [9.1].

The effect of the noise sources considered can be described in the model by means of a *single* noise generator which emits the noise current $n(t)$ with a time-dependent power spectrum $L_n = L_{\text{stat}} + L_{\text{inst}}$. As $n_{\text{stat}}(t)$ and $n_{\text{inst}}(t)$ are not mutually correlated, the noise power spectrum of the total noise current is given by

$$L_n(f, t) = L_{\text{stat}}(f, 0) + L_{\text{inst}}(0, t) \tag{9.17}$$

Hence the detection noise power defined by eqn (2.69) is also time dependent:

$$N_d(t) = \int_{-\infty}^{+\infty} L_n(f, t)|H_E(f)|^2 \, df \tag{9.18}$$

9.2 Error probability with signal-dependent noise

The calculation of the error probability of an optical digital system differs from that in Chapter 3 essentially because of the presence of signal-dependent noise. By analogy with eqn (3.37) we obtain in this case the mean error probability

$$p_M = 2^{-(n+v+1)} \sum_{i=1}^{2^{n+v+1}} Q\left\{\frac{|\tilde{d}_i(T_D) - E|}{\sigma_i(T_D)}\right\} \tag{9.19}$$

This equation is valid for the assumption of a nonredundant binary transmitter signal and Gaussian noise. v and n denote the number of precursors and postcursors of the basic detection pulse, and E denotes the decision value. $\tilde{d}_i(T_D)$ is the useful detection sample value for the ith eye line, and $\sigma_i(T_D) = \{N_d(T_D)\}^{1/2}$ the corresponding effective noise value. When the noise is signal independent, the $\sigma_i(T_D) =$ constant are the same for all symbol sequences, i.e. they are independent of i and also of the detection time T_D (see eqns (2.69) and (3.37)). In this case the optimum decision value $\mathring{E}$ lies in the center of the eye (see eqn (3.26)).

However, in the case of an optical transmission system the rms noise values $\sigma_i(T_D)$ differ for individual values of i because of the signal-dependent shot noise. Hence we obtain from eqns (9.15)–(9.18)

$$\{\sigma_i(T_D)\}^2 = N_d(T_D) = \int_{-\infty}^{+\infty} \{F(f)L_{\text{th}} + eS_{\text{PD}}M_{\text{PD}}{}^{2+x}y_i(T_D)\}|H_E(f)|^2 \, df \tag{9.20}$$

where $y_i(t)$ is the optical received signal which produces the useful detection signal $\tilde{d}_i(t)$. The rms noise value is thus larger for a larger optical received signal at time T_D.

This means that in an optical transmission system not only the useful signal but also the statistical parameters of the noise signal depend on the symbol sequence and the detection time. Here the "inner contours" of the eye pattern do not necessarily correspond to the worst-case symbol sequences, i.e. the sequences with the greatest error probability. In addition the optimum decision value $\mathring{E}$ is not in the center of the eye as it is in a system with signal-independent noise.

The decision value is an optimum if the (mean) error probability defined by eqn (9.19) has its minimum value. This optimization cannot generally be performed analytically but only by means of simulation. The problem is simplified significantly if the optimization is based on the worst-case error probability p_U (eqn (3.21)) rather than on p_M. For simplicity, it is also assumed in the following that the inner eye contours represent the worst cases with regard to the error probability.

This approximation is permissible in almost all practical applications. If, as in Fig. 3.4, we denote the upper and lower boundaries of the eye opening at the detection time T_D by $\tilde{d}_{up}(T_D)$ and $\tilde{d}_{lo}(T_D)$ and the corresponding rms noise values by $\sigma_{up}(T_D)$ and $\sigma_{lo}(T_D)$, we obtain the error probabilities of the two worst-case symbol sequences as follows:

$$p_{up} = Q\left\{\frac{\tilde{d}_{up}(T_D) - E}{\sigma_{up}(T_D)}\right\} \tag{9.21a}$$

$$p_{lo} = Q\left\{\frac{E - \tilde{d}_{lo}(T_D)}{\sigma_{lo}(T_D)}\right\} \tag{9.21b}$$

$\sigma_{up}(T_D)$ and $\sigma_{lo}(T_D)$ can be calculated from eqn (9.20) when the relevant signal values $y_{up}(T_D)$ and $y_{lo}(T_D)$ are substituted for $y_i(T_D)$.

The worst-case error probability p_U is equal to the maximum value of either p_{up} or p_{lo}. p_U is a minimum if $p_{up} = p_{lo}$. Hence the optimum decision value can be calculated from eqn (9.21):

$$\mathring{E} = \frac{\sigma_{lo}(T_D)\tilde{d}_{up}(T_D) + \sigma_{up}(T_D)\tilde{d}_{lo}(T_D)}{\sigma_{lo}(T_D) + \sigma_{up}(T_D)} \tag{9.22}$$

Inserting this value in eqn (9.21) yields the worst-case symbol error probability for the optimum decision value:

$$p_U = p_{up} = p_{lo} = Q\left\{\frac{\tilde{d}_{up}(T_D) - \tilde{d}_{lo}(T_D)}{\sigma_{up}(T_D) + \sigma_{lo}(T_D)}\right\} \tag{9.23}$$

If the error probability is sufficiently small, p_U is a good approximation to the mean symbol error probability p_M. As a nonredundant binary source is assumed in this chapter, the mean bit error probability p_B is also equal to p_M. Hence we obtain the sink signal-to-noise ratio of a nonredundant binary optical system from eqn (7.4):

$$\rho_v = \frac{\{\tilde{d}_{up}(T_D) - \tilde{d}_{lo}(T_D)\}^2}{\{\sigma_{up}(T_D) + \sigma_{lo}(T_D)\}^2} \tag{9.24}$$

The special case of signal-independent noise is obtained by putting $\sigma_{up}(T_D) = \sigma_{lo}(T_D)$ in these general equations. Comparison of eqn (7.19) with eqn (9.24) shows that all previous statements also apply to signal-dependent noise if the following value is used for the detection noise power:

$$N_d(T_D) = \tfrac{1}{4}\{\sigma_{up}(T_D) + \sigma_{lo}(T_D)\}^2 \tag{9.25}$$

9.3 Optimum optical transmission system and system efficiency

The system efficiencies defined in Section 7.2 are again used for optimization and comparison of optical digital systems. It is assumed that the peak value of the transmitter signal, i.e. the instantaneous optical power, is limited ($0 \leqslant s(t) \leqslant s_{max}$) so that the system efficiency $\eta_A = \rho_v/\rho_A$ given by eqn (7.45) represents a suitable optimization criterion.

ρ_A defines the maximum sink signal-to-noise ratio under the co-condition of peak-value limitation for the calculation of which an optimum transmitter, an ideal channel and an optimum receiver are assumed (see eqn (7.15)). As in the following ρ_A is merely used as a normalizing value for the actual sink signal-to-noise ratio ρ_v, the technological limitations need not be considered. Therefore the components of the optimum optical transmission system are determined by analogy with Section 7.2.

(a) *Optical transmitter* The basic transmitter pulse $g_s(t)$ is an NRZ rectangular pulse of amplitude $\hat{g}_s = s_{max}$. The optical base power is vanishingly small ($s_0 \approx 0$).

(b) *Ideal channel* The channel frequency response $H_K(f)$ given by eqn (9.9) is unity ($a_0 = 0$, $\Delta t_K = 0$). This implies an attenuation- and dispersion-free optical waveguide and lossless optical coupling and decoupling.

(c) *Optimum receiver* It is also possible in principle to use an optimum Viterbi receiver, as described in Chapter 6, in an optical transmission system. However, it was shown in Section 8.3 that with an optimum transmitter and an ideal channel, a threshold detector produces the same (maximum) sink signal-to-noise ratio if the receiver filter $\mathring{H}_E(f) = H_S{}^*(f) = \mathrm{si}(\pi f T)$ is suitably dimensioned.

With assumptions (a) and (b) we obtain $y_{lo}(T_D) = 0$ and $y_{up}(T_D) = s_{max}$. From assumption (c) and eqn (2.62) it follows that the basic detection pulse is a triangular pulse of amplitude $g_d(0) = S_{PD} M_{PD} s_{max}$. For $|t| \geqslant T$, $g_d(t) = 0$, i.e. $g_d(t)$ is a Nyquist pulse. From this we obtain

$$\tilde{d}_{lo}(0) = 0 \tag{9.26a}$$

$$\tilde{d}_{up}(0) = S_{PD} M_{PD} s_{max} \tag{9.26b}$$

for the lower and upper boundaries of the eye opening at time $\mathring{T}_D = 0$. The corresponding rms noise values $\sigma_{lo}(\mathring{T}_D = 0)$ and $\sigma_{up}(\mathring{T}_D = 0)$ are still needed to calculate the maximum sink signal-to-noise ratio ρ_A (see eqn (9.24)). Both these values can be determined from eqn (9.20). They are a minimum if both the (thermal) noise factor $F(f) = 1$ and the shot noise exponent $x = 0$ have their theoretically smallest values. For $y_{lo}(\mathring{T}_D = 0) = 0$ and $R = 1/T$ we obtain

$$\sigma_{lo}{}^2(0) = \int_{-\infty}^{+\infty} L_{th} |\mathring{H}_E(f)|^2 \, df = L_{th} \int_{-\infty}^{+\infty} \mathrm{si}^2(\pi f T) \, df = L_{th} R \tag{9.27}$$

for the noise power of the lower eye boundary at time $\mathring{T}_D = 0$. By analogy it follows from eqn (9.20) that with $y_{up}(\mathring{T}_D = 0) = s_{max}$

$$\sigma_{up}^2(0) = (L_{th} + eS_{PD}M_{PD}^2 s_{max})R \tag{9.28}$$

Inserting both these equations into eqn (9.24) results in the following sink signal-to-noise ratio under the present assumptions:

$$\rho_v = \frac{S_{PD}^2 M_{PD}^2 s_{max}^2}{R\{L_{th}^{1/2} + (L_{th} + eS_{PD}M_{PD}^2 s_{max})^{1/2}\}^2} \tag{9.29}$$

Now let us consider the dependence on the gain M_{PD} of the avalanche photodiode. As the shot noise exponent x is zero for the optimum system, ρ_v is a maximum when $M_{PD} \to \infty$. Thus the maximum sink signal-to-noise ratio ρ_A of an optical digital system under peak-value limitation to s_{max} is given by

$$\rho_A = \lim_{M_{PD} \to \infty} \rho_v = \frac{S_{PD} s_{max}}{eR} \tag{9.30}$$

This means that, if the shot noise exponent x is zero, the perturbing effect of the thermal noise L_{th} can be made negligibly small by increasing the mean avalanche gain. However, if x is greater than zero the optimum avalanche gain $\mathring{M}_{PD}$ has a finite value (see Section 9.4).

Equation (9.30) shows further that the maximum sink signal-to-noise ratio ρ_A increases in proportion to the responsivity S_{PD} of the photodiode. However, for a given optical wavelength λ_L the responsivity cannot be arbitrarily increased. We obtain the possible maximum value $S_{PD,max} = e/E_{ph}$ from eqn (9.14) for the quantum efficiency $\eta_Q = 1$. Inserting this value in eqn (9.30) gives the maximum sink signal-to-noise ratio of an optical transmission system:

$$\rho_A = \frac{s_{max}}{E_{ph}R} \tag{9.31}$$

In this case we assumed an optimum transmitter, an ideal channel and an optimum receiver. If the transmitter, channel or receiver deviate from the ideal, then the sink signal-to-noise ratio ρ_v is less than ρ_A. Therefore, in the case of an optical transmission system also the system efficiency $\eta_A = \rho_v/\rho_A$ (where $0 \leqslant \eta_A \leqslant 1$) is a suitable measure of the transmission quality.

In an optimum optical transmission system the useful detection power is the same as the peak value s_{max} of the transmitter power $s(t)$ (see eqn (9.31)). In this case the noise power is calculated as $E_{ph}R$ where E_{ph} is the energy of a photon. For $\lambda_L = 850$ nm (1300 nm) E_{ph} is 2.34×10^{-19} W s (1.53×10^{-19} W s). This means that in an optical transmission system the minimum noise power per transmitted bit corresponds to the energy of a photon. Comparison with the corresponding eqn (7.42) makes it clear that in an optical system the energy E_{ph} of a photon has the same fundamentally limiting effect on data transmission as the thermal noise power density L_{th} does in an electrical (lower-frequency) system.

9.4 Optimization of transmitter and receiver parameters

In this section the optimum parameters of an optical transmission system are calculated on the basis of the block diagram in Fig. 9.2. The digital source is binary and nonredundant. If the decision value E given by eqn (9.22) is chosen optimally, as is assumed here, we obtain an approximation to the sink signal-to-noise ratio from eqn (9.24):

$$\rho_v = \frac{\{o(T_D)\}^2}{4N_d(T_D)} = \frac{\{o(T_D)\}^2}{\{\sigma_{up}(T_D) + \sigma_{lo}(T_D)\}^2} \tag{9.32}$$

$o(T_D)$ is the vertical eye opening of the useful detection signal, which can be calculated from the basic detection pulse $g_d(t)$ using eqn (3.32) (for systems without DFE) or eqn (5.12) (for systems with ideal DFE). Hence the absolute eye opening can be written in terms of the normalized eye opening (eqn (3.25)):

$$o(T_D) = s_{max} S_{PD} M_{PD} H_K(0) H_E(0) o_{norm}(T_D) \tag{9.33}$$

In this equation it is assumed that $H_I(f) = H_K(f) S_{PD} M_{PD} H_E(f)$ for an optical transmission system.

Under the assumption, which is generally valid, that not only the signal-dependent shot noise $n_{inst}(t)$ but also the signal-independent thermal noise $n_{stat}(t)$ has a frequency-independent power spectrum, we obtain the noise power for both worst-case symbol sequences from eqn (9.20) as follows:

$$\sigma_{lo}^2(T_D) = \{FL_{th} + eS_{PD}M_{PD}^{2+x}y_{lo}(T_D)\} \int_{-\infty}^{+\infty} |H_E(f)|^2 \, df \tag{9.34a}$$

$$\sigma_{up}^2(T_D) = \{FL_{th} + eS_{PD}M_{PD}^{2+x}y_{up}(T_D)\} \int_{-\infty}^{+\infty} |H_E(f)|^2 \, df \tag{9.34b}$$

F is the mean noise factor of the thermal noise (see eqn 2.71)) and $y_{lo}(T_D)$ and $y_{up}(T_D)$, which depend on the detection time T_D, are the instantaneous values of the optical received signal $y(t)$ for both worst-case symbol sequences. For simplicity these two values are abbreviated below to y_{lo} and y_{up}.

Inserting eqns (9.33) and (9.34) into eqn (9.32) gives the sink signal-to-noise ratio of an optical transmission system (Fig. 9.2) as follows:

$$\rho_v = \frac{s_{max}^2 S_{PD}^2 M_{PD}^2 H_K^2(0) o_{norm}^2(T_D)}{\Box f_E\{(FL_{th} + eS_{PD}M_{PD}^{2+x}y_{lo})^{1/2} + (FL_{th} + eS_{PD}M_{PD}^{2+x}y_{up})^{1/2}\}^2} \tag{9.35}$$

This equation is generally valid under the assumption of an optimum decision value. $\Box f_E$ is the equivalent noise bandwidth of the receiver filter according to eqn (2.39).

9.4.1 Optimization of the mean avalanche gain

ρ_v depends, among other things, on the mean avalanche gain M_{PD} which gives an optimum (depending on the other system parameters in eqn (9.35)). If we set the differential ratio $\mathrm{d}\rho_v/\mathrm{d}M_{\text{PD}}$ to zero we obtain a quadratic equation for the optimum avalanche gain $\mathring{M}_{\text{PD}}$ with the real solution

$$\mathring{M}_{\text{PD}}{}^{2+x} = \frac{FL_{\text{th}}(y_{\text{up}} + y_{\text{lo}})}{2eS_{\text{PD}}y_{\text{up}}y_{\text{lo}}} \left\{ 1 + \frac{16(1+x)}{x^2} \frac{y_{\text{up}}y_{\text{lo}}}{(y_{\text{up}} + y_{\text{lo}})^2} - 1 \right\}^{1/2} \tag{9.36}$$

$\mathring{M}_{\text{PD}}$ is plotted against the ratio $y_{\text{lo}}/y_{\text{up}}$ for the two excess noise exponents $x = 0.5$ and $x = 1$ in Fig. 9.3(a). The optimum mean avalanche gain $\mathring{M}_{\text{PD}}$ increases with decreasing x.

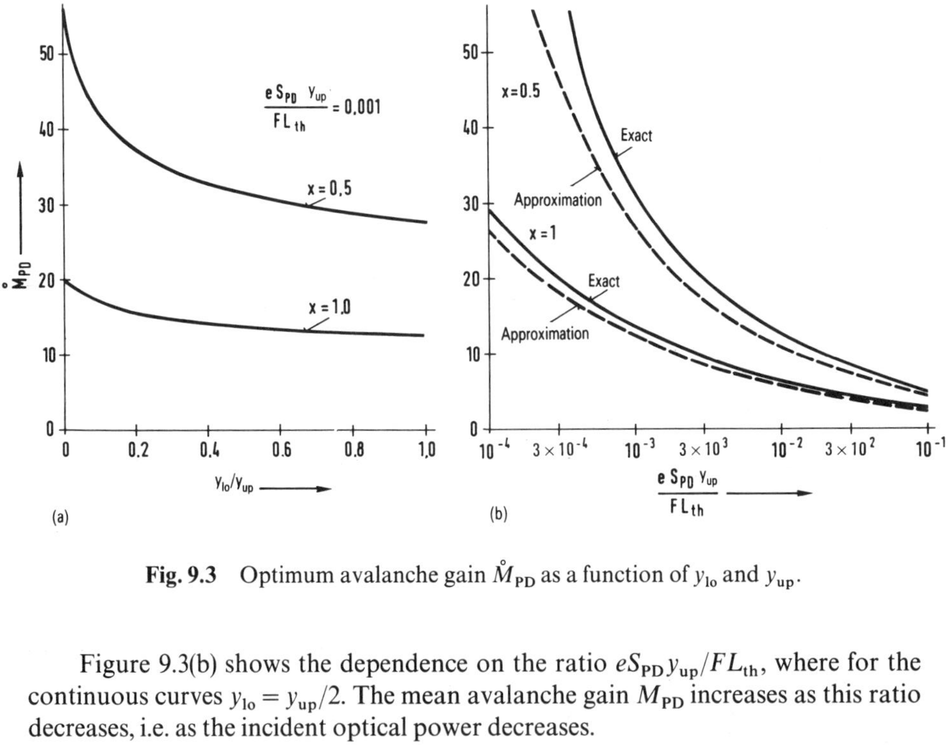

Fig. 9.3 Optimum avalanche gain $\mathring{M}_{\text{PD}}$ as a function of y_{lo} and y_{up}.

Figure 9.3(b) shows the dependence on the ratio $eS_{\text{PD}}y_{\text{up}}/FL_{\text{th}}$, where for the continuous curves $y_{\text{lo}} = y_{\text{up}}/2$. The mean avalanche gain M_{PD} increases as this ratio decreases, i.e. as the incident optical power decreases.

For $y_{\text{lo}} \ll y_{\text{up}}$ the following approximation can be obtained from eqn (9.36):

$$\mathring{M}_{\text{PD}} \approx \left\{ \frac{4(1+x)}{x^2} \frac{FL_{\text{th}}}{eS_{\text{PD}}y_{\text{up}}} \right\}^{1/(2+x)} \tag{9.37}$$

The condition $y_{\text{lo}} \ll y_{\text{up}}$ is very well satisfied if, for example, the optical base power s_0 is negligibly small and the eye of the optical received signal $y(t)$ is very wide open.

This occurs, for example, with very short lengths of fiber and for optical waveguides with low dispersion (single-mode fibers).

Inserting the optimum avalanche gain $\mathring{M}_{PD}$ given by eqn (9.37) into eqn (9.34) we obtain for the equivalent detection noise power of eqn (9.25)

$$N_d(T_D) = \tfrac{1}{4}\{\sigma_{up}(T_D) + \sigma_{lo}(T_D)\}^2 \approx \square f_E F \left(1 + \frac{1}{x}\right)^2 L_{th} \tag{9.38}$$

This means that if the condition $y_{lo} \ll y_{up}$ is satisfied and, in addition, the mean avalanche gain given by eqn (9.37) is optimally designed, the signal-dependent, and hence also time-dependent, shot noise can be replaced with sufficient accuracy by a time-independent noise source. The additional perturbing effect of the shot noise relative to the thermal noise is thus defined by an increase in the noise figure F by a factor $(1 + 1/x)^2$.

Example As in Section 9.3 we now consider an optical transmission system with an optimum transmitter, an ideal channel (i.e. attenuation and dispersion free) and an optimum equalizer. Thus $y_{lo} = 0$ and $y_{up} = s_{max}$. The normalized eye opening $o_{norm}(\mathring{T}_D) = 1$ and the equivalent noise bandwidth $\square f_E = R$. In contrast with Section 9.3, in which the excess noise component $x = 0$ was assumed, x is greater than zero in this case. Hence from eqns (9.32), (9.33) and (9.38) we have for the sink signal-to-noise ratio with optimum avalanche gain given by eqn (9.37)

$$\rho_v = \left\{ \frac{x^{2x}}{2^{2x}(1+x)^{2+2x}} \frac{S_{PD}{}^{2+2x} s_{max}{}^{2+2x}}{e^2 F^x L_{th}{}^x R^{2+x}} \right\}^{1/(2+x)} \tag{9.39}$$

It follows from this, together with eqn (9.31), that the system efficiency under peak-value limitation is

$$\eta_A = \left[\left\{ \frac{x}{2(1+x)^{1+1/x}} \right\}^2 \frac{e^2 s_{max}}{E_{ph} L_{th}} \right]^{x/(2+x)} \tag{9.40}$$

where the quantum efficiency η_Q and the noise figure F are both assumed to be unity. This value defines the loss in signal-to-noise ratio occurring in principle, which is due to the excess noise exponent $x \geqslant 0$ with otherwise ideal assumptions. For $x = 0$, η_A is by definition equal to unity. With increasing x, η_A reduces monotonically. For the following numerical values

$$\begin{aligned} s_{max} &= 1\,\text{mW} \qquad E_{ph} = 1.53 \times 10^{-19}\,\text{W s}\ (\lambda_L = 1300\,\text{nm}) \\ L_{th} &= 1.6 \times 10^{-22}\,\text{A}^2\,\text{Hz}^{-1}\ (G_{th} = 20\,\text{mS},\ \theta = 290\,\text{K}) \end{aligned} \tag{9.41}$$

we obtain $\eta_A = 0.35$ (for $x = 0.5$) and $\eta_A = 0.25$ (for $x = 1$).

The approximation $y_{lo} \ll y_{up}$ is permissible only with very short fibers. In contrast, it can often be assumed with length-optimized systems that the eye of the optical received signal $y(t)$ at the end of the fiber is almost closed. If y_{lo} is larger than

$y_{up}/2$ the following approximation for the optimum avalanche gain (see eqn (9.36)) is valid:

$$\mathring{M}_{PD} \approx \left\{ \frac{4FL_{th}}{xeS_{PD}(y_{up} + y_{lo})} \right\}^{1/(2+x)} \tag{9.42}$$

Inserting this approximation into eqn (9.34), we obtain the noise power for both the worst-case symbol sequences:

$$\sigma_{lo}{}^2(T_D) = \square f_E F L_{th} \left(1 + \frac{4}{x} \frac{y_{lo}}{y_{up} + y_{lo}}\right) \tag{9.43a}$$

$$\sigma_{up}{}^2(T_D) = \square f_E F L_{th} \left(1 + \frac{4}{x} \frac{y_{up}}{y_{up} + y_{lo}}\right) \tag{9.43b}$$

It should be remembered here that y_{lo} and y_{up} depend on the detection time T_D. For the equivalent detection noise power of eqn (9.25) we obtain, again assuming $y_{lo} \geqslant y_{up}/2$,

$$N_d(T_D) = \frac{1}{4}\{\sigma_{up}(T_D) + \sigma_{lo}(T_D)\}^2 \approx \left(1 + \frac{2}{x}\right) \square f_E F L_{th} \tag{9.44}$$

This means that under these conditions, which are very close to reality, the signal-dependent shot noise can be represented by a signal-independent noise current source with the power spectrum $(2/x)FL_{th}$. An optimum mean avalanche of the photodiode gain as given by eqn (9.42) is assumed.

Under this assumption ($M_{PD} = \mathring{M}_{PD}$) it follows that the sink signal-to-noise ratio of an optical transmission system (see eqns (9.35) and (9.44)) is given by

$$\rho_v = \left\{ \frac{1}{4^x(2+x)^2(1+2/x)^x} \frac{(S_{PD}s_{max}H_K(0))^{2+2x}}{\gamma^2 e^2 (FL_{th})^x} \right\}^{1/(2+x)} \frac{o_{norm}{}^2(T_D)}{\square f_E} \tag{9.45}$$

For abbreviation the coefficient

$$\gamma = \frac{y_{up} + y_{lo}}{H_K(0)s_{max}} \tag{9.46}$$

is used here. Its significance and calculation are described in the following section.

9.4.2 Optical system with Gaussian receiver filter

The results given above will now be clarified by means of an example. Let us consider an optical digital system with a Gaussian transmitter pulse

$$g_s(t) = \hat{g}_s \exp\left\{ -\pi \left(\frac{t}{\Delta t_s}\right)^2 \right\} \tag{9.47}$$

This is a sensible assumption at high bit rates and with the use of a laser diode. $\hat{g}_s$ is the amplitude (in milliwatts) and Δt_s is the equivalent duration of the basic transmitter pulse. Hence with the pulse response $h_K(t) \circ\!\!-\!\!\bullet H_K(f)$ of the optical waveguide given by eqn (9.9) we obtain for the optical received pulse

$$g_y(t) = g_s(t) * h_K(t) = \hat{g}_s \frac{\Delta t_s}{(\Delta t_s{}^2 + \Delta t_K{}^2)^{1/2}} \exp\left(-2a_0 - \pi \frac{t^2}{\Delta t_s{}^2 + \Delta t_K{}^2}\right) \quad (9.48)$$

a_0 is the direct signal attenuation value of the optical waveguide and is expressed in nepers (see eqn (9.10)). For the equivalent duration of the optical received pulse, which is also Gaussian, we have, with the dispersion parameter Δt_K of the optical waveguide given by eqn (9.11),

$$\Delta t_y = (\Delta t_s{}^2 + \Delta t_K{}^2)^{1/2} \quad (9.49)$$

An avalanche photodiode is used to convert the received optical power into an electrical current. This is completely described by the parameters S_{PD}, M_{PD} and x defined in Section 9.1. The receiver filter $H_E(f)$ is a Gaussian low-pass filter with cut-off frequency f_E. This filter, which is necessary for noise-power limitation, broadens the received pulses. Hence the basic detection pulse is given by

$$g_d(t) = S_{PD} M_{PD} \{g_y(t) * h_E(t)\} = \hat{g}_s S_{PD} M_{PD} \frac{\Delta t_s}{\Delta t_d} \exp\left\{-2a_0 - \pi\left(\frac{t}{\Delta t_d}\right)^2\right\} \quad (9.50)$$

The equivalent duration of this pulse is calculated using $\Delta t_E = 1/2f_E$ as follows:

$$\Delta t_d = (\Delta t_s{}^2 + \Delta t_K{}^2 + \Delta t_E{}^2)^{1/2} \quad (9.51)$$

The sink signal-to-noise ratio ρ_v of this transmission system can be calculated from eqn (9.35) (for an arbitrary avalanche gain M_{PD}) or from eqn (9.45) (for the optimum avalanche gain $\mathring{M}_{PD}$). Here $|H_K(0)| = \exp(-2a_0)$. The equivalent noise bandwidth of the Gaussian receiver filter is given by $\square f_E = 2^{1/2} f_E$ (Appendix, Table 3). Because of the binary and unipolar amplitude coefficients the vertical eye opening of a system without DFE is given by

$$o(T_D) = g_d(T_D) - \sum_{\nu \neq 0} |g_d(T_D - \nu T)| \quad (9.52)$$

where the basic detection pulse of eqn (9.50) has to be used. As $g_d(t)$ is symmetrical, the optimum detection time $\mathring{T}_D$ is zero. If the equivalent duration Δt_d of the basic detection pulse is less than the symbol duration T, the eye is sufficiently wide open and the normalized eye opening (eqn (3.25)) is approximately

$$o_{norm}(\mathring{T}_D = 0) \approx \frac{\hat{g}_s}{s_{max}} = \frac{\hat{g}_s}{s_0 + \hat{g}_s} \quad (9.53)$$

However, if Δt_d is greater than T, the eye opening decreases very rapidly with increasing Δt_d. The following is a good approximation for a system without DFE in the range $T < \Delta t_d < 2T$ (see Fig. 9.4):

$$o_{\text{norm}}(\mathring{T}_{\text{D}}=0)=\frac{\hat{g}_s}{s_0+\hat{g}_s}\left(\frac{2T}{\Delta t_{\text{d}}}-1\right) \tag{9.54}$$

For $\Delta t_d \geqslant 2T$ the eye is closed in this case.

The signal values $y_{\text{lo}}(T_{\text{D}})$ and $y_{\text{up}}(T_{\text{D}})$ for the two worst-case symbol sequences are still required to calculate the sink signal-to-noise ratio ρ_v (eqn (9.35)). As the basic detection pulse $g_d(t) \geqslant 0$, the worst-case symbol sequence relative to the lower boundary of the eye opening (symbol O) is the sequence ... LLOLL ... . In this case the distance of the useful signal from the threshold is the least and, in addition, the disturbing effect of the shot noise is the greatest in comparison with all other lower eye lines. Correspondingly we obtain from eqns (9.2) and (9.48)

$$\begin{aligned} y_{\text{lo}}(\mathring{T}_{\text{D}}=0) &= H_{\text{K}}(0)s_0+\sum_{v\neq 0} g_y(vT) \\ &\approx \exp(-2a_0)\left\{s_0+\frac{\Delta t_s}{T}\hat{g}_s\left(1-\frac{T}{\Delta t_y}\right)\right\} \end{aligned} \tag{9.55}$$

This approximation results from the fact that the infinite summation of all sample values $g_y(vT)$ of the optical received pulse is equal to $\hat{g}_s \exp(-2a_0)\Delta t_s/T$.

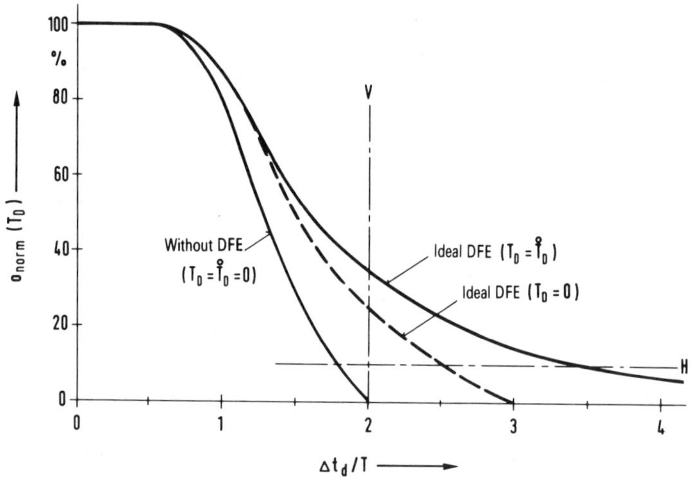

Fig. 9.4 Normalized eye opening for a Gaussian basic detection pulse with equivalent duration Δt_d.

The determination of the worst-case symbol sequence for the upper boundary of the eye opening (symbol L) is somewhat more difficult. In this case the inner eye line comes from the symbol sequence ... OOLOO ... . This sequence is not the worst case for the shot noise. However, in general it can be assumed that the perturbing effect of the intersymbol interference outweighs that of the shot noise, so we have for the upper value

$$y_{up}(\mathring{T}_D = 0) = H_K(0)s_0 + g_y(0) = \exp(-2a_0)\left(s_0 + \frac{\Delta t_s}{\Delta t_y}\hat{g}_s\right) \qquad (9.56)$$

This completes the determination of all the parameters needed to calculate the signal-to-noise ratio and the system efficiency. The optimizable system parameters are the mean avalanche gain M_{PD}, the equivalent transmitter pulse duration Δt_s and the parameter $\Delta t_E = 1/2f_E$ of the Gaussian receiver filter. The decision value E is assumed to be optimum as defined by eqn (9.22).

The dependence of the system efficiency η_A on these three parameters is illustrated in Fig. 9.5 [9.11]. The values assumed for other system parameters are given below the figure. Figure 9.5(a) shows the dependence of the system efficiency on the duration Δt_s of the transmitter pulse when the receiver parameter Δt_E has its optimum value of $T/2$. The smaller the chosen transmitter pulse is, the smaller is the amplitude of the basic detection pulse. In contrast, both the intersymbol interference and the shot noise increase with increasing Δt_s. Therefore there is an optimum value of Δt_s, which in the present case is $\Delta \mathring{t}_s \approx 0.6T$. However, Fig. 9.5(a) also shows that the loss in the signal-to-noise ratio is less than 2 dB if Δt_s varies between $0.3T$ and $0.9T$.

We now consider the effect of the cut-off frequency f_E of the receiver filter which can be obtained from Fig. 9.5(b). The optimum value is $\Delta \mathring{t}_E \approx T/2$, which corresponds to the optimum cut-off frequency $\mathring{f}_E = R$. Δt_E can also be varied over a wide range without significantly degrading the system efficiency.

The dependence of the mean avalanche gain M_{PD} is illustrated in Fig. 9.5(c). The optimum value $\mathring{M}_{PD}$ agrees exactly with the value calculated using eqn (9.36). The

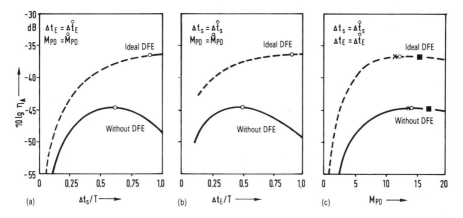

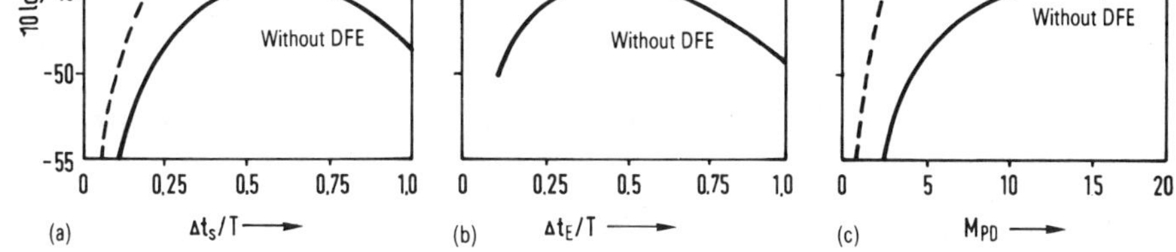

Fig. 9.5 System efficiency $10\lg\eta_A$ as a function of the (normalized) parameters (a) Δt_s, (b) Δt_E and (c) M_{PD}. The following assumptions are made: $s_0 = 0.25\,\text{mW}$; $\hat{g}_s = 1\,\text{mW}$; $a_0 = 20\,\text{dB}$; $\Delta t_K = 2^{1/2}T$; $x = 1$; $\eta_Q = 60\%$ ($\hat{=}\ S_{PD} = 0.63\ \text{AW}^{-1}$ with $\lambda_L = 1300$ nm); $L_{th} = 1.6 \times 10^{-22}\ \text{A}^2\,\text{Hz}^{-1}$; $F = 5$; $E = \mathring{E}$ given by eqn (9.22).

approximations of eqns (9.37) (full squares) and (9.42) (crosses) are also plotted. The figure shows that the system efficiency remains approximately unchanged if M_{PD} lies in the range determined by the two approximations. If this requirement is satisfied, the system efficiency $\eta_A = \rho_v/\rho_A$ can be determined from eqns (9.31) and (9.45). The normalized eye opening is given approximately by eqn (9.54). The following expression for the factor γ (eqn (9.46)) is obtained for a system (without DFE) with a Gaussian receiver filter (see eqns (9.55) and (9.56)):

$$\gamma = \frac{y_{up} + y_{lo}}{s_{max} \exp(-2a_0)} = 2\frac{s_0}{s_{max}} + \frac{\Delta t_s}{T}\frac{\hat{g}_s}{s_{max}} \tag{9.57}$$

For example, $\gamma = 0.88$ when $s_0 = \hat{g}_s/4$ and $\Delta\mathring{t}_s = 0.6T$.

Figure 9.6 shows the optimum values for Δt_s and Δt_E as a function of the dispersion parameter Δt_K of the optical waveguide for various values of the optical base components s_0. In the systems without DFE the optimum values of Δt_s and Δt_E decrease with increasing Δt_K as the equivalent duration Δt_d of the Gaussian basic detection pulse cannot exceed the value $2T$ (see eqn (9.51) and Fig. 9.4).

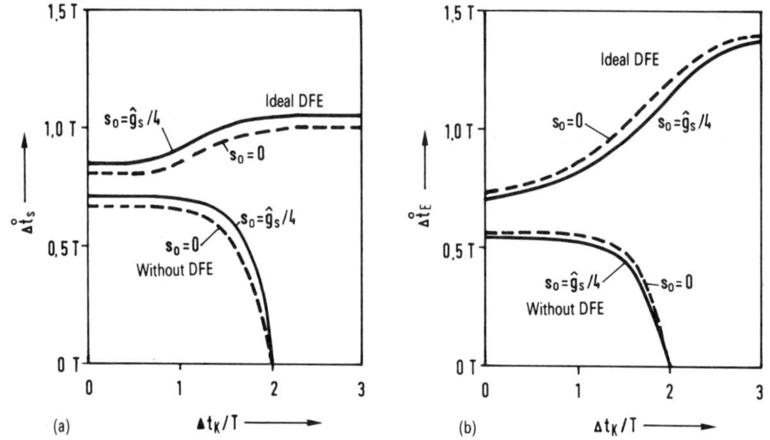

Fig. 9.6 Optimum values for (a) the transmitter pulse duration and (b) the receiver filter time constant $\Delta t_E = 1/2f_E$ as a function of Δt_K (under the same assumptions as in Fig. 9.5).

9.4.3 Optical transmission system with decision feedback equalization

Whereas for a system without DFE the eye is closed for $\Delta t_d \geqslant 2T$, the eye opening for a system with (ideal) DFE is still sufficiently large (see Fig. 9.4). For example, for $\Delta t_d = 2T$ the normalized eye opening is about 25% if the detection time T_D is zero (which is not the optimum). By additional optimization of T_D the eye can be opened to 34% (compare the systems along the vertical line V in Fig. 9.4).

Comparison of the systems along the horizontal line H shows that the normalized eye opening is still only 10% in a system without DFE for $\Delta t_d = 1.8T$. However, in a system with ideal DFE Δt_d can be almost twice as large for the same eye aperture of 10% at the decision device [9.4].

An additional difference from the last section is that different worst-case symbol sequences emerge for a system with (ideal) DFE than for a system without DFE. The worst-case symbol sequence for the lower boundary of the eye opening (symbol O) is, as in a system without DFE, ... LLOLL ..., so that the signal value $y_{lo}(\mathring{T}_D)$ can be calculated from eqn (9.55). However, it must be remembered that, in general, $\mathring{T}_D$ is nonzero for a system with DFE.

Let us now consider the worst-case symbol sequence for the upper boundary of the eye opening (symbol L) which, for a system without DFE and the basic detection pulse corresponding to eqn (9.50), is ... OOLOO ... . As the pulse postcursors are compensated in a system with DFE, the useful signal for the symbol sequence ... LLLOO ... is just as large. However, the contribution of this sequence to the shot noise is larger than that of the sequence ... OOLOO ..., so that instead of eqn (9.56) we have for a system with DFE

$$y_{up}(\mathring{T}_D) = H_K(0)s_0 + \sum_{\nu=0}^{+\infty} g_y(\mathring{T}_D + \nu T) \tag{9.58}$$

Hence the quantity γ defined by eqn (9.46), which is a measure of the shot noise, is rather higher than in a system without DFE. With the numerical values considered in the example we obtain $\gamma \approx 1.5$.

Figure 9.5 shows that with the dispersion parameter $\Delta t_K = 2^{1/2}T$ considered here the improvement in signal-to-noise ratio owing to DFE is about 8 dB. The optimum values for Δt_s and Δt_E are rather higher, as the intersymbol interference is less marked than in a system without DFE (see Fig. 9.6). As the shot noise has a somewhat larger effect here, M_{PD} has a smaller value than is the case without DFE.

Note Optical transmission systems which have optimum Nyquist equalization (see Section 8.3) are discussed in refs 9.9 and 9.11. The signal-to-noise ratio can also be calculated from eqn (9.45) when $o_{norm}(\mathring{T}_D) = 1$ is used. Both worst-case symbol sequences ... LLOLL ... and ... LLLLL ... are determined here exclusively by the shot noise. The factor γ can thus be calculated in an analogous manner to eqn (9.57) [9.11].

Bibliography

Textbooks

1.1 Abramowitz, M. and Stegun, A. *Handbook of Mathematical Functions*, Dover Publications, New York, 1970.

1.2 Bennett, W. and Davey, J. *Data Transmission*, McGraw-Hill, New York, 1965.

1.3 Bocker, P. *Datenübertragung*, Vols I and II, Springer, Berlin, 1976.

1.4 Bronstein, I. and Semendjajew, K. *Taschenbuch der Mathematik*, Harri, Frankfurt, 1980.

1.5 Doetsch, G. *Anleitung zum praktischen Gebrauch der Laplace-Transformation*, Oldenbourg, Munich, 1961.

1.6 Fano, R. M. *Informationsübertragung*, Oldenbourg, Munich, 1966.

1.7 Fetzer, V. *Einschwingvorgänge in der Nachrichtentechnik*, Porta, Munich, 1958.

1.8 Gallager, R. G. *Information Theory and Reliable Communication*, Wiley, New York, 1968.

1.9 Grau, G. *Optische Nachrichtentechnik*, Springer, Berlin, 1981.

1.10 Herter, E., Röcker, W. and Lörcher, W. *Nachrichtentechnik: Übertragung, Vermittlung und Verarbeitung*, Hauser, Munich, 2nd edn, 1981.

1.11 Hölzler, E. and Holzwarth, H. *Pulstechnik*, Vols I and II, Springer, Berlin, 1975 and 1976.

1.12 Kaden, H. *Theoretische Grundlagen der Datenübertragung*, Oldenbourg, Munich, 1968.

1.13 Kress, D. *Theoretische Grundlagen der Übertragung digitaler Signale*, Akademie, Berlin, GDR, 1979.

1.14 Küpfmüller, K. *Einführung in die theoretische Elektrotechnik*, Springer, Berlin, 1973.

1.15 Lucky, R., Saltz, J. and Weldon, E. *Principles of Data Communications*, McGraw-Hill, New York, 1968.

1.16 Lüke, H. D., *Signalübertragung*, Springer, Berlin, 1975.

1.17 Marko, H. *Methoden der Systemtheorie*, Springer, Berlin, 1st edn, 1977; 2nd edn, 1982.

1.18 McElice, R. J. *Theory of Information and Coding*, Addison-Wesley, Reading, MA, 1977.

1.19 Müller, R. *Rauschen*, Springer, Berlin, 1979.

1.20 Oberhettinger, F. *Tabellen zur Fourier-Transformation*, Springer, Berlin, 1957.

1.21 Papoulis, A. *Probability, Random Variables, and Stochastic Processes*, McGraw-Hill, New York, 1965.

1.22 Ryshik, I. M. and Gradstein, I. S. *Summen-, Produkt- und Integral-Tafeln*, VEB Deutscher Verlag der Wissenschaften, Berlin, GDR, 1963.

1.23 Schüssler, H. W. *Netzwerke, Signale und Systeme*, Vol. I, *Systemtheorie linearer elektrischer Netzwerke*, Springer, Berlin, 1981.

1.24 Stein, S. and Jones, J. *Modern Communication Principles*, McGraw-Hill, New York, 1967.

1.25 Steinbuch, K. and Rupprecht, W. *Nachrichtentechnik*, Springer, Berlin, 2nd edn, 1973.

1.26 Thomas, J. B. *An Introduction to Statistical Communication Theory*, Wiley, New York, 1969.

1.27 Tröndle, K. *Nachrichtentechnische Systeme*, Skriptum zur Vorlesung.

1.28 Tröndle, K. and Weiß, R. *Einführung in die Puls-Code-Modulation*, Oldenbourg, Munich, 1974.

1.29 Unbehauen, R. *Systemtheorie*, Oldenbourg, Munich, 2nd edn, 1980.

1.30 Unger, H. G. *Optische Nachrichtentechnik*, Elitera, Berlin, 1976.

1.31 Van Trees, H. L. *Detection, Estimation, and Modulation Theory*, Wiley, New York, 1968.

1.32 Wozencraft, J. M. and Jacobs, I. M. *Principles of Communication Engineering*, Wiley, New York, 1965.

Chapter 2

2.1 Abrahamsen, B., Holte, N. and Röste, T. A 560 Mbit/s digital transmission system over coaxial cable, *Proc. Int. Zürich Seminar*, 1976, A4.

2.2 Appel, U., Tröndle, K. and Weiß, R. Messung und Simulation von Impulsstörungen bei PCM-Übertragung, *Nachrichtentech. Z.*, **26**, 15–17, 1973.

2.3 Bauch, H., Jungmeister, H. G. and Möhrmann, K. H. Übertragung von 565 Mbit/s-Signalen über Koaxialkabel, *Nachrichtentech. Z.*, **32**, 608–611, 1979.

2.4 Bennett, W. R. Statistics for regenerative digital transmission, *BSTJ*, **37**, 1501, 1958.

2.5 Byrne, C. J., Karafin, B. J. and Robinson, D. B. Systematic jitter in a chain of digital regenerators, *BSTJ*, **42**, 2679–2714, 1963.

2.6 Catchpole, R. J., Norman, P. and Waters, D. B. Digitale Übertragung mit 565 Mbit/s auf Koaxialkabeln, *Elektr. Nachrichtenwes.*, **54**, 39–47, 1979.

2.7 CCITT Recommendations G. 742, G. 744 and G. 751, *Orange Book*, Vol. III-2, ITU, Geneva, 1977.

2.8 Dirndorfer, H. and Warkotsch, A. Simulation und Aufbau eines experimentellen 140 Mbit/s-Binärsystems für Koaxialkabel, *AEÜ, Arch. Elektron. Übertragungstech.*, **38**, 79–84, 1984.

2.9 Helm, G. and Plontke, J. Einfluß der Längs- und Querverluste auf das Übertragungsverhalten des Koaxialkabels, *Nachrichtentech. Elektron.*, **25**, 82–84, 1976.

2.10 Huber, J., Tröndle, K. and Lutz, E. Einfluß unsymmetrischer Empfangsimpulsformen auf die Phasenjitterakkumulation. *Nachrichtentech. Z., Arch. 1*, 187–192, 1979.

2.11 Irmer, T. PCM-Übertragungssysteme bei der Deutschen Bundepost. *Fernmelde-Ing.*, **27** (10, 12), 1973.

2.12 Irmer, T., Kersten, R. and Schweizer, L. Begriffe der Digital-Übertragungstechnik, *Frequenz*, **32**, (8–10), 1978.

2.13 Jessop, A., Norman, P. and Wahters, D. B. 120 megabit per second coaxial system demonstrated in the laboratory, *Electr. Commun.*, **48**, 79–92, 1973.

2.14 Kaufhold, B. Empfangsfilteroptimierung bei PCM-Übertragung für ausgewählte Leitungscodes, *Nachrichtentech. Elektron.*, **32**, 161–166, 1982.

2.15 Kersten, R. Signalarten und Signalformen bei der Übertragung von PCM-Signalen auf symmetrischen Fernsprechkabeln. *AEÜ, Arch. Elektron. Übertragungstech.*, **22**, 461–471, 1968.

2.16 Kress, D. Zur Übertragung digitaler Signale auf Kabeln, *Habilitation*, Technische Hochschule Ilmenau, 1975.

2.17 Kohlschmidt, R. Jitterstörungen bei der Übertragung von PCM-Signalen über Kabel, *Habilitation*, Technische Hochschule Ilmenau, 1976.

2.18 De Lange, O. E. The timing of high-speed regenerative repeaters, *BSTJ*, 1455–1485, 1958.

2.19 Lutz, E. and Tröndle, K. Einfluß von Schwellendrift und Verstimmung auf die Taktrückgewinnung mit PLL-Schaltungen, *Frequenz*, **34**, 164–169, 1980.

2.20 Lutz, E. and Tröndle, K. Mittelwert, Effektivwert und Wahrscheinlichkeitsdichte des Taktjitters in digitalen Übertragungssystemen, *AEÜ, Arch. Elektron. Übertragungstech.*, **34**, 104–110, 1980.

2.21 Lutz, E., Huber, J., Söder, G. and Tröndle, K. Einfluß des Impulsformers und der Taktrückgewinnung auf die Akkumulation des Phasenjitters, *NTG-Fachber.*, **65**, 382–387, 1978.

2.22 Marko, H. Eine allgemeine Spektraltransformation, die Fourier-Transformation and Laplace-Transformation gleichzeitig umfaßt, *AEÜ, Arch. Elektron. Übertragungstech.*, **31**, 363–370, 1977.

2.23 Marko, H. Planungsprinzipien digitaler Weitverkehrssysteme, *Nachrichtentech. Z.*, **27**, 2–8, 1974.

2.24 Morgenstern, G. Zur Berechnung der spektralen Leistungsdichte von digitalen Basisband-Signalen, *Fernmelde-Ing.*, **33** (12), 1979.

2.25 Peters, U., Tröndle, K. and Wellhausen, H. W. Zur Untersuchung der Übertragungseigenschaften und Störungen bei Ortskabeln hinsichtlich einer Digitalübertragung oberhalb von 2,048 Mbits/s, *Nachrichtentech. Z., Arch. 2*, 111–116, 1980.

2.26 Röste, T., Holte, N. and Knapskog, S. J. A 140 Mbit/s digital transmission system for coaxial cable using partial response class 1 line code with quantized feedback, *IEEE Trans. Commun.*, **28**, 1425–1430, 1980.

2.27 Sperlich, J. PCM-Systeme und ihr hierarchischer Aufbau, *Nachrichtentech. Z.*, **27**, 182–186, 1974.

2.28 Sunde, E. D. Self-timing regenerative repeaters, *BSTJ*, 891–936, 1957.

2.29 Thomas, R. Einschwingverhalten der Taktphase und des Entscheidungsvorgangs bei regenerativen digitalen Übertragungssystemen, *Dissertation*, Hochschule der Bundeswehr München, 1981.

2.30 Tröndle, K., Söder, G. and Lutz, E. Entstehung des Phasenjitters in regenerativen digitalen Übertragungssystemen und seine statistischen Kenngrößen, *Nachrichtentech. Z.*, **31**, 613–614, 1978.

2.31 Tröndle, K., Söder, G. and Lutz, E. Akkumulation des Phasenjitters in regenerativen digitalen Übertragungssystemen, *AEÜ, Arch. Elektron. Übertragungstech.*, **32**, 341–349, 1978.

2.32 Tufts, D. and Berger, T. Optimum pulse amplitude modulation, Part II, Inclusion of timing jitter, *IEEE Trans. Inf. Theory*, **13**, 209–216, 1967.

2.33 Wellhausen, H. W. Beitrag zur Übertragung digitaler Basisband-Signale über Koaxialkabel. *Fernmelde-Ing.*, **26** (1), 1972.

2.34 Wellhausen, H. W. Dämpfungsverhalten von Koaxialpaaren 1,2/4,4 und 2,6/9,5 mm oberhalb 60 MHz, *Technischer Bericht FI der DBP, 454 TBr 25*, April 1975.

2.35 Wellhausen, H. W. Dämpfung, Phase und Laufzeiten bei Weitverkehrs-Koaxialpaaren, *Frequenz*, **31**, 23–28, 1977.

2.36 Wellhausen, H. W. Digitale Übertragung großer Informationsflüsse über Kabel des Fernliniennetzes. *Mitte. Forschungsinst. Dtsch. Bundespost*, No. 3, 1980.

2.37 Wellhausen, H. W. Entwicklungstendenzen der Digitalübertragung über metallische Leiterpaare, *Frequenz*, **35**, 128–137, 1981.

2.38 Zapf, H. Über den Einfluß der Taktrückgewinnung mittels Resonanzkreisen auf das Jitterverhalten von Regeneratoren—eine zusammengefaßte Darstellung, *Technischer Bericht FI der DBP, 4444 TBr 7*, November 1978.

Chapter 3

3.1 Aaron, M. R. PCM transmission in the exchange plant, *BSTJ*, **41**, 99–141, 1962.

3.2 Antreich, K., Hauk, W. and Welzenbach, M. Zur Auslegung der Entzerrernetzwerke für die Übertragung von PCM-Signalen auf Kabeln, *AEÜ, Arch. Elektron. Übertragungstech.*, **25**, 109–116, 1971.

3.3 Cariolaro, G. L. Error probability in digital fiber communication systems, *IEEE Trans. Inf. Theory*, **24**, 213–221, 1978.

3.4 Glave, F. E. An upper bound on the probability of error due to intersymbol interference for correlated digital signals, *IEEE Trans. Inf. Theory*, **18**, 356–363, 1972.

3.5 Jeng, Y. C., Liu, B. and Thomas, J. B. Probability of error in PAM-systems with intersymbol interference and additive noise, *IEEE Trans. Inf. Theory*, **23**, 575–582, 1977.

3.6 Larsen, G. Näherung der Verteilungsfunktion der Nachbarimpulsstörungen, *AEÜ, Arch. Elektron. Übertragungstech.*, **33**, 403–406, 1979

3.7 Lugananni, R. Intersymbol interferences and probability of error in digital systems, *IEEE Trans. Inf. Theory*, **15**, 682–688, 1969.

3.8 Mathews, J. W. Sharp error bounds for intersymbol interference, *IEEE Trans. Inf. Theory*, **18**, 440–447, 1972.

3.9 Saltzberg, B. R. Error probabilities for a binary signal perturbed by intersymbol interference and Gaussian noise, *IEEE Trans. Commun. Syst.*, **12**, 117–120, 1964.

3.10 Saltzberg, B. R. Intersymbol interference error bounds with application to ideal band-limited signaling, *IEEE Trans. Inf. Theory*, **14**, 563–568, 1968.

3.11 Shimbo, O. and Celebiler, M. The probability of error due to intersymbol interference and Gaussian noise in digital communication systems, *IEEE Trans. Commun. Syst.*, **19**, 113–119, 1971.

3.12 Söder, G. Optimierung und Vergleich binärer und mehrstufiger digitaler Übertragungssysteme mit und ohne quantisierte Rückkopplung, *Dissertation*, Technische Universität München, 1981.

Chapter 4

4.1 Appel, U. Vergleich von Codes für die Übertragung digitaler Signale in regenerativen Kanälen, *Dissertation*, Technische Universität München, 1971.

4.2 Appel, U. and Tröndle, K. Zusammenstellung und Gruppierung verschiedener Codes für die Übertragung digitaler Signale, *Nachrichtentech. Z.*, **23**, 11–16, 1970.

4.3 Appel, U. and Tröndle, K. Vergleich verschiedener Codes für die Übertragung digitaler Signale, *Nachrichtentech. Z.*, **23**, 189–196, 1970.

4.4 Bertelsmeier, M. Blockcodes für digitale Basisbandübertragung—Vergleich einiger 4B3T-Blockcodes, *NTG-Fachber.*, **64**, April 1978.

4.5 Bertelsmeier, M. Digitale Signalübertragung mit einem 4B3T-Code—Untersuchungen an einem Codec für 139, 264 Mbit/s, *Technischer Bericht 44 TBr 65*, September 1978.

4.6 Buchner, J. B. Ternary line codes, *Philips Telecommun. Rev.*, **34**, 72–86, 1976.

4.7 Calfish, M. and Müller, K. H. Leistungsspektrum von Blockcodes, *Nachrichtentech. Z.*, **27**, 219–224, 1974.

4.8 Cariolaro, G. L. and Tronca, G. P. Correlation and spectral properties of multilevel (M, N) coded digital signals with applications to pseudoternary (4, 3) codes, *Alta Freq.*, **18**, 2–15, 1974.

4.9 Catchpole, R. J. Digital line codes and their effect on repeater spacing. *Telecommun. Trans.*, 91–94, 1975.

4.10 Huber, J. Codierung für gedächtnisbehaftete Kanäle, *Dissertation*, Hochschule der Bundeswehr München, 1982.

4.11 Jessop, A. and Waters, D. B. 4B3T, an efficient code for PCM coaxial line systems. *Proc. 17th Int. Scientific Congr. on Electronics, Rome, 1970*, pp. 275–283.

4.12 Krick, W. and Baack, C. Vergleich von "Nyquist-" und "Partial-Response-" Übertragungsverfahren für digitale Lichtwellenleitersysteme, *AEÜ, Arch. Elektron. Übertragungstech.*, **35**, 265–274, 1981.

4.13 Marko, H. Das optimale Sendesignal (Leitungscode) für digitale Übertragungssysteme, *Nachrichtentech. Z.*, **28**, 7–12, 1975.

4.14 Morgenstern, G. Calculation of the power density of rectangular multilevel and AMI coded signals, *Nachrichtentech. Z.*, **31**, 210–211, 1978.

4.15 Morgenstern, G. Vergleich der Leistungsdichtespektren verschiedener binärer Basisbandsignale, *Technischer Bericht, 44 TBr 71*, July 1978.

4.16 Peterson, W. W. *Prüfbare und korrigierbare Codes*, Oldenbourg, Munich, 1967.

4.17 Rocks, M. Calculation of duobinary transmission systems with optical waveguides, *IEEE Trans. Commun.*, **30**, 2464–2470.

4.18 Roth, D. Wahrscheinlichkeitstheoretische Eigenschaften und Spektralverhalten des duobinären Codes im Basisband, *Frequenz*, **27**, 98–103, 1973.

4.19 Ruopp, G. Entwurf und Eigenschaften von 4B3T-Codes, *AEÜ, Arch. Elektron. Übertragungstech.*, **31**, 481–488, 1977.

4.20 Sauer, K. Untersuchung des Phasenjitters in digitalen Übertragungssystemen unter Berücksichtigung des Übertragungscodes, *Diplomarbeit*, Lehrstuhl für Nachrichtentechnik, Technische Universität München, 1979.

4.21 Tröndle, K. Die Verbesserung der Geräuschschwelle pulscodemodulierter Signale durch Anwendung fehlererkennender und fehlerkorrigierender Codes, *Dissertation*, Technische Universität München, 1970.

4.22 Tröndle, K. Codier- und Decodiermethoden zur Fehlerkorrektur digitaler Signale, *Habilitation*, Technische Universität München, 1974.

4.23 A new code of the 4B-3T type: the FOMOT Code, *CCITT, GM/CNC No. 50-E*, January 1975.

Chapter 5

5.1 Andexser, W. Untersuchung von Datenübertragungssystemen mit adaptiver Quantisierter Rückkopplung, *Dissertation*, Universität Stuttgart, 1977.

5.2 Belfiore, C. A. and Park, J. H. Decision feedback equalization, *Proc. IEEE*, **67** (8), 1143–1156, 1979.

5.3 Bennett, W. R. Synthesis of active networks, *Proc. Polytech. Inst. Brooklyn Symp. Ser.*, **5**, 45–61, 1955.

5.4 Gibson, B. Equalization design for a 600 MBd quantized feedback PCM repeater, *IEEE Trans. Commun.*, **27**, 134–142, 1979.

5.5 Hauser, W. and Heidner, D. Ein neuer Impulsregenerator mit quantisierter Rückkopplung für das PCM-Digitalverstärker-Hybridsystem für 560 Mbit/s, *Nachrichtentech. Z.*, **30**, 712–717, 1977.

5.6 Höge, H. Berechnung der Fehlerwahrscheinlichkeit in regenerativen Systemen mit Quantisierter Rückkopplung, *AEÜ, Arch. Elektron. Übertragungstech.*, **30**, 130–135, 1976.

5.7 Marko, H. Kann man über die Nyquistrate übertragen? (Grenzen und Möglichkeiten der Quantisierten Rückkopplung), *AEÜ, Arch. Elektron. Übertragungstech.*, **36**, 238–244, 1982.

5.8 Marko, H., Warkotsch, A. and Dirndorfer, H. The maximum achievable repeater spacing for high speed digital coaxial systems, *Int. Conf. on Communications, Denver, CO, 1981*.

5.9 McColl, L. A. *U.S. Patent 2,056,284*, 1936.

5.10 Milnor, J. W. *U.S. Patent 1,717,116*, 1929.

5.11 Monsen, P. Feedback equalization for fading dispersive channels, *IEEE Trans. Inf. Theory*, **17**, 56–64, 1971.

5.12 Roza, E. A practical design approach to decision feedback receivers with conventional filters, *IEEE Trans. Commun.*, **26**, 679–689, 1978.

5.13 Tamburelli, G. Decision feedback and feedforward receiver (for rates faster than Nyquist's), *Alta Freq.*, **45**, 224–231, 1976.

5.14 Thaller, F. X. Realisierungsformen und Fehlerfortpflanzung der Quantisierten Rückkopplung, *Diplomarbeit*, Lehrstuhl für Nachrichtentechnik, Technische Universität München, 1981.

5.15 Waldhauer, F. Quantized feedback in an experimental 280 Mb/s digital repeater for coaxial transmission, *IEEE Trans. Commun.*, **22**, 1–5, 1974.

5.16 Warkotsch, A. and Goßlau, A. Pilottonzusatz bei digitaler Basisbandübertagung über Koaxialkabel, *AEÜ, Arch. Elektron. Übertragungstech.*, **38**, 79–84, 1984.

5.17 Warkotsch, A. Ein experimentelles 140 Mbit/s Binärsystem mit Quantisierter Rückkopplung und Pilottonzusatz für Koaxialkabel mit sehr hoher Streckendämpfung, *Dissertation*, Technische Universität München, 1983.

5.18 Wellhausen, H. W. Redundanzfreie Codes für digitale Basisbandübertragung, *NTG-Fachber*, **65**, 185–189, 1978.

5.19 Wellhausen, H. W. and Fahrendholz, J. Über die Verwendung binärer, redundanzfreier Basisbandsignale für die Digitalübertragung, *Technischer Bericht, 442 TBr 59*, January 1978.

5.20 Zador, P. L. Error probabilities in data system pulse regenerators with dc restoration, *BSTJ*, **45**, 979, 1966.

Chapter 6

6.1 Beare, C. The choice of the desired impulse response in combined linear–Viterbi algorithm equalizers, *IEEE Trans. Commun.*, **26**, 1301–1307, 1978.

6.2 Burkhardt, H. and Barbosa, L. C. Contributions to the application of the Viterbi algorithm, *Res. Rep. RJ 3377 (40413)*, Communications IBM Research Laboratory, San Jose, CA, January 22, 1982.

6.3 Cantoni, A. and Kwong, K. Further results on the Viterbi algorithm equalizer, *IEEE Trans. Inf. Theory*, **20**, 764–767, 1974.

6.4 Desblache, A. Optimal short desired impulse response for maximum likelihood sequence estimation, *IEEE Trans. Commun.*, **25**, 735–738, 1977.

6.5 Falconer, D. and Magee, F. Adaptive channel memory truncation for maximum likelihood sequence estimation, *BSTJ*, **52**, 1541–1562, 1973.

6.6 Forney, G. D. Lower bounds on error probability in the presence of large intersymbol interference, *IEEE Trans. Commun.*, **20**, 76–77, 1972.

6.7 Forney, G. D. Maximum likelihood sequence estimation of digital sequences in the presence of intersymbol interference, *IEEE Trans. Inf. Theory*, **18**, 363–378, 1972.

6.8 Forney, G. D. The Viterbi algorithm, *Proc. IEEE*, **61**, 268–278, 1973.

6.9 Foschini, G. A reduced state variant of maximum likelihood sequence detection attaining optimum performance for high signal-to-noise ratios, *IEEE Trans. Inf. Theory*, **23**, 605–609, 1977.

6.10 Fredricsson, S. Optimum transmitting filter in digital PAM systems with a Viterbi detector, *IEEE Trans. Inf. Theory*, **20**, 834–838, 1974.

6.11 Gitlin, R. and Ho, E. A null-zone decision feedback equalizer incorporating maximum likelihood bit detection, *IEEE Trans. Commun.*, **23**, 1243–1250, 1975.

6.12 Heller, J. and Jacobs, I. M. Viterbi decoding for satellite and space communication, *IEEE Trans. Commun.*, **19**, 835–848, 1971.

6.13 Helstrom, C. W. *Statistical Theory of Signal Detection*, Pergamon, Oxford, 1968.

6.14 Kaltenbacher, H. Der Viterbi Algorithmus. Ein Verfahren zur Decodierung rekurrenter Codes und zur Entzerrung digitaler Signale, *Diplomarbeit*, Institut für Nachrichtentechnik, Technische Universität München, 1976.

6.15 Kawas-Kaleh, G. Double decision feedback equalizer, *Frequenz*, **33**, 146–149, 1979.

6.16 Lee, W. U. and Hill, F. S. A maximum-likelihood sequence estimator with decision-feedback equalization, *IEEE Trans. Commun.*, **25**, 971–979, 1977.

6.17 Magee, F. and Proakis, J. Adaptive maximum-likelihood-sequence estimation for digital signaling in the presence of intersymbol interference, *IEEE Trans. Inf. Theory*, **19**, 1128–1154, 1973.

6.18 McLane, P. J. A residual intersymbol interference error bound for truncated-state Viterbi detectors, *IEEE Trans. Inf. Theory*, **26**, 548–553, 1980.

6.19 Peters, U. Detection with prediction and decision feedback, *Signal Processing II*, pp. 543–546, EUSIPCO, Erlangen, 1983.

6.20 Peters, U. Empfangsstrategien für Digitalsignalübertragung im Basisband, *Dissertation*, Hochschule der Bundeswehr, München, 1983.

6.21 Qureshi, S. and Newhall, E. An adaptive receiver for data transmission over time-dispersive channels, *IEEE Trans. Inf. Theory*, **19**, 448–457, 1973.

6.22 Schwartz, M. Abstract vector spaces applied to problems in detection and estimation theory, *IEEE Trans. Inf. Theory*, **12**, 327–336, 1966.

6.23 Tröndle, K. Optimierung digitaler Übertragunssysteme. NTG-Fachtagung "Neue Aspekte der Informations- und Systemtheorie", *NTG Fachber.*, **84**, 61–71, 1983.

6.24 Tröndle, K. and Peters, U. Prädiktionsdetektion mit Quantisierter Rückkopplung, *Nachrichtentech. Z. Arch.*, **5**, 63–75, 1983.

6.25 Ungerboeck, G. Nonlinear equalization of binary signals in Gaussian noise, *IEEE Trans. Commun.*, **19**, 1128–1137, 1971.

6.26 Ungerboeck, G. Adaptive maximum-likelihood receiver for carrier-modulated data-transmission system, *IEEE Trans. Commun.*, **29**, 886–890, 1981.

6.27 Vachula, G. M. and Hill, F. S. On optimal detection of band-limited PAM signals with excess bandwidth, *IEEE Trans. Commun.*, **29**, 886–890, 1981.

6.28 Vermeulen, F. L. and Hellman, M. E. Reduced state Viterbi decoders for channels with intersymbol interference, *Proc. 10th Int. Conf. on Communication*, pp. 37B1–37B4, 1974.

6.29 Viterbi, A. Error bounds for convolutional codes and an asymptotically optimum decoding algorithm, *IEEE Trans. Inf. Theory*, **13**, 260–269, 1976.

Chapter 7

7.1 Ballweg, A. Ein Beitrag zur Kanalkapazität bei Leistungs- und Amplitudenbegrenzung, *Diplomarbeit*, Lehrstuhl für Nachrichtentechnik, Technische Universität München, 1984.

7.2 Benzel, R. The capacity region of a class of discrete additive degraded interference channels, *IEEE Trans. Inf. Theory*, **25**, 228–231, 1979.

7.3 Färber, G. Die Kanalkapazität allgemeiner Übertragungskanäle bei begrenztem Signalwertbereich, beliebiger Signalübertragungszeit sowie beliebiger Störung, *AEÜ, Arch. Elektron. Übertragungstech.*, **21**, 565–574, 653–660, 1976.

7.4 Färber, G. and Appel, U. Bestimmung der Kanalkapazität signalwertbegrenzter Analogkanäle durch nichtlineare Programmierung, *Nachrichtentech. Z.*, **21**, 341–348, 1968.

7.5 Pierce, J. R. Information rate of a coaxial cable with various modulation systems, *BSTJ*, **45**, 1197–1207, 1966.

7.6 Raisback, S. Optimal distribution of signal power in a transmission link whose attenuation is a function of frequency, *IRE Trans. Inf. Theory*, **4**, 129–130, 1958.

7.7 Shannon, C. E. A mathematical theory of communication, *BSTJ*, **27**, 379–423, 623–656, 1948.

7.8 Shannon, C. E. and Weaver, W. *The Mathematical Theory of Communication*, University of Illinois Press, Urbana, IL, 1949.

7.9 Söder, G. and Ballweg, A. Die Kanalkapazität als Grenze für die Digitalsignalübertragung, *Frequenz*, **39** (2), 1985.

Chapter 8

8.1 Aaron, M. R. and Tufts, D. W. Intersymbol interference and error probability, *IEEE Trans. Inf. Theory*, **12**, 26–34, 1966.

8.2 Achilles, D. Impulsformung für die Datenübertragung, *AEÜ, Arch. Elektron. Übertragungstech.*, **24**, 186–193, 1970.

8.3 Aein, J. M. and Haucock, J. C. Reducing the effects of intersymbol interference with correlation receivers, *IEEE Trans. Inf. Theory*, **9**, 167–175, 1963.

8.4 Berger, T. and Tufts, D. W. Optimum pulse amplitude modulation, Part I, Transmitter–receiver design and bounds from information theory, *IEEE Trans. Inf. Theory*, **13**, 196–208, 1966.

8.5 Berghaus, T. Equalization of a coaxial cable for digital transmission, *TH-Rep. 78-E-80*, University of Eindhoven.

8.6 Bitzer, E. Optimierung digitaler Übertragungssysteme mit Partial-Response-Codes, *NTG Fachber.*, **65**, 276–281, 1978.

8.7 Brandes, M. Untersuchungen zur Optimierung von Parametern der PCM-Übertragung auf Leitungen für verschiedene Kodes bei praxisnahen Bedingungen und Toleranzen, *Dissertation*, Technische Hochschule Ilmenau, 1979.

8.8 Dippold, M. and Dirndorfer, H. Reduktion des Impulsnebensprechens durch quantisierte Rückkopplung bei optimiertem Phasenverlauf des Detektionssignals, *NTG-Fachber.*, **84**, 99–106, 1983.

8.9 Dirndorfer, H. Simulation und Analyse hochratiger, digitaler Übertragungssysteme unter Berücksichtigung der Quantisierten Rückkopplung, *Dissertation*, Technische Universität München, 1983.

8.10 Franks, L. E. A method for optimizing performance of pulse transmission systems, *IEEE Int. Conv. Rec.*, Part 5, pp. 279–290, 1964.

8.11 Gerst, I. and Diamond, J. The elimination of intersymbol interference by input signal shaping, *Proc. IRE*, **49**, 1195–1203, 1961.

8.12 Gibby, R. and Smith, J. W. Some extensions of Nyquist's telegraph transmission theory, *BSTJ*, **44**, 1487–1510, 1965.

8.13 Graf, K. Theoretische Untersuchungen für Digitalsignalübertragung über symmetrische Kabel, *Diplomarbeit*, Lehrstuhl für Nachrichtentechnik, Technische Universität München, 1984.

8.14 Hänsler, E. Entwurf optimaler Impulse für Puls-Amplituden-Modulationssysteme mit statistisch schwankendem Abtastzeitpunkt, *AEÜ Arch. Elektron. Übertragungstech.*, **31**, 349–354, 1977.

8.15 Irber, A. and Tröndle, K. Codierung zur vollständigen Beseitigung der empfangsseitigen Impulsinterferenz bei Digitalsignalübertragung, *AEÜ, Arch. Elektron. Übertragungstech.*, **38**, 153–156, 1984.

8.16 Kaufhold, B. Zur Optimierung des Empfangsfilters in PCM-Regeneratoren, *Nachrichtentech. Elektron.*, **28**, 235–238, 1978.

8.17 Kaufhold, B. Varianten zur Optimierung des Empfangsfilters in PCM-Regeneratoren mit hoher Bitrate, *Nachrichtentech. Elektron.*, **30**, 48–53, 1980.

8.18 Kreß, D. Zur optimalen Sendeimpulsform bei digitaler Übertragung, *Nachrichtentech. Elektron.*, **25**, 292–293, 1975.

8.19 Kreß, D. and Irmer, R. Optimierungsprinzipien bei der Entzerrung von PCM-Signalen, *Nachrichtentech. Elektron.*, **20**, 185–188, 1970.

8.20 Kreß, D. and Krieghoff, M. Elementare Approximation und Entzerrung bei der Übertragung von PCM-Signalen über Koaxialkabel, *Nachrichtentech. Elektron.*, **23**, 225–227, 1973.

8.21 Marko, H. Comparison of binary and multilevel digital transmission systems in the presence of intersymbol interference and thermal noise, *Nachrichtentech. Z.*, **27**, 239–242, 1974.

8.22 Marko, H. Optimale und fast optimale binäre und mehrstufige digitale Übertragungssysteme, *AEÜ, Arch. Elektron. Übertragungstech.*, **28**, 402–414, 1974.

8.23 Marko, H. and Tröndle, K. Der Grenzverstärkerabstand digitaler Übertragungssysteme für koaxiale Kabel bei gaußförmigem Übertragungsfaktor, *Nachrichtentech. Z.*, **28**, 160–165, 1975.

8.24 Marko, H., Tröndle, K. and Söder, G. Vergleich binärer und mehrstufiger Regenerativverstärkersysteme für koaxiale Kabel bei symmetrischer Impulsform unter Berücksichtigung der Toleranzen, *Nachrichtentech. Z.*, **29**, 601–608, 1976.

8.25 Marko, H., Tröndle, K. and Söder, G. Vergleich optimaler, binärer und mehrstufiger Regenerativverstärkersysteme mit quantisierter Rückkopplung und unsymmetrischer Impulsform, *Nachrichtentech. Z.*, **30**, 316–323, 1977.

8.26 Marko, H., Tröndle, K. and Söder, G. Comparison of optimized digital transmission systems for coaxial cables, *Proc. Eurocon.*, 1977, pp. 2.7.3.1–2.7.3.6.

8.27 Nyquist, H. Certain factors affecting telegraph speed, *BSTJ*, **3**, 324, 1924.

8.28 Nyquist, H. Certain topics in telegraph transmission theory, *Trans. Am. Inst. Electr. Eng.*, *Part 1*, **47**, 617–644, 1928.

8.29 Peters, U. Vergleich verschiedener Empfangsstrategien bei Digitalsignalübertragung, in the press.

8.30 Smith, J. W. The joint optimization of transmitted signal and receiving filter for data transmission systems, *BSTJ*, **44**, 2363–2392, 1965.

8.31 Söder, G. Optimierung und Vergleich digitaler Koaxialkabelsysteme, *AEÜ, Arch. Elektron. Übertragungstech.*, **36**, 451–464, 1982.

8.32 Stojanovic, V. S. and Dirndorfer, H. Über den Entwurf eines hochratigen digitalen Übertragungssystems mit Hilfe der Pol-Nullstellenverteilung, *Frequenz*, **37**, 155–161, 1983.

8.33 Stoll, R. Untersuchung der optimalen Sendeimpulsform bei regenerativen digitalen Übertragungssystemen, *Diplomarbeit*, Lehrstuhl für Nachrichtentechnik, Technische Universität München, 1978.

8.34 Tröndle, K. and Irber, A. Codeoptimierung für digitale Übertragungssysteme, *Frequenz*, **39** (2), 1985.

8.35 Tröndle, K., Krstic, D. and Dirndorfer, H. Methoden zur Bestimmung breitbandiger Funktionsersatzschaltungen für Transistoren, *AEÜ, Arch. Elektron. Übertragungstech.*, **31**, 436–438, 1977.

8.36 Tröndle, K. and Peters, U. Optimierungskriterien und Grenzen digitaler Übertragungssysteme, *Frequenz*, **39** (2), 1985.

8.37 Tufts, D. W. Nyquist's problem—the joint optimization of transmitter and receiver in pulse amplitude modulation, *Proc. IEEE*, **53**, 248–259, 1965.

8.38 van Etten, W. The joint optimization for transmitter and receiver in single pulse transmission, *AEÜ, Arch. Elektron. Übertragungstech.*, **31**, 93–97, 1977.

Chapter 9

9.1 Asatani, K., Sato, K. and Maki, K. Fibre optic analogue transmission experiment for high-definition television signals using semiconductor laser diodes, *Electron. Lett.*, **16**, 536–538, 1980.

9.2 Baack, C. Grenzrepeaterabstand digitaler, optischer Übertragungssysteme hoher Bitrate, *Nachrichtentech. Z.*, **30**, 65, 1977; *Nachrichtentech. Z. Forschungsdienst 1–77*.

9.3 Baack, C. Optimierung eines optischen, digitalen Übertragungssystems für 1,12 Gbit/s, *Frequenz*, **32**, 16–21, 1978.

9.4 Dippold, M. Empfänger für schnelle digitale Glasfaserübertragung mit quantisierter Rückkopplung. *AEÜ, Arch. Elektron. Übertragungstech.*, **37**, 15–24, 1983.

9.5 Geckeler, S. Übertragungseigenschaften von Glasfasern, *Siemens Forsch. Entwicklungsber.*, **2** (4), 1973.

9.6 Geckeler, S. Berechnungsverfahren für die Lichtausbreitung in Glasfasern, *Siemens Forsch. Entwicklungsber.*, **6** (3), 1977.

9.7 Gloge, D. and Marcatili, E. A. J. Multimode theory of graded-core fibers, *BSTJ*, **52**, 1563–1578, 1973.

9.8 Kersten, R. Th. *Einführung in die optische Nachrichtentechnik*, Springer, Berlin, 1983.

9.9 Klein, M. V. *Optics*, Wiley, New York, 1970.

9.10 Lutz, E. and Tröndle, K. *Systemtheorie der optischen Nachrichtentechnik*, Oldenbourg, Munich, 1983.

9.11 Marko, H. and Söder, G. Die Vergrößerung des Regeneratorabstandes bei Glasfasersystemen durch quantisierte Rückkopplung, *Proc. Professorenkonferenz 1981 im FTZ*, pp. 163–192.

9.12 Personick, S. D. Receiver design for digital fibre optic communication systems, *BSTJ*, **52**, 843–886, 1973.

9.13 Personick, S. D. Baseband linearity and equalization in fibre optic digital communication systems, *BSTJ*, **52**, 1175–1194, 1973.

9.14 Rugemalira, R. M. A. Optimum linear equalization of a digital fibre optic communication system, *Opt. Quantum Electron.*, **13**, 153–163, 1981.

9.15 Schnepf, L. Untersuchung und Optimierung von digitalen Übertragungssystemen mit Lichtwellenleitern, *Diplomarbeit*, Lehrstuhl für Nachrichtentechnik, Technische Universität München, 1980.

9.16 *Telcom Report 6, Beiheft "Nachrichtenübertragung mit Licht"*, 1985.

Appendix

Table A1 Mathematical functions

Row	*Definition*	*Sketch*	*Notes*
1	Step function $\sqsubset(x) = \begin{cases} 1 & \text{for } x > 0 \\ 1/2 & \text{for } x = 0 \\ 0 & \text{for } x < 0 \end{cases}$	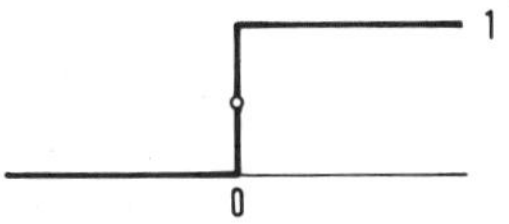	$\sqsubset(0) = \frac{1}{2}$
2	Rectangular function $\text{rec}(x) = \begin{cases} 1 & \text{for } \lvert x\rvert < 1/2 \\ 1/2 & \text{for } \lvert x\rvert = 1/2 \\ 0 & \text{for } \lvert x\rvert > 1/2 \end{cases}$	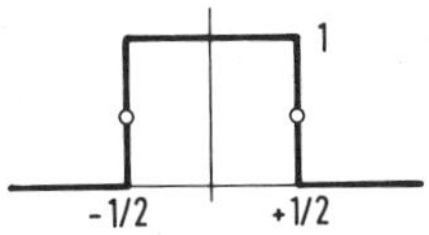	$\text{rec}(x) =$ $\sqsubset(x + \frac{1}{2}) - \sqsubset(x - \frac{1}{2})$
3	Triangular function $\wedge(x) = \begin{cases} 1 - \lvert x\rvert & \text{for } \lvert x\rvert \leqslant 1 \\ 0 & \text{for } \lvert x\rvert \geqslant 1 \end{cases}$	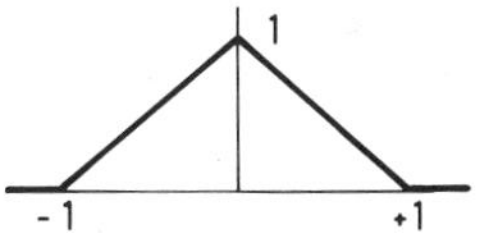	$\wedge(x) =$ $\text{rec}(x) * \text{rec}(x)$
4	Dirac "function" $\delta(x) = \begin{cases} +\infty & \text{for } x = 0 \\ 0 & \text{for } x \neq 0 \end{cases}$		$\lim_{\varepsilon \to 0} \int_{-\varepsilon}^{+\varepsilon} \delta(x)\,dx = 1$
5	si function $\text{si}(x) = \dfrac{\sin x}{x}$	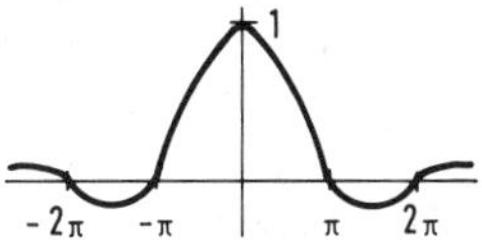	$\text{si}(x) \approx 1 - \dfrac{x^2}{3!} + \dfrac{x^4}{5!}$
6	si integral function $\text{Si}(x) = \int_0^x \text{si}(u)\,du$	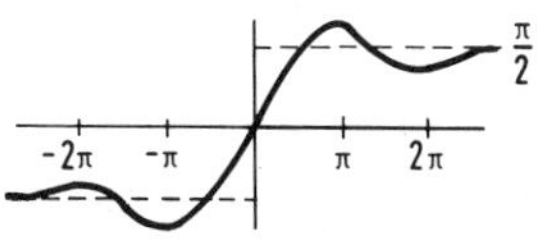	$\text{Si}(-x) = -\text{Si}(x)$ $\text{Si}(0) = 0$ $\text{Si}(+\infty) = \pi/2$
7	Gaussian error integral $\phi(x) = \dfrac{1}{(2\pi)^{1/2}} \int_{-\infty}^{x} \exp\left(\dfrac{u^2}{2}\right) du$	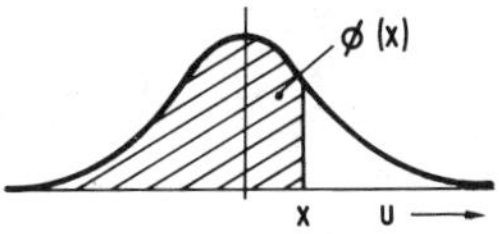	$\phi(x) = 1 - Q(x)$ $\phi(-x) = Q(x)$
8	Complementary Gaussian error integral $Q(x) = \dfrac{1}{(2\pi)^{1/2}} \int_{x}^{+\infty} \exp\left(\dfrac{u^2}{2}\right) du$	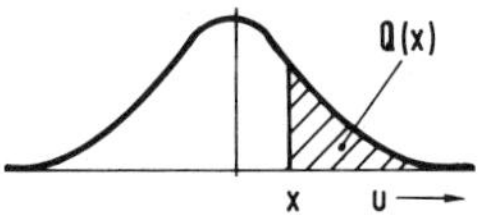	$Q(x) = 1 - \phi(x)$ for $x \geqslant 1$ $Q(x) \approx \dfrac{\exp(-x^2/2)}{(2\pi)^{1/2} x}$

Table A2 Table of values of the (complementary) Gaussian error integrals

x	$20 \lg x$	$\varphi(x)$	$\phi(x) = 1 - Q(x)$	$Q(x) = 1 - \phi(x)$ *Exact*	$Q(x) = 1 - \phi(x)$ *Approximate*
0.0	0.0	0.39894	0.50000	0.50000	0.00000
0.1	−20.0	0.39695	0.53983	0.46017	3.9695
0.2	−14.0	0.39104	0.57926	0.42074	1.9552
0.3	−10.5	0.38139	0.61791	0.38209	1.2713
0.4	−8.0	0.36827	0.65542	0.34458	0.92068
0.5	−6.0	0.35207	0.69146	0.30854	0.70413
0.6	−4.4	0.33322	0.72575	0.27425	0.55537
0.7	−3.1	0.31225	0.75804	0.24196	0.44608
0.8	−1.9	0.28969	0.78814	0.21186	0.36211
0.9	−0.9	0.26609	0.81594	0.18406	0.29565
1.0	0.0	0.24197	0.84134	0.15866	0.24197
1.1	0.8	0.21785	0.86433	0.13567	0.19805
1.2	1.6	0.19419	0.88493	0.11507	0.16182
1.3	2.3	0.17137	0.90320	0.96801×10^{-1}	0.13182
1.4	2.9	0.14973	0.91924	0.80757×10^{-1}	0.10695
1.5	3.5	0.12952	0.93319	0.66807×10^{-1}	0.86345×10^{-1}
1.6	4.1	0.11092	0.94520	0.54799×10^{-1}	0.69326×10^{-1}
1.7	4.6	0.94049×10^{-1}	0.95543	0.44565×10^{-1}	0.55323×10^{-1}
1.8	5.1	0.78950×10^{-1}	0.96407	0.35930×10^{-1}	0.43861×10^{-1}
1.9	5.6	0.65616×10^{-1}	0.97128	0.28717×10^{-1}	0.34535×10^{-1}
2.0	6.0	0.53991×10^{-1}	0.97725	0.22750×10^{-1}	0.26995×10^{-1}
2.2	6.8	0.35475×10^{-1}	0.98610	0.13903×10^{-1}	0.16125×10^{-1}
2.4	7.6	0.22395×10^{-1}	0.99180	0.81975×10^{-2}	0.93311×10^{-2}
2.6	8.3	0.13583×10^{-1}	0.99534	0.46612×10^{-2}	0.52242×10^{-2}
2.8	8.9	0.79155×10^{-2}	0.99744	0.25552×10^{-2}	0.28269×10^{-2}
3.0	9.5	0.44318×10^{-2}	0.99865	0.13500×10^{-2}	0.14773×10^{-2}
3.2	10.1	0.23841×10^{-2}	0.99931	0.68720×10^{-3}	0.74503×10^{-3}
3.4	10.6	0.12322×10^{-2}	0.99966	0.33698×10^{-3}	0.36242×10^{-3}
3.6	11.1	0.61190×10^{-3}	0.99984	0.15915×10^{-3}	0.16997×10^{-3}
3.8	11.6	0.29195×10^{-3}	0.99993	0.72372×10^{-4}	0.76828×10^{-4}
4.0	12.0	0.13383×10^{-3}	0.99997	0.31686×10^{-4}	0.33458×10^{-4}
4.2	12.5	0.58943×10^{-4}	0.99999	0.13354×10^{-4}	0.14034×10^{-4}
4.4	12.9	0.24942×10^{-4}	0.99999	0.54170×10^{-5}	0.56687×10^{-5}
4.6	13.3	0.10141×10^{-4}	1.00000	0.21146×10^{-5}	0.22045×10^{-5}
4.8	13.6	0.39613×10^{-5}	1.00000	0.79435×10^{-6}	0.82527×10^{-6}
5.0	14.0	0.14867×10^{-5}	1.00000	0.28710×10^{-6}	0.29734×10^{-6}
5.2	14.3	0.53610×10^{-6}	1.00000	0.10310×10^{-6}	0.10310×10^{-6}
5.4	14.6	0.18574×10^{-6}	1.00000	0.34396×10^{-7}	0.34396×10^{-7}
5.6	15.0	0.61826×10^{-7}	1.00000	0.11040×10^{-7}	0.11040×10^{-7}
5.8	15.3	0.19773×10^{-7}	1.00000	0.34092×10^{-8}	0.34092×10^{-8}
6.0	15.6	0.60759×10^{-8}	1.00000	0.10126×10^{-8}	0.10126×10^{-8}

Table A2 (cont.)

x	$20\lg x$	$\varphi(x)$	$\phi(x) = 1 - Q(x)$	$Q(x) = 1 - \phi(x)$	
				Exact	*Approximate*
6.2	15.8	0.17938×10^{-8}	1.00000	0.28932×10^{-9}	0.28932×10^{-9}
6.4	16.1	0.50881×10^{-9}	1.00000	0.79502×10^{-10}	0.79502×10^{-10}
6.6	16.4	0.13867×10^{-9}	1.00000	0.21010×10^{-10}	0.21010×10^{-10}
6.8	16.7	0.36310×10^{-10}	1.00000	0.53396×10^{-11}	0.53396×10^{-11}
7.0	16.9	0.91347×10^{-11}	1.00000	0.13050×10^{-11}	0.13050×10^{-11}
7.5	17.5	0.24343×10^{-12}	1.00000	0.32458×10^{-13}	0.32458×10^{-13}
8.0	18.1	0.50523×10^{-14}	1.00000	0.63153×10^{-15}	0.63153×10^{-15}
8.5	18.6	0.81662×10^{-16}	1.00000	0.96073×10^{-17}	0.96073×10^{-17}
9.0	19.1	0.10280×10^{-17}	1.00000	0.11422×10^{-18}	0.11422×10^{-18}
9.5	19.6	0.10078×10^{-19}	1.00000	0.10608×10^{-20}	0.10608×10^{-20}
10.0	20.0	0.76946×10^{-22}	1.00000	0.76946×10^{-23}	0.76946×10^{-23}

$$\varphi(x) = \frac{1}{(2\pi)^{1/2}} \exp\left(-\frac{x^2}{2}\right) \qquad \phi(x) = \int_{-\infty}^{x} \varphi(u)\,\mathrm{d}u \qquad Q(x) = \int_{x}^{+\infty} \varphi(u)\,\mathrm{d}u$$

Table A3 Frequency, pulse and step responses of some low-pass filters

Row	*Name and sketch*	*Frequency response* $H_I(f)$	*Pulse response* $h_1(t) \circ\!\!-\!\!\bullet H_I(f)$	*Step response* $c_I(t) = \int_{-\infty}^{t} h_1(x)\,dx$	*Cut-off frequency* f_I *Noise bandwidth* $\square f_I$ *Roll-off factor* r_I
1	First order low-pass Absolute value 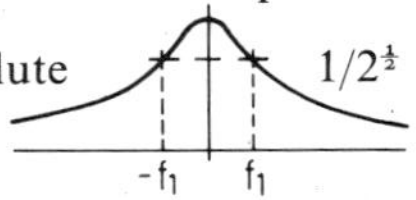	$\frac{1}{1+jf/f_1}$ (complex)	$2\pi f_1 \exp(-2\pi f_1 t)\,\text{ʃ}(t)$ $\text{ʃ}(t)$: step function	$\{1-\exp(-2\pi f_1 t)\}\,\text{ʃ}(t)$	$f_I = \frac{\pi}{2} f_1$ $\square f_I = \pi f_1 = 2 f_I$
2	Second-order low-pass Absolute value 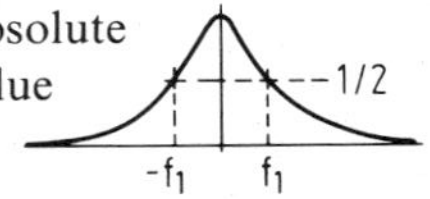	$\frac{1}{(1-jf/f_1)^2}$ (complex)	$4\pi^2 f_1{}^2 t \exp(-2\pi f_1 t)\,\text{ʃ}(t)$ $\text{ʃ}(t)$: step function	$\{1-(1+2\pi f_1 t)\exp(-2\pi f_1 t)\}\,\text{ʃ}(t)$	$f_I = 0$ $\square f_I = \frac{\pi}{2} f_1$
3	Rectangular (brick-wall) low-pass 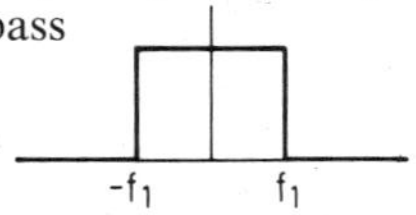	1 for $\lvert f\rvert < f_1$ 1/2 for $\lvert f\rvert = f_1$ 0 for $\lvert f\rvert > f_1$	$2f_1\,\mathrm{si}(2\pi f_1 t)$ $\mathrm{si}(x) = \frac{\sin x}{x}$	$\frac{1}{2} + \frac{1}{\pi}\mathrm{Si}(2\pi f_1 t)$ $\mathrm{Si}(x) = \int_0^x \mathrm{si}(u)\,du$	$f_I = f_1$ $\square f_I = 2f_1 = 2f_I$
4	si low-pass 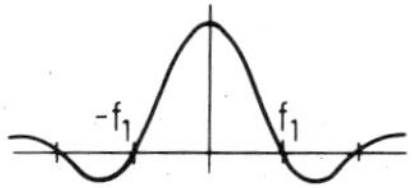	$\mathrm{si}\left(\pi \frac{f}{f_1}\right)$ $\mathrm{si}(x) = \frac{\sin x}{x}$	f_1 for $\lvert t\rvert < 1/2f_1$ $f_1/2$ for $\lvert t\rvert = 1/2f_1$ 0 for $\lvert t\rvert > 1/2f_1$	0 for $t \leqslant -1/2f_1$ $\frac{1}{2} + f_1 t$ otherwise 1 for $t \geqslant 1/2f_1$	$f_I = f_1/2$ $\square f_I = f_1 = 2f_I$

Table A3 Frequency, pulse and step responses of some low-pass filters

Row	*Name and sketch*	*Frequency response* $H_\mathrm{I}(f)$	*Pulse response* $h_1(t) \circ\!\!-\!\!\bullet H_\mathrm{I}(f)$	*Step response* $c_\mathrm{I}(t) = \int_{-\infty}^{t} h_1(x)\,\mathrm{d}x$	*Cut-off frequency* f_I *Noise bandwidth* $\square f_\mathrm{I}$ *Roll-off factor* r_I
5	si² low-pass 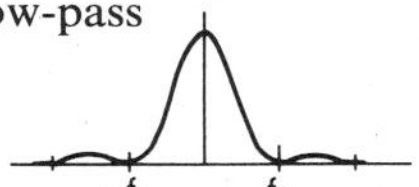	$\mathrm{si}^2\left(\pi \frac{f}{f_1}\right)$ $\mathrm{si}(x) = \frac{\sin x}{x}$	$f_1(1 - f_1\|t\|)$ for $\|t\| \leqslant 1/f_1$ 0 for $\|t\| \geqslant 1/f_1$	0 for $t \leqslant -1/f_1$ $\frac{1}{2} + \frac{1}{2}f_1{}^2 t(t)$ otherwise 1 for $t = 1/f_1$	$f_\mathrm{I} = f_1/2$ $\square f_\mathrm{I} = \frac{2}{3}f_1 = \frac{4}{3}f_\mathrm{I}$
6	Triangular low-pass 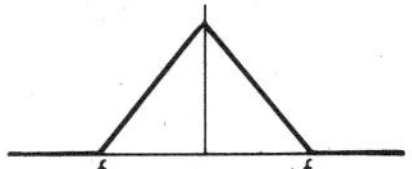	$1 - \frac{\|f\|}{f_1}$ for $\|f\| \leqslant f_1$ 0 otherwise	$f_1\,\mathrm{si}^2(\pi f_1 t)$ $\mathrm{si}(x) = \frac{\sin x}{x}$	$\frac{1}{2} + \frac{1}{\pi}\mathrm{Si}(2\pi f_1 t) -$ $-\frac{1}{\pi}\mathrm{si}(\pi f_1 t)\sin(\pi f_1 t)$ $\mathrm{Si}(x) = \int_0^x \mathrm{si}(u)\,\mathrm{d}u$	$f_\mathrm{I} = f_1/2$ $\square f_\mathrm{I} = \frac{2}{3}f_1$
7	Gaussian low-pass 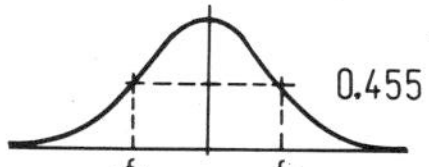	$\exp\left\{-\pi\left(\frac{f}{2f_1}\right)^2\right\}$	$2f_1\exp\{-\pi(2f_1 t)^2\}$	$\phi\{2(2\pi)^{1/2}f_1 t\}$ $\phi(x) = 1 - Q(x) =$ $\frac{1}{(2\pi)^{1/2}}\int_{-\infty}^{x}\exp\left(-\frac{u^2}{2}\right)\mathrm{d}u$	$f_\mathrm{I} = f_1$ $\square f_\mathrm{I} = 2^{1/2}f_1$

Table A3 Frequency, pulse and step responses of some low-pass filters

Row	*Name and sketch*	*Frequency response* $H_I(f)$	*Pulse response* $h_1(t) \circ\!\!-\!\!\bullet H_I(f)$	*Step response* $c_I(t) = \int_{-\infty}^{t} h_1(x)\,dx$	*Cut-off frequency* f_I / *Noise bandwidth* $\square f_I$ / *Roll-off factor* r_I
8	Cosine low-pass 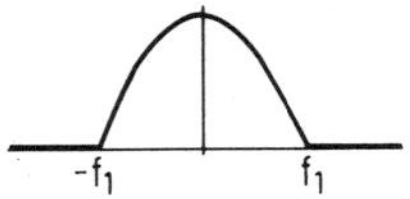	$\cos\left(\frac{\pi}{2}\frac{f}{f_1}\right)$ for $\lvert f\rvert \leqslant f_1$; 0 otherwise	$\frac{4f_1\cos(2\pi f_1 t)}{\pi(1-16{f_1}^2t^2)}$	Numerical integration over $h_I(t)$	$f_I = \frac{2f_1}{\pi}$; $\square f_I = f_1 = \frac{\pi}{2}f_I$
9	Trapezoidal low-pass 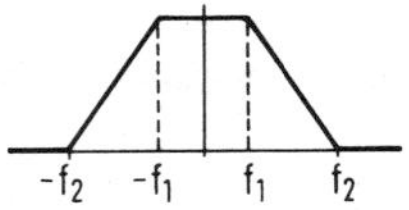	1 for $\lvert f\rvert \leqslant f_1$; 0 for $\lvert f\rvert \geqslant f_2$; $\frac{f_2-\lvert f\rvert}{f_2-f_1}$ otherwise	$2f_I\sin(2\pi f_I t)\,\mathrm{si}(2\pi f_I r_I t)$	$\frac{1}{2}+\frac{1+r_I}{2\pi r_I}\times$ $\times\ \mathrm{Si}\{2\pi f_I(1+r_I)t\} -$ $-\frac{1-r_I}{2\pi r_I}\mathrm{Si}\{2\pi f_I(1-r_I)t\} -$ $-\frac{1}{\pi}\mathrm{si}(2\pi f_I r_I t)\sin(2\pi f_I t)$	$f_I = \frac{f_1+f_2}{2}$; $\square f_I = \frac{4}{3}f_1 + \frac{2}{3}f_2$; $r_I = \frac{f_2-f_1}{f_2+f_1}$
10	Cosine roll-off low-pass 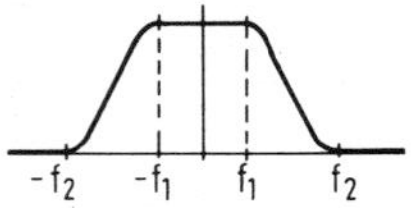	1 for $\lvert f\rvert \leqslant f_1$; 0 for $\lvert f\rvert \geqslant f_2$; $\cos^2\left(\frac{\lvert f\rvert - f_1}{f_2-f_1}\frac{\pi}{2}\right)$ otherwise	$\frac{2f_I}{1-(4r_I f_I t)^2}\mathrm{si}(2\pi f_I t)\times$ $\times\cos(2\pi r_I f_I t)$	Numerical integration over $h_I(t)$	$f_I = \frac{f_1+f_2}{2}$; $\square f_I = \frac{5}{4}f_1 + \frac{3}{4}f_2$; $r_I = \frac{f_2-f_1}{f_2+f_1}$

Glossary

$c = 3 \times 10^8\,\mathrm{m\,s^{-1}}$	Lichtgeschwindigkeit	velocity of light
$e = 1.6 \times 10^{-19}\,\mathrm{A\,s}$	Elementarladung	elementary charge
$h = 6.62 \times 10^{-34}\,\mathrm{W\,s^2}$	Planck'sches Wirkungsquantum	Planck's constant
$k_B = 1.38 \times 10^{-23}\,\mathrm{W\,s\,K^{-1}}$	Boltzmann-Konstante	Boltzmann's constant

$a_K(f)$	Dämpfungsmaß (des Kanals)	attenuation constant (of the channel)
a_0	Gleichsignaldämpfung	direct signal attenuation value
a_*	charakteristische Kabeldämpfung (bei der halben Bitrate)	characteristic attenuation value (at 50% bit rate)
a_ν	Amplitudenkoeffizient (sendeseitig)	amplitude coefficient (at the transmitter)
a_ν'	Amplitudenkoeffizient (empfangsseitig)	amplitude coefficient (at the receiver)
$\langle a_\nu \rangle$	zeitliche Folge der Amplitudenkoeffizienten	temporal sequence of amplitude coefficients
$\{a_\mu\}$	Menge der möglichen Amplitudenkoeffizienten	set of possible amplitude coefficients
B_K	einseitige Bandbreite (des Kanals)	one-sided bandwidth (of the channel)
$\mathring{B}_K$	optimale (einseitige) Bandbreite (des Kanals)	optimum (one-sided) bandwidth (of the channel)
$b_K(f)$	Phasenmaß (des Kanals)	phase constant (of the channel)
C	Kanalkapazität	channel capacity
C_A	Kanalkapazität bei Amplitudenbegrenzung	channel capacity under peak-value limitation
C_L	Kanalkapazität bei Leistungsbegrenzung	channel capacity under power limitation
$c(t)$	Codersignal	coded signal
$c_I(t)$	Sprungantwort (des Impulsformers)	step response (of the pulse shaper)
c_ν	Codesymbol	coded symbol
$\langle c_\nu \rangle$	Codesymbolfolge	coded symbol sequence

D_U	ungünstigster Detektionsabstand	worst-case detection distance
D_i	Detektionsabstand (der i-ten Augenlinie)	detection distance (of the ith eye line)
$D_{i,lo}$	unterer Detektionsabstand	lower detection distance
$D_{i,up}$	oberer Detektionsabstand	upper detection distance
$d(t)$	Detektionssignal	detector input signal (detection signal)
$\tilde{d}(t)$	Detektionsnutzsignal	useful detector input signal (useful detection signal)
$\overset{\times}{d}(t)$	Detektionsstörsignal	detector input noise signal (detection noise signal)
$d_{DFE}(t)$	Korrektursignal	correction signal
$d_E(t)$	Detektionssignal (bei empfängerseitiger Codierung)	detection signal (coded at the receiver)
$d_S(t)$	Detektionssignal (bei sendeseitiger Codierung)	detection signal (coded at the transmitter)
$\tilde{d}_i(t)$	i-te Augenlinie (des Detektionnutzsignals)	ith eye line (of the useful detection signal)
$\tilde{d}_{lo}(t)$	untere Begrenzung der Augenöffnung	lower boundary of the eye opening
$\tilde{d}_{up}(t)$	obere Begrenzung der Augenöffnung	upper boundary of the eye opening
d_ν	Detektionsabtastwert	detection sample value
$\tilde{d}_\nu$	Detektionsnutzabtastwert	useful detection sample value
$\overset{\times}{d}_\nu$	Detektionsstörabtastwert	detection noise sample value
$\Delta d_i(t)$	lineare Fehlergröße	linear error quantity
$\Delta\tilde{d}_{kj}(t)$	Nutzdifferenzsignal	useful difference signal
Δd_μ	Werteintervall(e) des Detektionssignals	detection signal interval(s)
E	Schwellenwert	decision value, threshold value
$\mathring{E}$	optimaler Schwellenwert	optimum decision value
$E_d^{(i)}$	Energie des i-ten Detektionsnutzsignals	energy of the ith useful detection signal
$E_e^{(i)}$	Energie des i-ten Empfangsnutzsignals	energy of the ith useful received signal
E_{ph}	Energie eines Photons	photon energy
E_μ	Schwellenwert(e) eines mehrstufigen Systems	decision value(s) of a multilevel system
$E_{\Delta d}^{(i)}$	Störenergie bei der gesendeten Quellensymbolfolge Q_i	noise energy due to transmission of the source symbol sequence Q_i
ΔE_{kj}	Energieabstand (zwischen k-tem und j-tem Detektionsnutzsignal	energy distance (between the kth and jth useful detection signals)
ΔE_{min}	minimaler Energieabstand	minimum energy distance
$e(t)$	Empfangssignal	received signal
$\tilde{e}(t)$	Empfangsnutzsignal	useful received signal

$\tilde{e}_i(t)$	Empfangsnutzsignal bei der gesendeten Quellensymbolfolge Q_i	useful received signal from transmission of the source symbol sequence Q_i
$\Delta e_i(t)$	Differenz(empfangs-)signal	difference (received) signal
F	Bandrauschzahl	average noise factor, noise figure
$F(f)$	Spektralrauschzahl	spectral noise factor
f	Frequenz	frequency
f_{DFE}	Grenzfrequenz des linearen QR-Netzwerks	cut-off frequency of the feedback network
f_{I}	Impulsformer-Grenzfrequenz	cut-off frequency of the pulse shaper
$\check{f}_{\mathrm{I}}$	minimale Inpulsformer-Grenzfrequenz	minimum cut-off frequency of the pulse shaper
$\mathring{f}_{\mathrm{I}}$	optimale Impulsformer-Grenzfrequenz	optimum cut-off frequency of the pulse shaper
f_{K}	Kanal-Grenzfrequenz	cut-off frequency of the channel
f_{N}	Nyquist-Frequenz	Nyquist frequency
$f_d(d)$	WDF des Detektionssignals	PDF of the detection signal
$f_{\tilde{d}}(\tilde{d})$	WDF des Detektionnutzsignals	PDF of the useful detection signal
$f_{\overset{\times}{d}}(\overset{\times}{d})$	WDF des Detektionstörsignals	PDF of the detection noise signal
$f_e(e)$	WDF des Empfangssignals	PDF of the received signal
f_{lo}	untere Grenzfrequenz eines Hochpasses	lower cut-off frequency of a high-pass filter
$f_n(n)$	WDF des Störsignals	PDF of the noise signal
$f_s(s)$	WDF des Sendesignals	PDF of the transmitted signal
f_{up}	obere Grenzfrequenz eines Tiefpasses	upper cut-off frequency of a low-pass filter
$\Box f_{\mathrm{E}}$	äquivalente Rauschbandbreite des Entzerrers	equivalent noise bandwidth of the equalizer
$\Box f_{\mathrm{I}}$	äquivalente Rauschbandbreite des Impulsformers	equivalent noise bandwidth of the pulse shaper
$\Box f_{\mathrm{K}}$	(äquivalente) Rauschbandbreite des Kanals	(equivalent) noise bandwidth of the channel
Δf_{K}	systemtheoretische Bandbreite des Kanals	system theory bandwidth of the channel
G_{A}	Störabstandsgewinn bei Amplitudenbegrenzung	signal-to-noise ratio gain under peak-value limitation
G_{L}	Störabstandsgewinn bei Leistungsbegrenzung	signal-to-noise ratio gain under power limitation
$G_s(f)$	Sendeimpulsspektrum	transmitter pulse spectrum
G_{th}	äquivalenter Rauschleitwert	equivalent noise conductance
$g_{\mathrm{DFE}}(t)$	QR-Grundimpuls	basic correction pulse
$g_{\mathrm{N}}(t)$	Nyquist-Impuls	Nyquist pulse

$g_{N_2}(t)$	Nyquist-2-Impuls	Nyquist-2 pulse
$g_d(t)$	Detektions-Grundimpuls	basic detection pulse
$g_{dE}(t)$	Detektions-Grundimpuls (bei empfangsseitiger Codierung)	basic detection pulse (coding at the receiver)
$g_{dS}(t)$	Detektions-Grundimpuls (bei sendeseitiger Codierung)	basic detection pulse (coding at the transmitter)
$g_e(t)$	Empfangs-Grundimpuls	basic received pulse
$g_k(t)$	korrigierter Detektions-Grundimpuls	corrected basic detection pulse
$g_m(t)$	Matched-Filter Grundimpuls	basic matched-filter pulse
$g_s(t)$	Sende-Grundimpuls	basic transmitted pulse
$\hat{g}_s$	Sendeimpulsamplitude	transmitter pulse amplitude
g_U	Summe der Beträge aller Vor- und Nachläufer d. Grundimpulses	sum of the absolute values of all precursors and postcursors
$g_v(t)$	Sinken-Grundimpuls	basic sink pulse
$g_y(t)$	optischer Empfangs-Grundimpuls	basic optical received pulse
g_0	Hauptwert des Grundimpulses	main value of the basic detection pulse
g_v	Detektions-Grundimpulswert(e)	basic detection pulse value(s)
$H_{BP}(f)$	Bandpaß-Frequenzgang	band-pass frequency response
$H_C(f)$	Coder-Frequenzgang	coder frequency response
$H_{DFE}(f)$	Frequenzgang des QR-Netzwerks	frequency response of the feedback network
$H_E(f)$	Entzerrer-Frequenzgang	equalizer frequency response
$H_E(p)$	komplexer Entzerrer-Frequenzgang	complex frequency response of the equalizer
$\|H_E(0)\|$	Gleichsignal-Übertragungsfaktor des Entzerrers	direct signal transmission factor of the equalizer
$H_{HP}(f)$	Hochpaß-Frequenzgang	high-pass frequency response
$H_I(f)$	Impulsformer-Frequenzgang	pulse-shaper frequency response
$\|H_I(0)\|$	Gleichsignal-Übertragungsfaktor des Impulsformers	direct signal transmission factor of the pulse shaper
$H_I(p)$	komplexer Impulsformer-Frequenzgang	complex frequency response of the pulse shaper
$H_K(f)$	Kanal-Frequenzgang	channel frequency response
$H_K'(f)$	auf die Rauschzahl $F(f)$ bezogener Kanal-Frequenzgang	channel frequency response based on the noise factor $F(f)$
$\|H_K(0)\|$	Gleichsignalübertragungsfaktor des Kanals	direct signal transmission factor of the channel
$H_K(f_0)$	Bezugswert bei Bandpaß-systemen	channel frequency response reference value for band-pass systems
$H_{K,max}(f)$	Maximalwert des Kanal-Frequenzgangs	maximum value of the channel frequency response

$H_{LP}(f)$	Tiefpaß-Frequenzgang	frequency response of low-pass filter
$H_{MF}(f)$	Matched-Filter-Frequenzgang	frequency response of matched filter
$H_N(f)$	Nyquist-Frequenzgang	Nyquist frequency response
$\mathring{H}_N(f)$	optimaler Nyquist-Frequenzgang	optimum Nyquist frequency response
$H_{N2}(f)$	Nyquist-2-Frequenzgang	Nyquist-2 frequency response
$H_S(f)$	Sender-Frequenzgang	transmitter frequency response
H_T	Synentropie (mittlerer Transinformationsgehalt eines Symbols)	synentropy (mean transinformation content of a symbol)
$H_T(f)$	Taktfilter-Frequenzgang	frequency response of a clock filter
$H_{TF}(f)$	Transversalfilter-Frequenzgang	frequency response of a transversal filter
$H_{WF}(f)$	Dekorrelationsfilter-Frequenzgang	frequency response of a whitening filter
$\|H_K(f)\|$	Amplitudengang des Kanals	amplitude response of the channel
H_e	Entropie des Empfangssignals	entropy of the received signal
H_n	Entropie des Störsignals	entropy of the noise signal
H_q	Entropie der Quelle	source entropy
$H_{q,\max}$	Nachrichtengehalt der Quelle	decision content of the source
H_v	Sinkenentropie	sink entropy
$H_{v\|q}$	Streuentropie	irrelevance
$h_{DFE}(t)$	Impulsantwort des QR-Netzwerkes	pulse response of the feedback network
$h_E(t)$	Impulsantwort des Entzerrers	pulse response of the equalizer
$h_I(t)$	Impulsantwort des Impulsformers	pulse response of the pulse shaper
$h_K(t)$	Impulsantwort des Kanals	pulse response of the channel
I_v	Informationsgehalt	information content
K	Korrekturgröße	correction term
$k(t)$	korrigiertes Detektionssignal	corrected detection signal
$\tilde{k}(\nu T)$	Abtastwert(e) des korrigierten Detektionsnutzsignals	sample value(s) of the corrected useful detection signal
k_λ	Filterkoeffizienten	filter coefficients
$L_a(f)$	spektrale Leistungsdichte (Leistungsspektrum) der Amplitudenkoeffizienten	power spectrum of the amplitude coefficients
$L_{\tilde{d}}(f)$	Leistungsspektrum des Detektionnutzsignals	power spectrum of the useful detection signal
$L_{\overset{\times}{d}}(f)$	Leistungsspektrum des Detektionstörsignals	power spectrum of the detection noise signal
$L_{\tilde{e}}(f)$	Leistungsspektrum des Empfangsnutzsignals	power spectrum of the useful received signal

$L_{eq}(f)$	Leistungsspektrum des äquivalenten Störsignals	power spectrum of the equivalent noise signal
$L^{\bullet}_{g_s}(f)$	Energiespektrum des Sende-Grundimpulses	energy spectrum of the basic transmitter pulse
$L_{inst}(f)$	instationärer Anteil des Störleistungsspektrum	nonstationary component of the noise power spectrum
$L_n(f)$	Störleistungsspektrum	noise power spectrum
$L_n(f,t)$	instationäres Störleistungsspektrum	time-dependent noise power spectrum
$L_s(f)$	Leistungsspektrum des Sendesignals	power spectrum of the transmitted signal
$\mathring{L}_s(f)$	optimales Sendeleistungsspektrum	optimum transmitter power spectrum
$L_{stat}(f)$	stationärer Anteil des Störleistungsspektrums	stationary noise power spectrum
L_{th}	thermische Rauschleistungsdichte (bei Widerstandsanpassung)	thermal noise power density (impedence matching assumed)
$L_{th,L}$	thermische Rauschleistungsdichte (im Leerlauf)	thermal noise power density (zero load assumed)
L_0	Rauschleistungsdichte (bei Weißem Rauschen)	noise power density (white noise assumed)
l	Regeneratorfeldlänge (Kabellänge)	regenerator-section length (cable length)
$l_a(\lambda)$	diskrete AKF der Amplitudenkoeffizienten	discrete ACF of the amplitude coefficients
$l^{\bullet}_{gd}(\tau)$	Energie-AKF des Detektions-Grundimpulses	energy ACF of the basic detection pulse
$l^{\bullet}_{g_s}(\tau)$	Energie-AKF des Sende-Grundimpulses	energy ACF of the basic transmitter pulse
$l_m(\lambda T)$	diskrete AKF von $m(t)$	discrete ACF of $m(t)$
$l_{\overset{\times}{m}}(\lambda T)$	diskrete AKF von $\overset{\times}{m}(t)$	discrete ACF of $\overset{\times}{m}(t)$
$l_s(\tau)$	AKF des Sendesignals	ACF of the transmitted signal
$l_{xy}(\tau)$	KKF von $x(t)$ und $y(t)$	CCF of $x(t)$ and $y(t)$
$l^{\bullet}_{xy}(\tau)$	Energie-KKF von $x(t)$ und $y(t)$	energy CCF of $x(t)$ and $y(t)$
M	Stufenzahl, Quellensymbolumfang	level number, source symbol set size
$\mathring{M}$	optimale Stufenzahl	optimum level number
M_{PD}	mittlere Lawinenverstärkung	mean avalanche gain
$M_{PD}{}^x$	Zusatzrauschfaktor	excess noise factor
M_c	Symbolumfang (Coder)	symbol set size (coder)
M_q	Symbolumfang (Quelle)	symbol set size (source)
MSE	mittlerer quadratischer Fehler	mean squared error
m_c	Blocklänge (Coder)	block length (coder)
$m_i(t)$	Ausgangssignal des i-ten Matched-Filters	output signal of the ith matched filter
m_q	Blocklänge (Quelle)	block length (source)

$\overset{\times}{m}_\nu$	Störanteile des abgetasteten Signals $m(t)$	noise components of the sampled signal $m(t)$
N	Ordnung des Codes	code order
N_B	Störleistung von bandbegrenztem Weißen Rauschen	power of band-limited white noise
N_d	Detektionsstörleistung	detector input noise power (detection noise power)
N_{dE}	Detektionsstörleistung (bei empfangsseitiger Codierung)	detection noise power (coding at the receiver)
N_{dS}	Detektionsstörleistung (bei sendeseitiger Codierung)	detection noise power (coding at the transmitter)
N_{norm}	normierte Störleistung	normalized noise power
N_v	Sinkenstörleistung	sink noise power
n	Anzahl der Nachläufer	number of postcursors
$n(t)$	(Empfangs-)Störsignal	noise signal
$n_{eq}(t)$	äquivalentes Störsignal	equivalent noise signal
$o(t)$	vertikale Augenöffnung	eye opening
$o(T_D)$	vertikale Augenöffnung zum Detektionszeitpunkt T_D	eye opening at detection time T_D
$o_{norm}(T_D)$	normierte Augenöffnung	normalized eye opening
$P_F(\xi_0 \mid \xi_{-i})$	mittlere Folgewahrscheinlichkeit	mean follow-up probability
P_T	Übergangswahrscheinlichkeit	transition probability
p_B	mittlere Bitfehlerwahrscheinlichkeit	mean bit error probability
p_G	günstigste mittlere Fehlerwahrscheinlichkeit	best-case mean error probability
p_{Glv}	obere Schranke für p_M nach Glave	upper boundary for p_M according to Glave's approximation
p_{Gr}	Grenzfehlerwahrscheinlichkeit	critical error probability
p_M	mittlere Symbolfehlerwahrscheinlichkeit, mittlere SFW	mean symbol error probability, mean SEP
p_{S_i}	SFW der i-ten Augenlinie	SEP of the ith eye line
p_{S_ν}	SFW des ν-ten Symbols	SEP of the νth symbol
p_U	ungünstigste Symbolfehlerwahrscheinlichkeit	worst-case symbol error probability
p_i	Auftrittswahrscheinlichkeit der i-ten Augenlinie	occurrence probability of the ith eye line
p_k	mittlere Fehlerwahrscheinlichkeit eines Systems mit k gleichen Regeneratorabschnitten	mean error probability of a system consisting of k equally designed regenerator sections
$p_{\nu\mu}$	Symbolwahrscheinlichkeit(en)	symbol probability(ies)
p_μ	statistisch unabhängige Symbolwahrscheinlichkeit(en)	statistically independent symbol probability(ies)

Q_N	Quellensymbolfolge bestehend aus N Symbolen	source symbol sequence consisting of N symbols
$Q(x)$	komplementäres Gauß'sches Fehlerintegral	complementary Gaussian error integral
$q(t)$	Quellensignal	source signal
$q_\delta(t)$	äquivalentes Diracquellensignal	equivalent Dirac source signal
q_v	Quellensymbol	source symbol
$\bar{q}_v$	Komplementärsymbol zu q_v	complementary symbol to q_v
$\langle q_v \rangle$	Quellensymbolfolge	source symbol sequence
$\{q_\mu\}$	Quellensymbolvorrat	source symbol set
R	(äquivalente) Bitrate Nachrichtenfluß	(equivalent) bit rate, maximum information rate per unit time
R_c	Bitrate nach Codierung	bit rate of the coded signal
R_q	Bitrate der Quelle	bit rate of the source signal
R_{th}	thermischer Widerstand	thermal impedance
RDS	laufende digitale Summe	running digital sum
r_{I}	roll-off-Faktor (des Impulsformers)	roll-off factor (of the pulse shaper)
$\mathring{r}_{\text{I}}$	optimaler roll-off-Faktor (des Impulsformers)	optimum roll-off factor (of the pulse shaper)
r_{N}	roll-off-Faktor (bei Nyquist-Frequenzgängen)	roll-off factor (of Nyquist frequency responses)
r_s	roll-off-Faktor (des Sendeimpulses)	roll-off factor (of the transmitter pulse)
r_c	relative Redundanz des Codes	relative redundancy of the coded signal
r_q	relative Redundanz der Quelle	relative redundancy of the source signal
rec(x)	Rechteck-Funktion	rectangular function
$\langle r_v \rangle$	regenerierte Symbolfolge	regenerated symbol sequence
S_{PD}	Steilheit einer Photodiode	responsivity (sensitivity) of the photodiode
S_d	Detektionsnutzleistung	useful detection power
S_e	Empfangsnutzleistung	useful received power
S_s	mittlere Sendeleistung	mean transmitter power
S_v	Sinkennutzleistung	useful sink power
$s(t)$	Sendesignal	transmitted signal
$s_{\max}$	Maximalwert des Sendesignals	maximum value of $s(t)$
$s_{\min}$	Minimalwert des Sendesignals	minimum value of $s(t)$
s_0	Sendesignal-Grundanteil	base component of $s(t)$
si(x)	$\sin(x)/x$, si-Funktion	$\sin(x)/x$, si function
$\{s(t)\}$	stochastischer Proze mit der Musterfunktion $s(t)$	stochastic process with representative function $s(t)$

Δs	Aussteuerbereich des Sendesignals	modulation range of $s(t)$
T	Symboldauer	symbol duration
T_D	Detektionszeitpunkt (für das nullte Symbol)	detection time instant (for the zeroth symbol)
$\mathring{T}_D$	optimaler Detektionszeitpunkt	optimum detection time
T_{eye}	zeitliche Augenöffnung	temporal eye opening
T_c	Symboldauer (Coder)	symbol duration of the coded signal
T_q	Symboldauer (Quelle)	symbol duration of the source signal
T_s	absolute Sendeimpulsdauer	absolute transmitter pulse duration
$1/T$	Symbolrate	symbol rate
t	Zeit	time
t_ν	Detektionszeitpunkt(e)	detection time instant(s)
Δt_K	Dispersionskonstante	dispersion constant
Δt_{Ma}	Materialdispersionskonstante	material dispersion constant
Δt_{Mo}	Modendispersionskonstante	modal dispersion constant
Δt_s	äquivalente Sendeimpulsdauer	equivalent transmitter pulse duration
$\Delta \mathring{t}_s$	optimale äquivalente Sendeimpulsdauer	optimum equivalent transmitter pulse duration
V_N	Sinkensymbolfolge bestehend aus N Symbolen	sink symbol sequence consisting of N symbols
V_{tol}	Verlustfaktor durch Toleranzen	loss factor due to tolerances
$V_\nu^{(l)}$	bedingt detektierte Teilsymbolfolge	tentatively detected partial symbol sequence
v	Anzahl det Vorläufer	number of precursors
$v(t)$	Sinkensignal	sink signal
$v_\nu^{(l)}$	bedingt detektiertes Symbol	tentatively detected symbol
$\langle v_\nu \rangle$	Sinkensymbolfolge	sink symbol sequence
$w_n(\tau)$	inverse AKF des Störsignals	inverse ACF of the noise signal
x	Zusatzrauschexponent	excess noise exponent
$y(t)$	optisches Empfangssignal	received optical signal
Z_{PD}	Anzahl von Sekundärelektronen	number of secondary electrons
$z(t)$	Taktsignal	clock signal
$\alpha_0, \alpha_1, \alpha_2$	Dämpfungskonstante	attenuation coefficients
$\alpha_\nu^{(k,j)}$	Amplitudenkoeffizienten des Differenznutzsignals	amplitude coefficients of the useful difference signal
β_1, β_2	Phasenkonstante	phase coefficients

$\gamma_{\mathrm{EP,DFE}}$	Fehlerfortpflanzungsfaktor durch QR	error propagation factor due to decision feedback equalization
$\gamma_{\mathrm{EP,code}}$	Fehlerfortpflanzungsfaktor durch Codierung	error propagation factor due to coding
γ_{pu}	Korrekturfaktor	correction factor
$\gamma_{\nu}^{(i)}$	ν-te Gesamtfehlergröße	νth total error quantity
$\mathring{\gamma}_{\nu}^{(i)}$	minimale Gesamtfehlergröße	minimum total error quantity
$\delta(t)$	Diracimpuls	Dirac pulse
$\varepsilon_i(t)$	quadratische Fehlergröße	square error quantity
$\varepsilon_{\nu}^{(i)}$	ν-te Fehlergröße der Symbolfolge $Q_N^{(i)}$	νth error quantity of the symbol sequence $Q_N^{(i)}$
$\varepsilon_{\nu}'^{(i)}$	modifizierte Fehlergröße	modified error quantity
η_{A}	Systemwirkungsgrad bei Amplitudenbegrenzung	system efficiency under peak-value limitation
η_{E}	Empfängerwirkungsgrad	receiver efficiency
η_{K}	Kanalwirkungsgrad	channel efficiency
η_{L}	Systemwirkungsgrad bei Leistungsbegrenzung	system efficiency under power limitation
η_{Q}	Quantenwirkungsgrad	quantum efficiency
η_{S}	Senderwirkungsgrad	transmitter efficiency
$\eta_{\mathrm{S,A}}$	Senderwirkungsgrad bei Amplitudenbegrenzung	transmitter efficiency under peak-value limitation
$\eta_{\mathrm{S,L}}$	Senderwirkungsgrad bei Leistungsbegrenzung	transmitter efficiency under power limitation
κ_{S}	Crestfaktor, Sendespitzenwertfaktor	transmitter peak-value factor (crest factor)
λ_{L}	Lichtwellenlänge	wavelength
ξ_{ν}	logische Fehlervariable	logic error variable
ρ_{A}	maximales Sinken-Signalstörleistungsverhältnis bei Amplitudenbegrenzung	maximum sink signal-to-noise ratio under peak-value limitation
ρ_{L}	maximales Sinken-Signalstörleistungsverhältnis bei Leistungsbegrenzung	maximum sink signal-to-noise ratio under power limitation
ρ_{U}	ungünstigstes Signalstörleistungsverhältnis	worst-case signal-to-noise ratio
$\rho_{\mathrm{U,tol}}$	ungünstigstes S/N-Verhältnis unter Berücksichtigung von Toleranz	worst-case signal-to-noise ratio with tolerance
ρ_{v}	Sinken-Signalstörleistungsverhältnis	sink signal-to-noise ratio

$\rho_{v,E}$	Sinken-Signalstörleistungsverhältnis bei optimalem Empfänger	sink signal-to-noise ratio with optimum receiver
$\rho_{v,KE}$	Sinken-Signalstörleistungsverhältnis bei optimalem Empfänger und idealem Kanal	sink signal-to-noise ratio with optimum receiver and ideal channel
$\rho_{v,SKE}$	Sinken-Signalstörleistungsverhältnis bei optimalem Empfänger, idealem Kanal und optimalem Sender	sink signal-to-noise ratio with optimum receiver, ideal channel and optimum transmitter
ρ_d	Detektions-Signalstörleistungsverhältnis	detection signal-to-noise ratio
ρ_e	minimales Empfangs-Signalstörleistungsverhältnis	minimum receiver signal-to-noise ratio
σ_n	Streuung (Standardabweichung) des Störsignals	root mean square (rms, standard deviation) of the noise signal
θ	absolute Temperatur	absolute temperature
τ_v	Taktjitterwert	clock jitter value
ϕ	Informationsfluß	information rate per unit time
ϕ_T	maximaler mittlerer Transinformationsfluß	maximum mean transinformation rate
$\phi(x)$	Gauß'sches Fehlerintegral	Gaussian error integral
$\varphi_T(t)$	Phase der Taktschwingung	clock oscillation phase
$\_\Gamma(t)$	Sprungsignal	step signal
$\Lambda(t)$	Dreieckssignal	triangular signal

Index